ROYAL AIR FORCE MUSEUM

BRITISH AVIATION
The Ominous Skies
1935-1939

HARALD PENROSE

LONDON
HER MAJESTY'S STATIONERY OFFICE

First published 1980

Other titles by the same author:

The Pioneer Years: 1903–1914

The Great War and Armistice: 1915–1919

The Adventuring Years: 1920–1929

Widening Horizons: 1930–1934

ISBN 0 11 290298 7

CONTENTS

Introduction

Previous volumes of *British Aviation* have extended from the days of that deductive genius of aerial experimentation, Sir George Cayley, through the pioneering years to the Great War, the adventuring decade of the 1920's, the high summer of flying to the mid-1930's, and now the threat of a second world war. Once again it is the dedication of the pioneer manufacturers and their quietly devoted teams which enabled their country to have the power of defeating aerial attack – for had they not clung to their vocation through difficult days there might have been no experienced aircraft industry on which to rely.

As the elderly Gaius Cornelius Tacitus wrote in his *Annals* between 115 and 117 AD: 'My purpose is not to relate everything at length, but only such things as were conspicuous for excellence: this I regard as the highest function of history – to let not worthy action be uncommemorated.'

By fortunate coincidence of time-span in the world of aeronautics I encountered almost all the figure-head pioneers, their successors, chief designers, and pilots, for our common meeting ground was the Air Ministry's great aircraft testing station on Martlesham Heath in Suffolk and later at Boscombe Down in Wiltshire. There were also the helpful researchers at Government Establishments such as Farnborough, Air Ministry officials in London, senior RAF administrators, industrialists, politicians, celebrities, enthusiasts – and scores of young pilots who sometimes held high rank or office later, as well as many who untimely died. My own sphere of work as test pilot continued throughout the period of this volume, through World War II, and beyond, spanning a quarter century, and for fourteen years more I dealt with world sales of the Westland Group in which Fairey, Saunders-Roe, and the Bristol Helicopter interests were merged – so I am well aware that a great story of war-time endeavour and subsequent peace-time development remains to be told.

Three hundred years ago that quaintly delightful antiquary, John Aubrey, aptly expressed my position: 'Considering that if I should not finish what I had begun, and not knowing Anyone that would undertake this Design while I live, I have tumultarily sticht up what I have many years since collected: I hope hereafter it may be an Incitement to some Ingenious and Publick-spirited young Man to polish and complete what I have delivered rough hewn.'

Meanwhile I would like to express indebtedness to the many who have sent information and photographs, and in particular offer profound thanks to Marjorie Nitch, herself a pilot, whose devoted and accurate typing and retyping, checking and indexing over the years has made the sequence of *British Aviation* possible.

List of Abbreviations

A&AEE	Aircraft & Armament Experimental Establishment
AC	Army Co-operation
ADGB	Air Defence of Great Britain Command
AEW	Association of Engineering Workers
AGM	Annual General Meeting
AID	Aircraft Inspection Department
AMRD	Air Member for Research & Development
AMSR	Air Member for Supply and Research
ANT	A. N. Tupolev
AOA	Aerodrome Owners Association
ARB	Air Registration Board
ARC	Aeronautical Research Committee
ARP	Air Raid Precautions
AST	Air Service Training Ltd.
BAC	British Aircraft Co. Ltd.
BALPA	British Airlines Pilots Association
BGA	British Gliding Association
CAG	Civil Air Guild
CFI	Chief Flying Instructor
CG	Centre of Gravity
CinC	Commander in Chief
CIGS	Chief of Imperial General Staff
CO	Commanding Officer
CSSAD	Committee for Scientific Survey of Air Defence
CTO	Chief Technical Officer
DAP	Director of Aircraft Production
DDO	Deputy Director of Organization
DDTD	Deputy Director of Technical Development
DGCA	Director General of Civil Aviation
DGP	Director General of Production
DGRD	Director General of Research and Development
DOR	Director of Operational Requirements
DSR	Director of Scientific Research
DTD	Director of Technical Development
EKW	Eidg. Konstructions Werkstätte
FAA	Fleet Air Arm
FTS	Fighter Training School
GAC	Gloster Aircraft Company
GAPAN	Guild of Air Pilots and Navigators

GP	General Purpose
ICAN	International Convention of Aerial Navigation
ITP	Instruction to Proceed
KLM	Royal Dutch Airline
M&AEE	Marine & Armament Experimental Establishment
MOS	Ministry of Supply
NACA	National Advisory Committee for Aeronautics
NFS	National Flying Services Ltd.
NPL	National Physical Laboratory
OC	Officer Commanding
PAA	Pan American Airlines
PRO	Public Relations Officer
PV	Private Venture
RAAF	Royal Australian Air Force
RAE	Royal Aircraft Establishment
RAeC	Royal Aero Club
RAeS	Royal Aeronautical Society
RAF	Royal Air Force
RAFVR	Royal Air Force Volunteer Reserve
RCAF	Royal Canadian Air Force
RTO	Resident Technical Officer
SBAC	Society of British Aircraft Companies
SNCA	Societé Nationale pour Constructiones Aerien
SNCC	Societé Nationale du Centre
SSR	Superintendent of Scientific Research
STD	Superintendent of Technical Development
STOL	Steep Take-off & Landing
VP	Variable Pitch
WRAF	Womens Royal Air Force
W/T	Wireless Telegraphy

CHAPTER I

1935: *Peaceful Rearmament*

'When my brother and I built and flew our first man-carrying flying machine, we thought we were introducing into the world an invention which would make further wars practically impossible.'

Orville Wright (1917)

I

On 1st January *The Times*, that bible of Tory opinion, had its 150th anniversary. As one reader said: 'The greatest feats of airmanship have never tempted its writers from a proper attitude of stately acclaim, and consequently our own people as much as foreign readers have always been able to get from its columns the sane English view of every happening.'

The current issue, announcing the New Year's Honours, revealed promotion for a number of senior RAF Officers including HRH Edward, Prince of Wales to Air Chief Marshal. Not unexpected was the KBE for Sir Macpherson Robertson Kt., 'for philanthropic services in the State of Victoria,' but in fact because of his initiative in establishing the great MacRobertson Air Race to Australia of the previous autumn for which he gave prizes totalling £15,000. The specially designed DH Comet had won the speed section, flown by tough Charles Scott and equally determined Tom Campbell Black. Outstanding though their relatively small twin-engined racer might be, the startling pointers to the future were two big American airliner entrants – the Douglas DC-2 and Boeing 247 D low wingers distinctive with smooth metal skinning for the entire structure.

British designers were certainly impressed by the great advances the USA had made, though it was Britain who had led far back as 1920 with Oswald Short's pioneering metal monocoques, leading to metal-skinned flying-boat hulls and the Schneider racers – but prevailing practice was diagonally braced metal girder structures covered with fabric over ribs and stringers for aerodynamic shape.

The previous year's governmental decision of a five-year plan to treble the RAF's number of home-based aeroplanes was already creating a resurgence of the aircraft industry as indicated by *the Times* report of the sixth AGM of Fairey

Aviation Ltd, showing a profit of £47,534 which, despite being a reduction of £69,148 from the preceding year, was regarded as satisfactory, though the directors reduced the dividend to 5 per cent. Said Fairey: 'There has been further shrinkage of business with the Air Ministry, and a considerable part of our activities has been export. Because of national policy we have changed to metal construction, which increases the development time of new types, but though prototypes have been designed for repetition construction, a long period of preparation is necessary to make the requisite tools and organize processes, and the Air Ministry is concerned at the time taken to introduce new types. However, we have received a preliminary production order for one of our two new types, and have hopes for the second; but two unfortunate accidents caused delay before trials were completed. Next year we shall be extremely busy on experimental work. In order to produce aircraft outside the metropolis, the company has bought the war-time Heaton Chapel factory at Stockport, affording more than twice the area at Hayes, but the move can only be made if the company is assured of Air Ministry support.'

By comparison, de Havilland Aircraft Co Ltd, dependent solely on civil P.V. designs, made £50,994 profit, as revealed by Mr T. P. Mills, acting as chairman on behalf of Alan Butler who was ill. He commented that the trail of the MacRobertson Comet had left its mark in the account books as it had been an expensive exercise; nevertheless sales had increased by 16 per cent and purchase of 600 acres at Hatfield had been completed. A dividend of 7½ per cent was approved and Geoffrey de Havilland and his brother-in-law works director Frank Hearle were re-elected.

From Hawker Aircraft Ltd there wafted an air of new opportunism, for at an extraordinary general meeting on 9th January, with Sopwith's one-time mechanic, now director, Fred Sigrist presiding, a resolution was passed increasing the capital to £900,000 by creating 400,000 ordinary shares of 5*s*.

Major British aircraft companies could see a promising financial future dawning. The index of production, representing national prosperity, had risen a little above the post-war high of 1929. The cost of living began to fall. There was a mild boom in the building trade as well as in new industries such as chemicals, rayon, radio, and increasing production of cars and lorries. But exports had declined, resulting in the continuation of depressed industrial areas exacerbated by the marked fall in world commodity prices. Conditions had forced France's aviation industry to take new shape, the major firms amalgamating into four groups – respectively Breguet, Morane, Wibault, Maubousinn; followed by Lioré et Olivier and Dewoitine; then Farman, Les Mureaux, and Blériot; and last, Henri Potez and Marcel Bloch who held an option on the Lorraine Motor Works owned by the bankrupt Société comprising the now non-existent firms of Hanroit, CAMS, Amiot and Nieuport already bought up by these groups.

To help the many unemployed of the British ship-building industry, Miss Irene Ward, MP for Wallsend, on her own initiative endeavoured to secure an alliance of Anthony Fokker with Swan Hunter and Wigham Richardson, in conjunction with Airspeed Ltd, to build Fokkers and Douglases in vacant Tyneside shipyards. The prospect was encouraging, for Swan Hunter money had floated the new Airspeed (1934) Ltd with capital of £220,000, and Norway and Hessell Tiltman had been joined on the Board by C. S. Swan, Sir Philip Wigham Richardson and his son George, and Arthur Townsley of Swan Hunter as general manager, with his colleague J. Liddell as company secretary. Fokker was appointed technical adviser with emoluments of 1 per cent of gross receipts and an option on preferred shares. Meanwhile Tiltman was designing a 15-seat high-winger, the AS 14 tentatively named Ambassador – but though an elegant mock-up was made no airline was prepared to venture. Energies therefore concentrated on the Envoy with two 240 hp Lynx IVC engines instead of 185 hp Wolseley AR 9s, and the first was ordered by R. K. Dundas, their sales agents, who engaged Flt Lieut H. C. Johnson, accompanied by Sir Alan Cobham, to fly it to India for demonstrations. Others were being built for North Eastern Airways, largely owned by Lord Grimthorpe, Airspeed's chairman. Even more encouraging was an order from the Air Ministry for a military version.

As potential rival to the American DC-2 airliner, Hollis Williams at General Aircraft was initiating design of a twin-engined low-winger, the ST-18, accomodating ten passengers at an estimated 190 mph with a range of 600 miles. It would have a retractable undercarriage, variable-pitch propellers and landing flaps. Tentative price was £8,000, and there was interest from Oceanic Airways Ltd of Australia who operated ten Monospar four-seaters.

Another, but much slower ten-seater was the last of the antique Vickers Vellore series, the Vellox, which Charles Ulm was trying to borrow for a long-distance attempt at Vickers expense but which instead was sold at low price to Imperial Airways as a freighter. That was the end of Vickers attempts at civil aircraft for the foreseeable future. The Viastra monoplane demonstrator was broken up. Attention concentrated on completing the G4/31 'geodetic' low-wing monoplane and the similarly constructed B9/32 twin-engined bomber whose specification had been undergoing much revision of bomb load and requisite range.

At this juncture the Vickers design organisation was joined by a slim thirty-six-year-old industrial engineer with thinning hair and the determined personality that makes success. He was George Edwards who recollected: 'I started work under George Stannard who was absolute solid gold – a group leader of the old type and a wonderful man. The first aeroplane I worked on was the G4/31 biplane. I had to alter the wing trailing edge. My next significant job was the B9/32, and my charge was design of the rudder. When I asked Rex Pierson, the chief designer, what it should look like he said the one on the Stranraer was all right and 'Mutt' Summers, the Vickers pilot, was very pleased with it. So we scaled the shape and worked from there, though eventually the right shape was very different. We just had to learn the hard way in the absence of what has since become well documented science.'

A new twin-engined low-winger was the Avro 652 which had been swiftly completed, and in the absence of Sam Brown in Chile was flown by his eager young assistant, F. B. Tomkins, on 7th January who reported easy and delightful handling qualities and a top speed of 195 mph, but found it tedious to wind the many turns of the hand-wheel for the manually-operated retractable undercarriage. Closely following it in the shops was the military version with gun-turret cupola as a rival to the DH 89M with Lewis gun and patent DH mounting on top of the rear fuselage – both having been ordered for evaluation as coastal reconnaissance aircraft.

2

Despite concern at the number of RAF accidents, they had become accepted as inevitable; but now there was disquiet at several inexplicable civil accidents. Early in January a Dragon of Highland Airways was forced to descend in the sea when power gradually faded. Luckily it alighted near the shore, pilot and passengers suffering only a few cuts and bruises, but when salvaged there seemed nothing wrong with the engines. Few realized that carburettor icing could easily occur due to cooling by precipitation of water when fuel evaporated, rendering intake heating essential, or the addition of 3 to 5 per cent alcohol to the fuel.

More serious was a KLM crash in Iraq with a Douglas DC-2. Wreckage seemed to point to failure of the tail unit in the air, rather like the 1930 catastrophe of the Junkers at Meopham, Kent. Much of the earlier suspicion of metal monocoque construction suddenly revived. There was a feeling that strength was sacrificed to get sufficiently light structure weight to enable an adequate payload to be carried. There were rumours that European technicians in several countries would not accept the validity of American stress calculations. Nothing could be resolved until an official inquiry.

There was also the first fatal accident in England with an Autogiro. Flying Officer L. W. Oliver was flying the RAF's latest acquisition at Old Sarum at 2,000 ft when it disappeared in cloud, reappeared, and dived at 45 degrees until it struck the ground. Why? Rough-tongued George Ellis, the able chief draughtsman of Cierva's, admitted the C 30 should not be dived at top speed to steeper than 10 degrees, though 30 degrees was permissible at slow speeds. To the coroner he evasively said: 'There has been a similar accident in France to a French-designed machine, and proof has been given that the pilot fainted.' Discussing this with Arthur Davenport, Westland's chief designer, I reminded him of my tentative dives with the Autogiro and my certainty that there was a critical speed after which the dive would increase uncontrollably due to blade-twist.

At Yeovil the designer of the ill-fated French Autogiro, M. Lepère, had been introduced to the recently appointed youthful technical director, W.E.W. 'Teddy' Petter, during somewhat apprehensive attempts at overcoming the rotor vibration of the big C 29 Westland Autogiro and was entertained at Petter's house where French was the usual language because his wife was Franco-Swiss. Teddy saw the advantage of allying himself with a long-established if not wholly successful French designer, and induced his father Sir Ernest, chairman of Westland, to agree they could build a small two-seat Autogiro based on Lepère's crashed prototype. Concurrently Bristol's great engine designer Roy Fedden paid several visits to inculcate in his authoritative manner the advantages of his latest productions and the necessity of continuing to interest the Air Ministry in orthodox military designs, for he had exercised his influence to ensure the use of radials for the new generation of bombers and fighters. Thus the new F 5/34 specification admitted a radial-engined fighter with eight .303-in Browning guns. Assured by Fedden's powerful backing, Petter discussed the possibilities with Davenport, and a deep-bellied low-winger design with wide chord NACA cowling over its Mercury IX engine began to shape up. Soon a mock-up was being built in a screened-off bay of the main erecting shop. But Air Ministry caution over Petter prevailed, and eventually the AMSR, Air Vice-Marshal Sir Hugh Dowding, placed monoplane contracts with Gloster for what was to be the last of Henry Folland's designs, and with Bristol for their Type 146 which, unlike the Type 133 fighter, had tapered wings and retractable undercarriage.

A decade later Air Marshal Sir Ralph Sorley, recording the evolution of the eight-gun and cannon fighters, wrote: 'At that time two aeroplanes were being designed experimentally by Hawkers and Supermarine around the new Rolls-Royce Merlin engine in order to test the powerplant and monoplane conception rather than as operational types. If these aircraft were successful, we envisaged four guns might be installed in the fuselage. Both mock-ups were reaching completion, and as the general layout accorded with specification F 5/34, their respective designers Sydney Camm and Reggie Mitchell were confronted with the eight-gun theory and soon were enthusiastic to adopt the principle.

'Wing installation involved departures from current thought and practice. Guns in the wings dispensed with the complication of interrupter gear previously necessary to fire through the propeller. This offered saving in weight and complication, and enabled the guns to fire freely at their own rate instead of a mechanically restricted rate. However the question of clearing stoppages and re-cocking the guns became controversial. Access by the pilot to his guns had long been customary, and the idea of guns out of reach led to much argument, though it seemed obvious that with the guns working continuously there were likely to be fewer failures, and even if one or two guns stopped there were sufficient left. So re-cocking and clearing of stoppages was waived aside. Then came the question of instantaneous firing of all guns by the pilot's thumb pressure. The old Bowden control always had a lag between pressing the trigger and the operation of the gun, and that was not permissible if the requisite lethal density was to be built up in the brief time required; so a further departure was to provide pneumatic firing by pressing a button on the pilot's stick, as evolved by Dunlops. The question as to whether sighting accuracy would be as good with the guns off-set from the fuselage centre-line was settled by aligning the trajectories to cross at 400 yards range.

'The Supermarine and Hawker fighters to revised specifications F 36/34 and F 37/34 differed in one important feature: the former had a thin wing, the latter a thick one. Gun installation in the Hawker was therefore somewhat easier, enabling four guns each side to be grouped together. In the Supermarine it necessitated installing the guns separately, with the outer of each four well towards the wingtip.'

More immediately practicable was the installation by Fairey designers not only of four Browning guns but a 20 mm Oerlikon cannon in their latest 860 hp Hispano-Suiza version of the revamped Firefly biplane intended for the Belgian Government's fighter competition scheduled for 15th July. Sorley and his colleagues were immensely interested in this gun, despite its slow rate of fire, but the problem of its supersensitivity to the flexibility of mounting seemed to restrict use to a single gun as devised by the French and adopted by Fairey.

Multi-engined bombers were less urgent because strategy was defence and not offence; but Grp Capt Saundby of DOR was pressing for similar gun turrets for the twin-engined B 9/32 aircraft – the Vickers Type 271, Handley Page HP 52, and Armstrong Whitworth AW 38. Nobody knew the Germans were thinking of four-engined strategic bombers which could reach North Scotland or the Urals, as proposed by Oberstleutnant Wilhelm Wimmer and Generalleutnant Walther Wever. Some months later, on visiting Germany, Roy Fedden discovered that Dornier and Junkers were building prototypes, but when he reported this, the Minister of Defence dismissed the possibility of long-range bombing by saying that Fedden should have his head examined!

Range was certainly regarded by the Air Ministry as important, so there was considerable interest in Major Mayo's proposal to resuscitate the idea of a large aeroplane launching a smaller machine carried above its wings as first devised at Felixstowe in 1916 by Commander John Porte using a Bristol Scout C in crutches on the centre-section of the long top wing of a Port Baby flying-boat. In 1928 Norman Macmillan, at that time Fairey's test pilot, obtained a patent for the similar launching of a target-glider or winged bomb. Nevertheless Mayo had secured Patent 400,292 in April 1932 for 'aircraft carrying other aircraft', and pushed this project by publicity and direct approach, currently resulting in an Air Ministry contract to Short Bros. Ltd for a pick-a-back high-wing seaplane powered with four Napier Rapier engines, and modification of one of the big C-class Empire flying-boats to carry it.

Said Mayo: 'Refuelling in the air is another method of dodging the problem of getting a heavily loaded long-range aircraft safely into the air, but it is only a fair weather method. Catapulting cannot be considered a satisfactory solution because of high cost, lack of mobility, and risk in releasing an aircraft low down in a state in which it could not alight safely should an engine fail or there be other trouble. The Composite Aircraft links for take-off two aircraft whose characteristics, though widely different, combine to ensure a normal take-off. My first idea was to place the load in the lower component, and use the additional area and power of the upper component to get the lower unit into the air – but it became evident that this is less efficient than loading the upper component. The problem is then more difficult because the two components will not naturally separate unless specific features are incorporated, but our solution ensures safe and effective separation at the appropriate time.'

Despite the new monoplane vistas, Britain's major air transport business, Imperial Airways, remained influenced by past adherence to biplanes, for the lightly loaded four-engined HP 42s were still serving them well. Latest acquisitions were two Boulton Paul twin-engined six-seat P 71A biplanes derived from the ill-fated P 64 mail carrier. Wings and fuselage were fabric-covered except for the re-structured cabin with cross-diagonal side braces and corrugated aluminium skin. But the machine was an anacronism. Making the best of things, a commentator said: 'Probably the commendable passion of British designers for reducing structural weight to the absolute minimum colours the views of John North, its designer. He knows that craft built to these regulations are as safe as human calculation can make them.' That did not eliminate other doubts: 'A noticeable change is the greater length and slimmer fuselage and the triple-rudder arrangement recommended by the RAE as a cure for the directional instability of the central fin and rudder of the mail carrier.' Nor was performance particularly encouraging for top speed, despite the 980 hp of the two Jaguar VIAs, was only 5 mph faster than the DH 89 with two 200 hp Gipsy VIs and the same seating capacity.

North, however ingenious with steel, was conservative compared with Dr Hugo Junkers of Germany who died aged seventy-six on 3rd February. Both had been equally long at the game, yet Junkers was a forty-nine-year-old professor at Aachen when he patented his pre-war, crudely schematic 'flying-wing', but it was in 1915 that he built his trend-setting first all-metal cantilever monoplane which was detailed by his assistants Reuter and Mader. Early post-war successes were due to Karl Plauth as chief designer, but he was killed in a flying accident in 1927 and was succeeded by Ernst Zindel who designed the now famous three-engined Ju 52 – yet until retirement in 1932 it was Junkers who formulated every design.

3

With growing pressure for rearmament came collateral examination of methods of keeping peace in Western Europe. A meeting was held at the beginning of February between Sir John Simon as Secretary of State for Foreign Affairs, Pierre Etienne Flandin the French Prime Minister, and his predecessor Georges Laval now Foreign Minister. The Foreign Office announced: 'The British and French Ministers have been impressed by the special dangers to peace created by modern developments in the air and have given consideration to provision against these dangers by reciprocal regional agreements. The signatories would undertake to give immediate assistance of their air forces to whichever might be victim of unprovoked aerial aggression. The British and French Ministers on behalf of their Governments found themselves in agreement that a mutual arrangement of this kind would go far to operate as a deterrent, and they resolved to invite Italy, Germany, and Belgium to consider with them whether such a convention might be promptly negotiated. The Governments of France and the UK declare themselves ready to resume consultations without delay after receiving the replies of other interested Powers.' Through a breach of confidence by the French official newsagency this *communiqué* had been circulated to the world's newspapers eighteen hours before the British Press received official permission. 'No wonder British sympathies are so seldom with the French people,' said one editor.

The Times of 9th February reported Germany's view that, because her central situation in Europe rendered her important centres particularly vulnerable, a rigid Western European Air Convention was no safeguard as she must take into account all possible opponents in the East as well as the West, especially Russia, which in the German view was acting in close co-operation with France.

The Socialist British Prime Minister Ramsay MacDonald then played his long-favoured card of a Royal Commission on Armaments 'to consider and report upon the practicability and desirability, both nationally and internationally, of the adoption of a prohibition on private manufacture of, and trade in, arms and munitions of war, and the institution of a State monopoly of such manufacture and trade.' Fiercely questioned in the Commons, he said that the Commission as a rule would take evidence in public, but sessions *in camera* could be held if the chairman agreed. The Commission would have power to ask for, but not demand, papers relating to the proceedings of private firms and also documents in Government Departments at the discretion of the Government. Bitterly a journalist commented: 'The history of the members of the committee alone is enough to show that in their opinion the private manufacture of and trade in arms and munitions should be prohibited.'

MacDonald made no attempt to prevent official purchase from the USA of a Northrop, metal-skinned, long-range light bomber with 650 hp Wright Cyclone. Currently it was flying with British roundels at Martlesham for handling and performance assessment, as well as inspection by designers. Though C. G. Grey, the monocled, caustic editor of *The Aeroplane*, described it sweepingly as 'an obsolescent formula', it was remarkably similar to the drawings of the Fairey P 27/32 low-winger except for its trousered fixed undercarriage – but the shapely Fairey was a full year from completion, so interesting features could still be incorporated, and I gained the impression that Marcel Lobelle, its designer, was quick to take advantage. I was very impressed with the Northrop when I obtained a flight as passenger and briefly tried the dual control, finding it easy to fly. Here was the shape, and methods, of things to come.

If America had many successes, there were disasters too. On 12th February, a little after midnight, came SOS signals from their Zeppelin-type naval airship *Macon*, the largest in the world. She was returning to base after manoeuvres with the US Pacific Fleet. For 24 hours she had been violently battling in the turbulence of a 60 mph gale. Two hours later, searchlights of the battleship *Pennsylvania* picked out her dim shape on the sea and seven of her inflatable life-boats, eighty-one of the eighty-three on board having been saved; but the airship sank in fifty fathoms.

At the Court of Inquiry, Commander Wiley, captain of the airship, said that after an order for 'left rudder' the airship vibrated as it nosed into a solid bank of cloud with a curtain of rain beneath, passed through considerable turbulence as it lifted 2,700 feet, then suddenly dropped 1,000 feet. At that, the gasbags under the fin failed and the *Macon* tilted bow upward, with the stern heavily vibrating. He dropped trimming ballast and tried to put her down near the cruisers, but could not see the surface, and the airship touched stern first without forward motion.

Like the earlier ill-fated *Akron*, the *Macon* carried six fighter aeroplanes within the hull, and in these Pacific manoeuvres they had flown off and hooked on a number of times, thus proving they could be the eyes of the Fleet instead of using conventional carrier ships. The internal stowage necessitated unconventional hull structuring and had been criticized as a source of weakness, and so had an auxiliary control room in the lower vertical fin from which fighter launching was supervised. A typical Press comment was: 'This country requires no further proof that the aeroplane is the only type of air vehicle worth concentrating upon' – but at least the prospect of alighting on the open ocean and saving crew and passengers was much better than with a big flying-boat.

Three days after the *Macon* accident, one of four Short Singapore IIIs on their way to rearm No. 205 Squadron at Singapore crashed when endeavouring to avoid a mountainside hidden by thick cloud, catching a wingtip, turning over and bursting into flames, killing the nine crew. The Italian Regia Aeronautica had insisted the flight took a difficult S-turn corridor across the tip of Sicily instead of directly down the Straits of Messina which was a prohibited area. The Italians endeavoured to compensate for the tragedy with a processional funeral led by high-ranking officers and officials following the gun-carriages through streets crowded with onlookers; but there was only the normal military funeral for two other young Nimrod pilots who lost their lives that week when they crashed into each other off Malta. Analysis of RAF accidents of the past five years showed they were more prevalent among *ab initio* pilots on joining their first squadron as responsible but inexperienced officers immersed in the work of the Unit, and this was further complicated if posted overseas where climate and flying conditions were strange.

The path to safety with rotating wings was still being pursued by Juan Cierva. In a lecture to the Royal Aeronautical Society on *New Developments of the Autogiro* he revealed the jump-start capabilities of the C 30 fitted with an inclined rotor hinge enabling the blade to swing sideways as well as up and down. On revving up they moved backward and decreased the incidence, enabling them to spin to maximum rpm some 25 per cent faster than previously. On declutching they moved forward under the inertia forces of the whirling rotor and normal incidence was restored, generating sufficient lift to pull the machine off the ground and allow transition to forward flight and climb away. Cierva forecast that when fully developed the Autogiro would jump high enough to clear small buildings and trees a few yards ahead, and even 60 to 100 ft might by possible, though for practical purposes 20 ft was the greatest height needed. He was convinced the Autogiro could eventually compete with the aeroplane in speed, but the fundamental development as a safe, slow flying vehicle must be achieved before attacking performance since speed was not the only criterion of utility. As an example he showed a film of his very experienced pilot, Reggie Brie, taking off and landing on a small platform on the Italian cruiser *Fiume* stationary at moorings.

That early pioneer pilot, Lieut-Col J. T. C. Moore-Brabazon, MP, a keen sailing man, devised new use for an Autogiro rotor. With the co-operation of O. Reder, a Cierva engineer, he replaced the sails and spars of his Redwing class sloop with a boomed, turnable 10-ft mast carrying a laterally mounted light two-bladed, 18-ft rotor. Rashly their first sally was in a wind of 30 mph. After giving the rotor a starting push it swiftly speeded, causing alarming out-of-balance forces initially coinciding with the natural harmonics of the mast, but settled smoothly at 250 rpm, drawing the boat forward. Manoeuvring proved satisfactory, except that in tacking the rotor lost speed on coming into wind, though jybing presented no difficulty. In very light winds the standard Bermuda-rigged Redwings were faster, but in winds above 15 mph the performance was remarkable and exceptionally close to wind, though with risk of the blade-tips touching the water when heeled, or decapitating the crew on coming up to or leaving moorings – but 'Brab' remained optimistic of survival.

The increasing optimism of the major aviation firms echoed at the twenty-first AGM of D. Napier & Son Ltd, chaired by Sir Harold Snagge, KBE, who explained that important developments in air politics had made the company revise its estimates and postpone the idea of paying off preference capital which now might be wanted at short notice for legitimate employment in the Works. 'The enervating process of marking time has ceased and the outlook is promising. The company is engaged on a further substantial order from the Air Ministry for Dagger engines with which to equip a squadron of aeroplanes for the RAF.'

This had bearing on a contract promised to Westland for a Dagger-engined version of the Hawker Hart – for there was now only a trickle of Wallaces going through the shops. With the prospect of full production ahead, Sir Ernest Petter began to see a more rosy future and was considering the possibility of capitalizing on prospects.

Thanks largely to vigorous backing from Bob Graham, the RTO at Westland since 1929, supplemented by Fedden's advocacy to DTD and DOR of Teddy Petter's ability, Westland was invited to tender for a general purpose machine to specification A 39/34 as an Audax replacement. Visits were made to Army Co-operation squadrons to study operational problems in order to decide the formula best satisfying the requirements of piloting, maintenance, and arming. There was no unanimity among the pilots except a general indication that they needed unobstructed forward and downward view, preferably from a high-winger which must be easy to handle and have ability to land in small spaces. At least this confirmed Westland's predilection in view of their unique monoplane experience ranging from the parasol Widgeon and Wizard of the mid 1920s, to the Wessex, Witch, F 20/27 fighter, F 29/27 and ill-fated PV 7. Although the all-up weight of the Witch was similar to the targeted figure for the new design, its huge span of 61 ft and area of 534 sq ft was outmoded. Re-consideration led to a relatively high aspect ratio wing of 50-ft span and half the area of the Witch, compensating for increased wing loading with a full-span automatic leading-edge slot and relatively long, slotted trailing-edge flap interlinked so that its angle automatically increased as the nose slot extended.

Petter had been impressed by the simplicity of the strut-braced single spar and ply torsion nose of my small sailplane *Pegasus*, and despite his extensive study of R & Ms was unaware of the tension field theory of Herbert Wagner (of the Rohrbach Works) and his concept of a metal D-section torsion nose and single spar, although the theory had been published by the NACA some four years earlier, and the ply and spruce version was common practice in sailplanes. The relatively high aspect ratio plan-form which Petter adopted had similarities with *Pegasus*, but inboard taper became necessary to give better lateral view for pilot and gunner. Lest the single duralumin main spar and torsion-box seemed too daring it was supplemented with rear triangulations fixed to a false spar, and torsion was relieved by an additional strut, forming a vee with the main strut. Petter was determined to make his mark with novelty. Encouraged by W. C. Devereux, the tough director of High Duty Alloys, he made extensive use of extrusions, including a hollow, square-section duralumin beam bent like a hairpin to make a cantilever undercarriage carrying internally sprung Dowty wheels, and the apex of each lift strut assembly was attached by a massive alloy lug to the top of the hairpin a foot outboard of its forged mounting in the fuselage frame.

Competing with the novice Petter was the long-experienced Frank Barnwell with a Bristol low-winger, which was the antithesis of the Westland design, and had the same span, wing and tail plan-form, and fuselage design as his Type 133 fighter but eliminated the centre-section anhedral and had a fully retracting undercarriage. It was certainly cleaner than the Westland, and the estimated top speed was 20 mph greater. Because it was 1,000 lb lighter, the combination of split trailing-edge flap interlinked with wingtip automatic slots was expected to give similar STOL performance to the Petter design.

4

Early in March, Anthony Eden and Sir John Simon on behalf of the British Government visited Hitler who confirmed that the Deutscher Luftsportverband (DLV) had officially become the Reichsluftwaffe (soon known as the Luftwaffe) and that it equalled the RAF in strike power. Generalleutenant Göring, formerly Air Minister, was made C-in-C, and Generalmajor Wever became Chief of Air Staff, Erhard Milch continuing as Secretary of State for Air and nominally in control – but Göring began to appoint his World War I Luftwaffe friends to senior posts, and had under his command 20,000 men and 1,888 aircraft to which 200 were being added each month. Nevertheless, a few months later, in a great speech to the Reichstag, he made a plea not only for restoration of the prestige of the German nation but outlawry of bombing, gas, incendiaries and explosives outside any battle zone.

This knowledge of German strength was too late to alter the Air Estimates for 1935–36, so the increase of £3,685,500 in the gross total seemed alarmingly small. Said C. G. Grey: 'Thanks to Messrs MacDonald, Baldwin, Simon and Co, and their imbecile policy of disarmament, rather than to the succession of Socialist Governments which the ineptitude of the Conservative Party inflicted on this country, the RAF has been reduced to the state of a starving man who cannot eat a full meal because his stomach cannot assimilate it.' But more sober reflection led him to explain that had more money been allocated, the Air Council could not have spent it because 'there are not enough trained designers or workmen to produce a large number of aeroplanes; there are not enough trained pilots to fly them, nor mechanics to keep them in order; and there are not enough aerodromes to house them. Consequently we have to start at the beginning by equipping and manning aircraft factories, including motor factories, finding more instructors to train more pilots, and making more aerodromes on which sheds for the machines and quarters for the officers and men can be built. All that will have to be done gradually.'

Skilfully Sir Philip Sassoon, Under Secretary of State for Air, explained the complex figures of the Estimates to the Commons, saying: 'In past years I have found myself during these Debates in the unhappy position of a modern San Sebastian, assailed by arrows from all sides, but lacking the comfortable assurance that in due course I would reap the rewards of martyrdom. From one side came the shafts of those who thought the provisions inadequate; from the other barbed bolts of those who would abolish all armed forces.' He said that inaccurate figures had been bandied about and an unduly black picture painted of our present weakness. Much confusion had been caused by varying methods of different nations in stating their programmes. Nevertheless our country remained only fifth of the world's air power in first-line strength, outnumbered by Italy and the USA by a relatively small margin, but France had 1,650 first-line aircraft, and it was believed that Russia exceeded 2,000. Sassoon said the British position had parallels with the USA where an independent commission appointed by the President reported that: 'The past year has been marked by incessant attack from our own people, using every medium of publicity, upon the quality of American military aircraft. It has been alleged they are hopelessly inferior to corresponding machines of European origin. But it is an interesting commentary on the state of the public mind that these charges are similar to those currently made in the very countries held up to the USA as examples to imitate. The Press of Great Britain has rung with assertions of the remarkable qualities of American aircraft and the inability of the available British military types even to keep pace with American commercial machines. In France a section of the Press has debated furiously the rumoured inferiority of French military aircraft to those of most of the rest of the world.' After a telling pause Sassoon added: 'A strenuous programme of re-equipment is ahead and special steps are being taken to shorten the time in bringing new types into production. We can be confident that the British industry will rise to the occasion. To show what foreign nations think of our aircraft, there are twenty-nine countries using British aeroplanes and thirty-three have adopted British motors, resulting in a rise of 31 per cent in our aircraft exports last year.'

As to Home Defence, there would now be eleven new squadrons formed in 1935 and the equivalent of one and a half added to the FAA. By the end of 1935 there would be fifty-four Home Defence squadrons, including thirteen Auxiliaries. That left twenty-one squadrons to complete the present programme of seventy-five for Home Defence. Meanwhile the first-line strength was 890 machines in regular squadrons and 130 in non-regular, and this would rise to 1,170 in 1935 and 1,310 in 1936.

Civil aviation, he said, had also shown striking progress with the extension of the Imperial air services from India to Australia and a doubled weekly service to South Africa and Calcutta. But some £2 million would be required by Imperial Airways for a new fleet which would require improved ground organization and night-flying facilities. To attempt internal services in Britain on the scale of the USA was impossible. Even the huge expenditure of £28 million between 1927 and 1933 left American air transport unsound economically, and their aircraft industry had lost three-quarters of the amount invested by the public, 'some £95 million vanishing like gossamer in the sun.'

Of the score of members participating in the Debate only Winston Churchill was lambasted by the Press. 'His was about the worst speech he has ever made,' said one report. 'He started on the assumption that Germany is inevitably an enemy, and juggled with hypothetical figures of air strength today and next year and babbled about what are first-line machines and what are reserves. He talked a lot of nonsense about the distance of the German frontier from London, just as if London were all England. In fact his whole tone was summed up in one of his later sentences: "We are not faced with the prospect of a new war, but with something very like the possibility of the resumption of the war which ended in 1918."'

Certainly the insularity of the British public made the thought of trouble with Germany or Russia seem unreal to almost all. The trade depression was barely remembered. Mass production ensured a multitude of consumer goods at prices within reach of most people. A suit of clothes could be bought for 50*s*. Food was cheap. But though small cars had become commonplace most workmen still walked to their factories or used a bicycle, and the younger members sported second-hand motor-bikes. Yet the depressed areas in the North, and to a lesser extent in South Wales, remained. Rising incomes of those fully employed contrasted with the hopelessness of long idleness for those with neither initiative nor sufficient money to move into areas offering the chance of employment. Yet for nine-tenths of the population these were days of cheerful placidity, and it was more interesting to know that the Zoo had the first chimpanzee ever born in captivity than read of the revolution in Spain, or that Paraguay had withdrawn from the League of Nations, and even the news on 10th March that Germany had introduced compulsory military service caused no stir except to diehards of the London Clubs.

The Press preferred exciting news such as the crash of the Bristol low-wing fighter Type 133 on 8th March and how 'Jock' Campbell, the company's chief assistant pilot and senior instructor of their Flying School, had his foot trapped by the control column before he could parachute from his uncontrollably spinning machine. He had taken the fighter for a half-hour's handling during which he made a right-hand spin at 14,000 ft with undercarriage down. As the company naïvely explained: 'It was decided in the interests of young and inexperienced Service pilots to experiment at a safe height with the wheels down, and so subject the monoplane to manoeuvres for which it was not designed.' There was an air of familiarity about that excuse, but whatever the reason, a flat spin developed; controls were completely ineffective; the engine stopped from fuel starvation under centrifugal force. Only at 2,000 ft did Campbell finally get free, and the machine continued spinning until it crashed into a field at Longwell Green, east of Bristol. 'Probably several lives have been saved by this deliberate experiment, which must have cost either the Bristol Company or the Air Ministry a lot of money. And a lot more might well be spent on similar experiments,' happily concluded C. G. Grey.

On the previous day at Copenhagen, tall, fair-complexioned John Tranum, that thirty-five-year-old Dane who was one of the world's most famous parachutists, met his end. Originally a film stunt man, he turned to parachute demonstration in 1927, becoming the first to establish the falling speed of a man as 120 mph, and also pioneered delayed drops. It was in an attempt at beating his 1933 record that disaster came. He was taken to 25,000 ft by a Danish Army pilot with the intention of broadcasting his sensations with a small radio set as he dropped – but he suddenly signalled to the pilot to descend, and on landing was found unconscious and died in a coma. Within a few months Sir Alan Cobham's skilled parachutist, twenty-seven-year-old Ivor Price was killed during the National Aviation Day at Avro's aerodrome at Woodford – his canopy failing to extend after he had fallen into it.

There was also the death on 15th March of one of the best liked personalities in British sporting aviation, Capt Hugh Spooner, better known as 'Tony', brother of the late Winnie Spooner, best of British women pilots. A cavalryman, he learned to fly early in 1927, and soon proved the axiom that horsemen make good pilots, becoming chief instructor to the Montreal Flying Club in 1930 and joining Misr Airwork in Egypt two years later as superintendent of flying. He was piloting one of their Dragons when an engine cut, and with full load in the hot atmosphere he was forced to land in the sand and nosed over, killing two of the passengers and was so severely injured himself that he died the same night –

just another of the long stream of accidents to civilian and Air Force pilots, their crews and passengers.

Almost escaping notice because its significance was beyond imagination, was a veiled question to Sir Philip Sassoon, who discreetly replied that a special committee had been appointed 'to investigate the possibilities of countering air attacks by utilizing the recent progress of scientific invention.' He named the members as H. T. Tizard, Rector of Imperial College who was chairman; A. V. Hill, Foulerton Research Professor of the Royal Society; young P. M. S. Blackett, Professor of Physics at London University; and H. E. Wimperis, the DSR.

Tizard was prickly, tough, and unpredictable, but had greater catholicity of knowledge than any man in the aircraft industry. Although chairman of the ARC since 1933, supervising thirteen sub-committees and four panels, he maintained his duties as Rector of Imperial College. He was master of the penetrating word or phrase to sustain discussion. Sandy-haired, with pale tense face, and fierce eyes glinting through metal-rimmed spectacles, he could seem formidable, and therefore was not universally liked, yet almost everybody admired and respected him.

Sir Philip Sassoon explained that this group would be known as the Committee for the Scientific Survey of Air Defence (CSSAD), and that the Air Ministry intended to invite other distinguished scientists to contribute evidence if, as investigations proceeded, they considered its work would be facilitated. In due course it would report to the Committee of Imperial Defence.

This was the beginning of radar – originally intended as an investigation of what the popular Press called a 'death-ray'. Tizard asked Wimperis to find a scientist who could investigate the possibility of such beams, even if only to eliminate it from further discussion, and forty-three-year-old Robert Watson-Watt was appointed, who in turn set the mathematical problem to a twenty-eight-year-old assistant in the radio research station at Slough. As Sir Robert Watson-Watt later said: 'While I modestly believe myself to be father of radar, I am convinced that Arnold Frederick Wilkins has a unique claim to the title of mother of radar, for nobody did so much to give the embryonic war-winner its start in life.'

In the light of existing knowledge the pair deduced that a beam of sufficient power to harm human beings was impossible, but drew attention to the possibility of radio detection of aircraft because radio signals reflected back from the ionized layer above the earth could have the elapsed time measured, and it was known that three years earlier, engineers at a Post Office radio station between St Albans and Hatfield had found reception disturbed by reflection from aircraft taking off from nearby de Havillands. Within a week of Watson-Watt's report a meeting was arranged with Air Marshal Sir Hugh Dowding and Wimperis. A week later a Heyford bomber was flown along the centre-line of a 50-metre radio beam transmitted from the BBC Overseas Station at Daventry, and a cathode ray oscilloscope linked to a receiver displayed measurable displacement as the machine passed within eight miles of the beam, and grew stronger as it drew nearer. This was tremendously exciting. Here was realization of the most momentous discovery of a defensive measure. It was top secret. £10,000 was allotted for further experiments at Orfordness, a shingle beach on the Suffolk coast chosen for its isolation. Erection began of a 75 ft high aerial and installation of transmitters and receivers.

By comparison there was an absurd to-do over Leslie Hore-Belisha's latest ideas as Transport Minister, of placing a 30 mph limit on urban roads and whether to install lighted yellow beacons, like footballs on posts, to mark crossings where pedestrian had right of way. There was a storm of protest from car drivers.

If attacking aircraft were going to be readily located by radar then speed was essential to them for escape and attack. Engine power was always the limitation, but became less important if aircraft resistance could be drastically reduced. This led to a lecture on *Speed* before the Royal Institution by Melvill Jones, Professor of Aeronautics at Cambridge, who was known as 'Bones' – an absent-minded, loosely broad-shouldered, dark-haired man who lucidly explained the dynamics, but admitted that only experiment could show whether there was good streamline flow because now that aeroplanes had better streamlined form, skin friction was the dominating contribution to resistance. This was controlled by the thin sheet of flow in which the velocity parallel to the surface increased rapidly from zero to that of the main stream. 'When the flow within this boundary layer is smooth, in the sense that sheets of fluid slip over one another without mixing, we say that it is laminar; but when this condition breaks down, the flow within the boundary layer becomes turbulent, and the distance from the leading edge of a wing, or bow of a ship, at which this occurs is controlled by speed of flow, shape and roughness of the solid surface, and the turbulence in the fluid before it reaches the body.' The importance of discovering how far back the boundary layer could be persuaded to remain smooth was that frictional forces exerted by a laminar boundary layer on a high speed aeroplane were one-eighth those exerted by a turbulent layer, and might result in nearly a two-fold increase of speed, but certainly 20 per cent was readily practicable. Melvill Jones was also confident that stratospheric flying afforded unpredictable possibilities of speed and would eventually provide the principal means of long-distance transport. He concluded: 'I cannot believe that man will fail to take advantage of the arrangement, so conveniently provided by nature, whereby the air is thick where he has to land and take off, and thin a little way above where he can travel at high speed with small resistance and without inconveniencing those who remain on the ground. With luck I expect to see stratospheric flying becoming general for long-distance work at just about the time when I am myself preparing to operate in a more rarified, and, I trust, even more interesting medium.'

The twentieth annual report of the NACA showed that the USA had already tested a highly lacquered and polished aeroplane in their full-scale wind-tunnel, and found the drag coefficient was reduced by 0.001, which would add only 2.9 mph to its top speed of 200 mph, or alternatively 4 per cent less power could be used. That was far from Melvill Jones's predictions, but there was no evidence that the design even approached good aerodynamic shape in the first place. Because considerable money was available, American programmes were more comprehensive, so they now proposed a 500 mph wind-tunnel at Langley Field. Their fastest tunnel was investigating 'compressibility burble', for they had found that flow over an aerofoil broke down at very high speeds because the air was being compressed. Interference between wing and fuselage had also been intensively investigated, and the NACA concluded that the best position for a monoplane's wing was between the mid and shoulder arrangements, but the latter was nearly as efficient with suitable fillets, and though low-wing positions were unfavourable, careful filleting could bring the efficiency near the best high-wing combination. Perhaps it was a lucky coincidence that for bomb stowage facilities the twin-engined bombers being designed by Vickers, Armstrong Whitworth and Handley Page had an almost ideal wing location, which was confirmed by their wind-tunnel tests. Meanwhile Farnborough was hot on the trail of American full-scale experiments, having completed a 24-ft wide, open-jet wind-tunnel capable of testing a mounted Gauntlet fighter.

The almost forgotten soldier under whose charge had been the management and policy of Farnborough in its earliest days, Maj-Gen Sir Richard Matthews Ruck, died in mid March, aged eighty-four. It was while he was Director of Fortifications and Works in 1904 that he authorized a visit to the Wright brothers by the Factory's representative, Colonel Capper, and gave encouragement to construction of 'heavier-than-air machines'. From 1912 until 1919 he was chairman of the Aeronautical Society, not then Royal, but on the outbreak of War he also became chief engineer of the Central Force for Home Defence, then Major-General in charge of Administration until 1916, becoming in 1919 Vice-Chairman of the Air Inventions Committee, and a member of the Civil Aerial Transport Committee until 1928.

His death was followed on 29th March by that of Lieut-Col John Barrett-Lennard who, with Frederick Handley Page, was joint managing director of the Aircraft Disposal Co Ltd shortly after the Armistice, then joined the Board of Handley Page Transport Ltd and subsequently Imperial Airways Ltd. As his obituary stated: 'He was one of the small group of pioneers whose foresight combined with business ability made air travel a practical possibility. His knowledge and experience was of the greatest value to the administration of Imperial Airways, and his enthusiasm for air travel made him a keen and competent critic of its development.'

Demonstrating speedy air transport, there was a great flight from Australia in his Miles Falcon by dogged H. L. Brook of Harrogate, who at the end of the month returned from participation in the MacRobertson Race in 7 days 19 hours 15 minutes from Darwin, having taken 26 days 20 hours to reach it during the race. Though this did not beat the record of Scott and Campbell Black with their Comet it was faster by a day than Jim Mollison's 1931 official record, and 13 hours less than James Melrose's flight. Inevitably, the MacRobertson race had made such flights a surfeit, so there were no official receptions other than a small party by the York County Aviation Club which had taught Brook to fly at Sherburn-in-Elmet.

A month after his return came piquant Jean Batten from Australia, flying the Gipsy Moth with which a year earlier she took the Antipodean record from Amy Johnson. Again it was a tremendous flight. She left Port Darwin just after dawn on 12th April, with a strong south-east wind lifting a haze of red dust, and reached Timor in less than four hours after a startling moment when the engine suddenly choked 250 miles across the sea – but ever cautious, she had been flying at 6,000 ft, and after dropping through cloud at 3,000 ft the engine picked up and she continued to Rambang somewhat shaken but utterly determined. Day by day the capitals of the East saw her touch down to refuel and rest – Rangoon, Calcutta, Jodhpur, then Baghdad, Damascus, Nicosia, Athens; but *en route* to Rome on 25th April she had a forced landing at Foggia when an oil pipe broke. Next day she reached Marseilles where bad weather and repairs delayed her, but she managed to reach Dijon on the 28th. Next day she left at dawn for Croydon, but the cold was so intense that she landed at Abbeville to warm up, finally reaching Croydon at 2.15 that afternoon and was quietly greeted by a few reporters ready to make headline news that she was the first woman to fly to Australia and back; but even more interest was taken in the new fashion she set by wearing trousers instead of skirt, and being warmly clad with a fur coat and black helmet.

Amy Mollison, heroine of earlier flights, was out of the running as a rival. Money was running short. She was in open disagreement with her feckless, no less famous airman husband Jim Mollison, and Courtenay urged them to leave Grosvenor House. Jim in mid January sailed for the USA, and nobody knew his plans. Amy went to Madeira, saying 'Jim has been my undoing.' Nobody was interested in employing her, but she managed to secure the British agency for the Beechcraft back-staggered biplane, equivalent to Roy Chadwick's Avro Commodore.

5

Handley Page found the springtime days of April jubilant, for there was settlement in his favour on liability for income tax on £30,000 he received from the Royal Commission on Awards to Inventors and £9,000 paid by the American Government for use of his slots. The case had dragged on for years. Initially sued by the Tax Commissioners, he had been ruled liable because he owned all the shares of his company. Handley Page took this to the Court of Appeal who reversed the decision. In turn the Inspector of Taxes appealed to the Lords where Lord Tomlin decided that the sum of £30,483 must be treated as paid by the company to Mr Handley Page and not assessable to tax as an award. That went far to offset his dismay with the HP 46 deck lander which had suffered constant snags on controls, slot behaviour and stability. Perhaps Handley Page had thought his test pilot Tom Harry England rather fussy over these complaints, so he had stopped all work a year ago and insisted that the RAE must advise. They too thought the machine hopeless, and the following November the Air Ministry reluctantly agreed to pay the contract price of £36,000 – yet it had cost Handley Page twice as much in the five years since design inception. Now, early in April, it was sent to the RAE by road and broken up.

Meanwhile his general purpose low-wing HP 47, which had been flying since November 1933, was longitudinally unstable and had a big change of trim when the flaps were lowered, though wind-tunnel tests had not revealed this defect. On 20th April it was flown to Martlesham by Major Jim Cordes, the assistant pilot, in vain hope that it would be acceptable. This so often happened. A firm's pilot, knowing a prototype was unsatisfactory, was overruled by his directors in the belief that the RAF would be less critical. Inevitably it came back for correction.

It was to establish a closer relationship between model results and full-scale ones that the new big RAE tunnel had been built, powered by a 2,000 hp electric motor driving a huge six-bladed wooden suction fan. On 5th April the Secretary of State for Air, the Marquess of Londonderry, attended by representatives from the Air Ministry, War Office, Admiralty, scientific societies, and Air Attachés, pressed the switch which started the fan motor for the first official occasion. As a visitor said: 'Sometimes the national genius for compromise is to be deplored. There are times when only the best can be good enough, and this seems to be one of them. Our American friends of the NACA began with a 20-ft wind-tunnel years ago, but decided the job must be done properly and built a 70-ft tunnel for full-scale experiments.'

Except for wooden propellers there was little compromise about the silvery Bristol Type 142 which grey-haired Cyril Uwins, the chief test pilot, took for its first flight on 12th April at Filton. At last the Americans had met their match in metal-skinned aircraft of this size, and every current British fighter was rendered obsolete, for it was 50 mph faster than the Gauntlet. Immediately the Air Ministry urged the earliest possible attendance at Martlesham.

Despite the new outlook on speed, Geoffrey de Havilland, Britain's most famous pioneer designer, lecturing to the RAeS on 15th April, cautiously warned: 'To use slower types, with their lower first cost per seat, has been the only possibility for British commercial flying, because it has been slowly working up traffic without subsidy and pioneering once-a-week services over desolate regions without mail contracts. Under such conditions fast transport must take a long time to develop because operators feeling their way cannot incur the large capital needed until they can foresee the amount of traffic likely to follow. Natural development in the USA has been towards fast, highly loaded, all-metal machines because their operational circumstances reduce obsolescence to proper proportions.' But in Britain: 'For machines of moderate and small size, wood has been primarily adopted for reasons of cost and speed of production, and has been proved safe, light and reliable. The time has not yet arrived when use of metal is warranted by mass-production. Few doubt that metal or possibly synthetic materials will eventually be used universally, but wooden construction still has a long life before it.'

He described the racing Comet as a scale model for a commercial aeroplane, for it had the high index value of 9.6 ton-miles per gallon for a 500-mile range at its operational height of 10,000 ft. If scaled to four times the power, which was still quite moderate, the fuselage would become a commodious cabin without change of proportion. An increase in structure weight would be incurred, but something could be saved on the motors, so an outstandingly fast aeroplane could be developed along these lines. He agreed that current regulations for take-off served a useful purpose, but a fast, highly laden aeroplane of the future could well be licensed at some greater weight for particular aerodromes having long enough runways – which would be more economical than limiting the load of range to comply with requirements which otherwise adversely affected the usefulness of the aeroplane.

Summing up the respective merits of monoplane and biplane he said that a biplane was lighter and the monoplane faster, and taking off with a large load was easier with a biplane because light loading was more readily obtainable, so it was well suited for operation from poorer aerodromes and at great heights. 'During their evolution in the past twelve years, commercial aeroplanes have broken sharply from military types,' he said. 'Civil development aims at utmost performance for every hp developed, whereas the military method is performance regardless of the amount of power. The transport performance of small aeroplanes is about that of a motor-car, say 20 miles to the gallon with three aboard, but much faster, and they have surpassed the speed of their single-seat fighter ancestry which had twice the power. Such evolution has been steady and healthy, and is why there is more unsubsidized flying in this country than anywhere else. This progress in small and continuous steps without too great a commercial risk was our best and surest way of leading to high speed transport. When ground organization has been thoroughly developed and there is a solid foundation of mail and other contracts on which to base services, all the rest will follow, for only then can rapid progress take place in design and research for which there is no substitute.'

Geoffrey de Havilland was still the great and active inspirer of his company, despite delegation of administrative and engineering design responsibility to the confident, intellectual-looking, bespectacled Arthur Hagg. Key man in building the company's prosperity was Francis St Barbe, whose youthful, vigorous appearance belied his long service since the war days of Airco. De Havilland said of him: 'St Barbe can be very exacting and, at times, rather ruthless in support of anything which he feels requires drastic action. His genius is shown in his business ability and in the world-wide sales organization he has built up.' He also had boundless energy and enthusiasm, as well as great judgment of men, yet he was only interested in design and construction as far as it affected the saleability of a product. Appositely, St Barbe said of de Havilland: 'Having chosen the team, he had complete reliance upon it and did not interfere, even with things which he did not quite believe in or did not understand – such as certain sales functions. In the early days there was no security – only faith. We were all friends at all levels; there was no sense of grading.' An executive added: 'D.H. was the same to all at all levels, whether an Air Marshal or labourer – courtly, natural and disarming; he made you feel completely relaxed.' Yet seen at an air meeting, this tall and distinctive man with down-turned brim to his trilby hat seemed remote and isolated, guarded by his intimate coterie of Walker, Hearle, St Barbe, with their test pilot Hubert Broad sometimes in more animated attendance.

At the RAeS Garden Party on 5th May many of the 1,600 visitors arrived by air, and their parked machines lined the southern boundary of Fairey's aerodrome in the open countryside of Heathrow near Harmondsworth. Close to the visitors' enclosure fronting the big flight shed were triple rows of demonstration aircraft ranging from the formidable Fairey Hendon night bomber, with its latest form of 'glass-house' cockpit enclosure and gun cupola, down to a novel Baynes auxiliary sailplane with Carden engine which, with propeller, could be retracted into the fuselage. Less obvious was the new Fairey Prince engine installed in a Fairey Fox. *The Aeroplane* commented: 'In throwing their party as the first occasion of this year, the Society have rather wiped the eye of everybody else in the aeronautical world, for most of the forthcoming season's exhibits have now been displayed in the best possible weather and most delightful surroundings.' There was all the usual fun of the fair in the form of displays by the industry's test pilots in their latest aircraft, as well as the serenely beautiful arrival of Eric Collins in his Rhönadler sailplane after a tow to 2,000 ft above Reading – but essentially this was a social occasion of chatter and tea and meeting people.

6

The promise of better times, after the economic depression of the early '30s, seemed crystallized throughout the Empire by celebrations on 6th May on the twenty-fifth anniversary of the accession of King George V, the bearded emblem of Imperial Majesty, and tall, dignified Queen Mary. Every town and village had its bunting, celebration banquets and concerts, Silver Jubilee tree plantings, and Jubilee holiday. There were Jubilee stamps and Jubilee newspaper supplements of silver-gilt. There was even a new London and North Eastern railway express named *Silver Jubilee*, which by travelling at 112 mph broke the steam-train speed record. And down the Mall, densely lined with cheering people behind the ranked Household Regiments, and onwards through the streets of London gaily decorated and equally crowded, the King in Field Marshal's panoply drove with the Queen in the open State Coach to St Paul's for a Service of Thanksgiving attended by all the Royal Family and the nation's dignitaries and nobility. Later they toured the East End where every street also had its Union Jacks and bunting and cheering pavement crowds. In his diary that night the King wrote: 'There were the greatest number of people in the streets that I have ever seen in my life.'

The grave of Karl Marx might be enshrined at Highgate, but popular support, even among the poorest, was for a King as the Father figure, perhaps a little remote, but guardian of his people. Certainly Lady Houston of Schneider fame was moved by the occasion and wrote to the Secretary of State for Air: 'To commemorate the Jubilee of the King, and in view of the dire necessity of the present moment, Lady Houston begs to renew her offer of £200,000 for the Air Defence of London, the capital of the King's Empire, which she believes is now utterly unprotected.'

C. G. Grey added to the thanksgiving a little homily: 'Happily this Jubilee year seems to have brought with it a change of spirit at the Air Ministry. King George is not likely to see any aeroplane of startling performance at his great Jubilee Review at Mildenhall and Duxford on 6th July, but before the next year of his reign is over the spearhead of his Fighting Forces may have the beginning of adequate performance in its equipment, besides having that mass of obsolescent aeroplanes at present being ordered. And we all hope that His Majesty may live long to see his Royal Air Force leading the world in every way, as Queen Victoria saw her Navy ruling the sea.'

On that week-end of the Jubilee, Ramsay MacDonald resigned as Prime Minister on the grounds of failing health and became Lord President of the Council. Stanley Baldwin took office in his place. Sir John Simon was replaced by Sir Samuel Hoare as Minister for Foreign Affairs, Simon becoming Minister for Home Affairs. Said a commentator: 'After the circuitous lawyer-like methods of Sir John Simon, the Chancelleries of Europe will doubtless be completely put off by having to deal with a plain, straightforward, English gentleman of indisputable Quaker descent.' Among other changes, Lord Londonderry was replaced by fifty-one-year-old Sir Philip Cunliffe-Lister as Secretary of State for Air, and became Lord Privy Seal. It was also erroneously rumoured that Lord Trenchard was to take charge of RAF expansion because he had just retired as Commissioner of the Metropolitan Police and was succeeded by Air Vice-Marshal Sir Philip Game.

The USA was also making new dispositions. Recently a War Department special committee, known as the Baker Board, had investigated the relative efficiency of the Army Air Corps for peace or war. As a result a General Headquarters Air Force administration was established to develop and utilize the mobility and range of air power. This was followed by a Federal Aviation Commission to consider the entire field of American aviation activities, and their report stated that because of rapid expansion and modernization of world-wide military air forces the United States programme made in 1926 was no longer adequate, and that nothing short of radical change in the international situation should interfere with the Navy's programme of 1,910 aeroplanes by 1941 and the Army's programme of 2,320 by 1938. The Commission leant towards formation of an independent air arm but considered such an upheaval would be uneconomic at present although it could solve 'many of the problems of co-ordination, economy of supply, and efficiency of personnel now troubling the United States.' There was a riposte from Admiral Standley, Chief of Naval Operations, who stated to the House of Representatives Naval Committee on Consolidation of Air Forces: 'Britain has virtually no air force in the Navy. I have just been there and I know. Operations from her aircraft-carriers are so elementary in type and character that she is now doing some of the things we did ten years ago' – but as one reporter said: 'I assume Admiral Standley did more talking than observing.'

The less ambitious British problem was to ensure equality with Germany's first-line aircraft. Certainly the *Daily Telegraph* diplomatic correspondent was convinced that the expansion of the German air force 'is likely to necessitate radical changes in the British Government's latest Air Force expansion programme.' That more was suspected than revealed in Parliament was indicated by the reply of the Secretary of State for Home Affairs in answer to the indefatigable Mr Mander: 'In accordance with the policy foreshadowed by my Rt Hon friend the Lord President of the Council on 30th July last year, a special department of the Home Office, under an assistant Under-Secretary of State, is being formed to act as the channel for communication to local authorities in England and Wales of the measures which it will be necessary for them to take for the purpose of organizing local services for safeguarding the civil population against the effects of air attack.' It was even significant that the Air Ministry issued a notice warning aviators that: 'Official notification has been received that UK aircraft have recently infringed regulations when flying over German territory, and pilots are warned that severe measures are likely to be taken in the case of any further infringements.' Of course the reciprocal applied to German pilots, for there were many forbidden areas in Great Britain, particularly where there were naval stores, docks, and secret establishments – yet any high flying foreign aircraft could photograph them unobserved using infra-red cameras.

Those concerned with a future war were visualizing many problems beyond the more straightforward question of the supply of guns, ammunition, and bombs. As Lieut-Col Outram said at the tenth annual AID dinner early in May: 'We are about to face the most strenuous twelve months of the AID's existence since the end of the War. We have added 100 men in a year and total 632 to-day but cannot go on expanding at that rate, for staff must be added slowly in order to train them properly. There is also the problem of the industry competing with the Directorate for the few good men available.' In fact aircraft firms were finding the greatest difficulty in getting the right kind of draughtsmen, works staff, and even workmen. There would have to be training schemes if production greatly increased, and that required time. More factory space would be required. Other trades must be brought into airframe and aero engine manufacture. Major Bulman, DTD's engine specialist, recommended to AMRD that car firms should be selected for this purpose, and that it would mean enormous outlay on machine tools, jigs, and gauges, which would have to be installed and functioning within two years of the start of any real output.

War was a romantic recollection to many on reading the headlined news that the eccentric, mysterious, forty-four-year-old T. E. Lawrence, one-time leader of the Arab Revolt against Turkish rule, had died on 19th May in Bovington Camp hospital, Dorset, after a crash with his powerful Brough Superior motor-bike when endeavouring to avoid a bicycling butcher's boy near his rhododendron-enclosed cottage at Clouds Hill not far from the Camp. Few people were more famous and no one, except royalty, Ministers of the Crown and great public figures, attained the five-column obituary which *The Times* gave him. *Revolt in the Desert* and *Seven Pillars of Wisdom* were world famous, yet he seemed an insignificant little man until one saw the power of his blue eyes and felt the strength of his personality. An admirer wrote: 'One of the most entertaining experiences was to see Lawrence in his little RAF uniform and Service boots clattering down the marble halls of Sir Philip Sassoon's beautiful house at Lympne and thereafter hold forth to an assembly of the most brilliant brains in the country.' But his task as an aircraftsman in India, and later in the Pathan country, was never fully revealed. Of his later work the *Evening Standard* published his own memoir: 'I was fortunate to share in a little revolution in boat design. When I went into the RAF in 1929, all boats were round bottomed Admiralty design, and progressed by pushing their own bulk of water aside. Now in 1935 we have chosen, selected, or derived our own types which have three times the speed, less weight, less cost, more room, more safety, more seaworthiness. As speed increases they rise and run over the waves, and they cannot roll, nor pitch. Pundits met them with fierce hostility; all the RAF sailors, and all the Navy, said they would break, sink, wear out, be unmanageable – but to-day the War Office is refitting Coast defences entirely with boats of our model. It has been five years of intense and co-ordinated progress.' Perhaps he should have mentioned that these speed-boats were basically devised by Hubert Scott-Paine who had founded the British Power Boat Co at Southampton after his quarrel with Commander James Bird over management of Supermarine Aviation Works Ltd.

Lawrence was buried on Empire Air Day, 23rd May – but despite an overcast sky and cold north-east wind, it was also a day of thrills for the air-minded. All over the country, civil aerodromes and RAF stations were opened to the public who came in thousands. Air Commodore Chamier, Secretary-General of the Air League, had every reason to feel justified with the success of his idea. 'One of the most striking things was the enormous trouble taken by all ranks to give their visitors an interesting and entertaining afternoon,' wrote a reporter. 'Not only were there static displays of aircraft, engines, and maintenance processes, but also flying displays by every squadron, whether fighters, reconnaissance aircraft, bombers, or flying-boats. Some aircraft factories also welcomed visitors, and at Hatfield there was great interest in an almost completed Comet for the French Government.' At Martlesham even the secret Handley Page HP 47 for the G.P. competition was on display, and at nearby Felixstowe, all the latest flying-boats were there: the four-engined Short Singapore III production version, and its stable mate the Knuckleduster monoplane making an eye-catching contrast with the equivalent biplane Saunders-Roe A 27 London, of which seven had been ordered, and the Vickers-Supermarine Stranraer which had not long ago made its first flight and was quickly a favourite. There was also its slightly earlier sister, the two-bay R 20/31 Scapa of which thirteen were being built. Evidently marine pilots of the RAF remained faithful to the Supermarine Southampton tradition which had nurtured them.

Methods of reducing the time between issue of specifications and delivery of prototypes were the subject of an Air Ministry Press Conference on Empire Air Day. Firms often took twice as long as their initial estimate, and the disaster of Handley Page's HP 46 biplane showed that eighteen months instead of a few weeks might easily be consumed in testing. Service trials took at least six months before a decision was reached, and if the prototype was accepted a development order was placed for enough machines to form a preliminary squadron so that modifications could be incorporated in early quantity production. The new procedure involved publication of outline operational requirements some months before issue of the specification so that designers could make feasibility studies quickly alterable to suit the final specification which would contain fewer details than previously and there would be a freer hand in choosing solutions. A 'mock-up conference' was to follow as quickly as possible, but though Air Ministry specialists would offer opinions, these would not be enforced. Martlesham tests were to be limited, and armament trials postponed until after the preliminary handling assessment. Service trials would be shortened, interim production abolished and full production risked so that complete jigging could be made. The scheme was welcomed by constructors, but with reservations that the testing time might even increase due to the greater complexity of equipment and systems, nor was there hope that engine snags could be more quickly overcome during development of new powerplants.

7

On 22nd May Lord Londonderry in the Lords and Baldwin in the Commons announced the Government's revised schemes for expansion of the RAF. By 31st March, 1937, the strength at home, irrespective of the FAA, would be 1,500 first-line machines. This would require an additional 22,500 personnel, including 2,500 pilots to add to the existing 2,700, together with 400 trainees and a Reserve of 1,200. Under the new programme seventy one new Home Defence squadrons would be formed in the current and next financial years instead of twenty two under the previous programme, and the number of stations would be increased from the envisaged eighteen to thirty one. His lordship stressed that 'risks will now be taken with money because every machine is obsolete from the moment of its adoption.' In the Commons, Baldwin added the rider: 'The Government is determined that in the necessary efforts of the next two years there shall be no profiteering,' and explained they were in consultation with Lord Weir who so successfully had controlled aircraft production in the latter part of the Great War.

Baldwin had used a phrase 'collective security', saying: 'Today we have Russia in the League of Nations; we have had Germany in, but we have lost her, I hope only for a time; the United States is still outside. And Japan, which from the beginning was a member of the League, has now left and I see no immediate prospect of her return to it.' Bitterly C. G. Grey commented: 'So now you see where we are. We are in a League of Nations consisting chiefly of Russia and France – the two nations who like one another most and who like us least.'

For the pioneer members of the SBAC the rearmament programme was manna from heaven, justified by their years of devotion and struggle before 1914 and throughout the post-war decade-and-a-half of lean times. Yet they were not merely profit seekers, even though they had become business men on whom great numbers of employees were now dependent. Taking office as chairman of the SBAC was Sir Robert McLean in succession to Herbert J. Thomas of Bristols, who had held that post for the past two years. Sir Robert, on retiring from twenty-two years of Government Service with Indian Railways, had joined Vickers in 1927, and since 1928 had been managing director of Vickers (Aviation) Ltd. He was keen, shrewd, and forcefully determined that any viewpoints he held should be seriously regarded not only by his staff, but also by his co-directors and the Air Ministry.

One expected result of increased orders was a steep rise in share prices 'never before attained in the annals of the companies concerned' despite warnings that investors must be prepared for at least a temporary low yield. As a survey of company accounts revealed, factory extensions and expenditure on machinery, tooling, and design, coupled with the unavoidably long time between production outlay and receipt of payments, involved big ploughing back of profits and maintenance of a high ratio of liquid resources. Concurrently Handley Page Ltd. held its AGM, and the chairman, still Mr S. R. Worley, ponderously assured shareholders that 'Handley Page Ltd intends to place the facilities of their factory at the disposal of the government in the present emergency in so far as they might care to avail themselves of them', and meanwhile declared a 10 per cent dividend less tax and 5 per cent bonus less tax. To ensure the most up-to-date manufacturing techniques, George Volkert, the chief designer, was sent to the USA to study production methods at the Curtiss and Boeing factories, and Handley Page appointed Dr Gustav Lachmann as acting chief of design.

To many enthusiasts it seemed that the golden age of sporting flying ended with the announcement by de Havillands at the beginning of June that the open-cockpit Gipsy Moth would no longer be manufactured. In the ten years since the first 60 hp Moth had been introduced as the new look in light aircraft, some 2,000 had been built, including Tiger Moths. For an uncountable host of *ab initio* pilots its open cockpit had afforded the first thrilling introduction to the joys of flying and the vast panoramas of the countries and seas of the world.

Nothing could really replace it, though the Hornet Moth during the past year had been satisfactorily developed and was now in modest production with mainplanes of even more pronounced taper than the experimental sets used for the three prototypes.

More importantly the Vickers Type 287 private venture G 4/31 monoplane of 74 ft 7 in span, made its first flight at Brooklands in the hands of 'Mutt' Summers on 19th June. Three weeks earlier Air Vice-Marshal Sir Hugh Dowding and Air Commodore Verney had visited Martlesham to inspect the other seven contenders – the Vickers Type 253 semi-geodetic biplane, the Clarke-designed Parnall biplane with Pegasus III, the second version of the Fairey biplane now with Tiger IV engine, the Blackburn B-7 biplane based on the Shark with Tiger IV, the HP 47 monoplane, and the Pegasus-powered Hawker PV 4 derivative of the Hart, which had arrived only three days earlier. The A & AEE pilots and technical staff still recommended the Westland PV 7, despite its crash, but this was rejected by the Air Staff. The Vickers Type 253 with Pegasus X was agreed next best, followed by the Fairey TSR II although not actually in the competition but considered suitable as the result of handling trials made between May and October the previous year. The HP 47 was next, and fourth was the Blackburn B-7, but the Hawker was considered too small though handling was faultless, and it was eventually purchased by the Air Ministry for use by Bristol on engine development. In due course Vickers received an order for 150 of their biplanes – but Summers, backed by Rex Pierson, their massive chief designer, was confident that the monoplane was greatly superior, and after a few more confirmatory flights Sir Robert McLean pressed the Air Ministry to cancel the biplane order and replace it with the monoplane; but decisions could not be made as quickly as that despite new impetus to save time by cutting corners.

Two days before the flight of the Vickers monoplane, Edgar Percival, flying his Gull, made a remarkable flight of 2,300 miles in a day, having left Gravesend, where his new factory was located, in the darkness of 1.30 am. By 8.40 he was at Oran in time for breakfast. After a stay of two hours twenty minutes he left for Croydon, arriving at 6.20 that evening, the double journey having taken fourteen hours thirty minutes at an average of almost 160 mph. As one approving report stated: 'It demonstrates that a Percival Gull, designed three years ago, is a touring vehicle which treats continents as a car treats counties.' It also indicated that the day of those daring the unconquered air in uncomfortably bulky clothing and flying helmets was past, for Percival emerged from his cabin aeroplane in a smart double-breasted suit, with soft brown trilby jauntily on his head. Even Geoffrey de Havilland had come to this point of view, judging by his concentration on the enclosed Hornet Moth. Nevertheless there were still many who preferred the heady joys of the open cockpit and its unobstructed view. The gay procession of week-end flying meetings retained a big quota of such machines, and the RAF still delighted in their open-cockpit Bulldogs, Gauntlets, Furies, Harts, and even viewed with some dismay the new Mk II Heyford with its glass-sided lid. Certainly instructors felt it essential that *ab initio* pilots should not be immured from the sensation of actual contact with the air.

George Bulman, *maestro* of test pilots, went further, for habitually he flew helmetless, his bald pate gleaming. At the Hawker AGM on 25th June he and Camm, together with Herbert Chandler, secretary of the company for the past eleven years, Herbert Keene Jones in charge of overseas sales, and old-timer Reginald William Sutton, responsible for Air Ministry contracts, were elected to the Board – the first occasion that either a designer or a test pilot became a director of his employer's company.

Nevertheless, members of the SBAC were startled by the subsequent announcement that Hawker Aircraft and Armstrong Siddeley were to merge, forming a new company, Hawker-Siddeley Development Co Ltd, of which the directors were T. O. M. Sopwith, Fred Sigrist, F. S. Spriggs, and Philip D. Hill, financial director of Covent Garden Properties Ltd, Timothy Whites Ltd, Beechams Pills Ltd, Eagle Star and British Dominions Insurance Co Ltd, and others. Capital was £2 million divided in one million 5 per cent cumulative preference shares of £1 and four million ordinary shares at 5*s*. each, but the public issue was limited to preference shares and one million ordinary 5*s*. shares at 15*s*. each. The Armstrong Siddeley Development Co Ltd was to be paid £1 million, and Sir John Siddeley would retire. As Sir Thomas Sopwith told me many years later: 'I was very conscious all that week-end of the £1 million cheque I had in my pocket. Never before had I such a sum in my possession.'

As with the earlier Gloster takeover, there was the burning question as to whether the Armstrong Whitworth designer John Lloyd would be subservient to the less experienced but recently successful Sydney Camm. Later Lloyd told me: 'About eight o'clock on a nice hot summer's Saturday morning I was asked to go at once to see Sir John at his home at Crackley in Kenilworth. As we walked around his garden he asked me if I had seen the morning papers. When I said I had not he told me he had sold out to Hawkers. I was so stunned that when he asked if I thought he had done the right thing I could only say "Yes, Sir John, if you have sold at the right price." He said "Don't worry, I am quite satisfied with the deal." My reaction that week-end was of great uncertainty of the future because I had never met Frank Spriggs who in the future would be my boss – though after meeting him I was quite happy, and he subsequently proved to be just the opposite to that ruthless autocratic J.D.' Virtually the three design teams of Hawker, Gloster, and Armstrongs continued individually, and each company retained its personality and had its own Board.

On the strength of his handsomely repaid share capital, Sir John Siddeley generously offered £10,000 to Cambridge University for aeronautical research. This was immediately opposed by a University clique who delivered a notice of *non placet* stating: 'It is clear that the money is intended to supplement grants from the Air Ministry and that the primary objective is to subsidize research of a military character and of pecuniary value to armament manufacturers. There is within the University, in this country and in the world generally, a large and increasing number of men and women who view with horror the increasing production of armaments and realize that unless immediate steps are taken to prevent it, war will sooner or later be declared and aerial attacks launched upon civilian populations. We are fully alive to the value of aviation for humane and useful purposes, and we support wholeheartedly any scheme for fostering aeronautical research in this University, but in our opinion the gift of Sir John Siddeley should be accepted only with the express proviso that the Aeronautical Department should use it solely for fundamental research and for the improvement of civil aviation.' This led to the refusal, at any rate for the time being, of Sir John's offer.

8

Hendon and its great RAF Display overshadowed all other matters on 29th June. Under the guidance of Air Commodore B. C. H. Drew, secretary of the Display for the past five years, there was the usual faultless timetable for the arrival and parking of the squadrons as well as for the public. However, the ever critical C. G. Grey said: 'Apart from the beautiful precision of flying the show was dull. There was nothing new or startling among the machines. We have seen Harts and Audaxes (described by a flight sergeant as "A 'Art with a 'ook") and Demons and Ospreys till we know them backwards, and the Hind and the Hardy look the same. The Autogiro has ceased to be a comic turn and has become a serious conveyance for peace-loving senior officers as a change from their official Staff car.' Others were more glowing. Mrs C. M. McAlery, C. G. Grey's great authority on the RAF and WRAF, described the Display as 'marking a definite step in the progress of Air Service pageantry.' But people missed the usual set piece, for which an impressive fly-past of seventy-seven aircraft including six of the latest flying-boats was substituted. Perhaps the biggest thrill was in the preliminary events when a demonstration of an instructor and pupil flying an Avro Tutor in crazily clambering flight ended suddenly by hitting the ground, and the machine collapsed with drooping wing – but the fire tender was there in seconds, covering the stricken Avro with fire foam, and the crew were uninjured.

As always the New Type Park interested technical members of the industry, but only seven of the fourteen aircraft on show were flown – the newly named Gloster Gladiator, Hawker Pegasus-powered G 4/31, the Handley Page HP 47, Vickers GP biplane carrying a torpedo, Avro 652 of which 150 were in production as the Anson, the massive AW 23 bomber-transport which first flew on 4th June, and the Supermarine Seagull V, now renamed Walrus – all flown by Martlesham pilots. Those that remained in the fenced enclosure were the new Vickers GP monoplane of which a reporter said it would 'take all the gilt off the gingerbread of the Vickers biplane'; the handsome Bristol Type 130 twin-engined high-wing bomber-transport which had been taken by Uwins on its maiden flight only a week earlier – but the HP 51 rival, though listed, was not there; the Westland Pterodactyl Mk V which could not be flown because both the Westland pilot and the selected RAF pilot, Flt Lieut Stainforth, were victims of influenza; the hitherto scarlet-painted Comet, winner of the MacRobertson Race, was there in silver and RAF tricolour; and finally there should have been the Queen Bee radio-controlled version of the Tiger Moth, developed by the RAE as a flying target for naval gunnery, which was alleged to have arrived pilotless and then 'somebody at Farnborough must have pressed the necessary buttons for it taxied out of the Park and flew away!'

Two days later numerous demonstration aircraft and several prototypes arrived at Hendon for the SBAC Show. The Fairey multi-gun Fantome was displayed by Chris Staniland with his usual high artistry and verve, rivalled by George Bulman flying the Hawker Super Fury which had its fuel system arranged for inverted flying, enabling him to make steep inverted climbs and upward rolls. Apart from the Blackburn Shark, most demonstrators were strictly civilian such as British Aircraft's Eagle cabin monoplane with retractable undercarriage, and their Pobjoy-powered Swallow: others were the twin Pobjoy-powered Short Scion; Saunders-Roe Cloud amphibian; Airspeed Envoy and Courier; the new Monospar ST-25 five-seater; the Percival Gull and latest version of the Mew Gull with landing flaps and longer fuselage. Within the nearest hangar were displays of British equipment to be duly admired by representatives of the forty-seven visitor countries. Some commented on the absence of Miles Aircraft, now the second largest producer of light aircraft, but they still were not members of SBAC and the rules were strict.

For those with consciences that summer, a sample house-to-house poll, known as the Peace Ballot, was conducted by the League of Nations' Union and revealed that 6¾ million were for collective military measures compared with two million against, and a million abstained. The view seemed supported by the enormous interest in the King's Jubilee Review of the RAF on 6th July at Mildenhall, where 350 aircraft were marshalled on the aerodrome in a great eight-rank chevron – but it was an Air Force of obsolete biplanes, for not even the Gauntlet, fastest of all at 231 mph, had yet reached the squadrons.

His Majesty, in the uniform of a Marshal of the RAF, arrived at 11.20 hours by road, accompanied by Air Chief Marshal the Prince of Wales, Air Vice-Marshal the Duke of York, and Sir Philip Cunliffe-Lister, and they were received by the Lord Lieutenant of Suffolk. After the Secretary of State had presented members of the Air Council and Air Vice-Marshal P. L. H. Playfair, who was in command of flying, the King inspected the guard of honour, and then, accompanied by the C-in-C and an equerry, toured the ranks of aeroplanes in an open Rolls-Royce, followed by another containing the two Princes and Air Vice-Marshal Playfair, while photographers in key positions photographed proceedings for newspapers eagerly scanned by an admiring public next morning. The King was then driven to nearby Duxford, where he was joined by the Queen. A spectator near the Royal Box wrote: 'The Queen looked her usual stately and charming self. And the King looked so well in his RAF uniform that one forgot it was the first time he had worn it. And the Duke and Duchess of York bowed sweetly to the people in the airmen's enclosure, and the Prince of Wales looked happy as he drove past the Press pen away from the official guests!'

When the massed sqadrons from Mildenhall duly hove in sight at the appointed time and flew overhead, they gave little conviction of air power: twenty Heyfords flying at 98 mph, fifty-four Harts four miles astern doing 115 mph, and four miles behind them were eighty-one mixed fighters also doing 115 mph. Said a disconsolate onlooker: 'I sat there longing for something startling to happen, like that incident which shook up and revolutionized the Royal Navy at Queen Victoria's Jubilee Review in 1897 when the little *Turbinia* came dashing in at enormous speed among the Fleet.' But not even the Comet appeared.

Tizard was determined that Britain should do better, and convened a three-day conference at the City and Guilds College, South Kensington, so that technicians of the aircraft industry, Air Ministry, and RAF could discuss common problems. Moore-Brabazon, in shirt sleeves and smoking a cigarette in a long holder, was chairman. 'Two responsible RAF officers had disrobed down to their braces. A couple more of high eminence were in the lightest of flannels. Those not smoking cigarettes were smoking pipes. Practically everybody of importance on the technical side of the aircraft trade and Air Ministry was there, so the atmosphere was such as to inspire confidence as long as one was able to breathe it.'

Tizard warned that industry relied too heavily on government research and must build up bigger teams and facilities. Though every major constructor except Hawker and Blackburn had a specialized aerodynamic section and a low-speed wind-tunnel which provided not unreasonable approximation to control balancing and stability, the snag was that firms were constantly referring problems to the RAE and NPL for solutions, and then complaining that they were kept waiting for results for months or even years. As one speaker said, what was wanted was a think department on a particular line of research necessary to satisfy RAF future requirements. 'Whatever the end results, there was no doubt the conference has been definitely a good thing,' concluded *The Aeroplane*. 'Everybody in aeronautics seemed to see everybody else in the game in a light in which they had never seen them before, and everybody certainly came away feeling encouraged by the freedom of speech in the discussion and by the obvious willingness of the official technicians to investigate new ideas.'

That July every householder received from the Home Office a circular explaining the action to be taken in the event of a bombing war. 'Provision must be made to minimize the consequences of such attacks if ever they should be delivered. The use of poison gas in war is forbidden by the Geneva Gas Protocol of 1925, but it is a possibility which cannot be disregarded, and plans must therefore include measures for protection against such attacks. Respirators and protective clothing will be issued to persons employed on air raid precautionary services whose duties require them to work in gassed areas. A civilian Gas School is to be established to train instructors who will be competent to give local training in their own districts.' Construction of bomb-proof shelters on an extensive scale was considered impracticable, but the circular stated that arrangements made by local authorities would be described in a series of memoranda from the Home Office, and a provisional summary gave the principal types of action required under the headings 'Action to be taken by the Government'; 'Action to be taken by Local Authorities'; 'Action by Employers in Industry and Commerce,' and 'Action by Householders and Members of the Public.' Newspapers and magazines featured these matters for a week. In a fortnight people had forgotten about it, for the average man still firmly believed there could never be war, and that politicians were merely creating a furore as a matter of personal publicity and maybe to increase their dividends from the aircraft and armaments industries. Nevertheless those living near Chatham and Sheerness Dockyards proved very willing to co-operate when it was arranged that all lights should be blacked out so that night-flying RAF aircraft could judge the increased difficulty of finding such targets.

A great Naval Review followed on 16th July. As the aeronautical artist Leonard Bridgman recorded: 'An early morning passage through the Fleet revealed a veritable armada. Warships were drawn up in ten lines, the longest six miles long, and on either side of the area were liners, merchant ships, fishing boats and yachts moored for the Review. Outside them were five areas for small boats. The whole of Spithead was alive with movement, busy ships' pinnaces and motor-boats, private craft of all kinds rubbernecking, tripper steamers taking crowds down the lines before the area closed at one o'clock, and liners of several nations packed with spectators passing to take up their moorings in readiness for the afternoon. Overhead came a continuous stream of civil aeroplanes, some on regular services from Portsmouth, others making tours of Spithead, but all well outside the Fleet.

'At 13.00 hours a gun was fired to clear the Review area. Shortly after 14.00 hours the two yellow funnels and three high masts of the *Victoria and Albert* could be seen coming out of Portsmouth Harbour. After the King had received the Flag Officers, Board of Admiralty, and representatives of the Merchant Navy and fishing fleets, the Review began. The *Victoria and Albert* left her moorings ahead of the *Queen Elizabeth* at 16.00 hours, and, preceded by the Trinity House yacht *Patricia*, followed by the Admiralty yacht *Enchantress* and other vessels carrying distinguished visitors, began her hour-and-a-quarter tour of the lines.'

At its conclusion the Fleet Air Arm began its display. A Seal dived past the moored Royal ship, followed by nine squadrons variously equipped with Blackburn Baffin, Fairey Seal, Fairey IIIF, Blackburn Shark, Hawker Osprey and Nimrod, each diving in salute as they passed.

More dramatic were two radio-controlled Queen Bee seaplanes, catapulted from the light cruiser *Achilles*. In vivid journalese the naval correspondent of the *Daily Telegraph* wrote: 'It seemed incredible that these machines should be up there alone without a soul on board. The *Rodney* opened rapid fire on the first with three anti-aircraft guns. Tongues of flame pierced the dun-coloured clouds of cordite smoke, and the sky was filled with balls of black cotton wool, clustered thickly above the target. For a time it seemed as if the Queen Bee bore a charmed life. Suddenly a ball of smoke developed just below her, and I saw a large fragment break from the lower wing. She began to turn, dropped into a nose dive, and struck the water with a crash, and by the time a destroyer had reached the scene not a vestige remained afloat.' But that was expenditure enough. The other Queen Bee was brought round in a splendid circle and neatly alighted on the smooth sea.

Six days later the annual Air Exercises of ADGB began. Southland was the aggressor, with a force of light and heavy bombers, comprising Harts, Heyfords, ancient Virginias and Hinaidis, commanded by Air Vice-Marshal Playfair who was reinforced with Gordons, Harts, Sidestrands and Overstrands commanded by Air Commodore H. R. Nicholl. Their objective was Northland defended by Demons, Bulldogs, Gauntlets, Furies, Nimrods, Ospreys, Harts and Audaxes all under the command of Air Vice-Marshal Joubert de la Ferté.

The previous year's operations had shown that the interception of approaching raiders by the fastest fighters stationed on the coast had been a failure. This year fighters were moved to a sector affording least warning of impending raids but with the hope of overcoming this by operating standing patrols warned by radio transmission of the approach of the enemy; however Southland still managed to avoid interception to a considerable extent whether going in or coming out, day or night. This resulted in the statement: 'No general conclusion can be drawn from the Exercise because of the many artificial limitations, but they have provided excellent training for the staffs and units concerned.' Yet there must have been the obvious inference that better methods of guiding fighters to bombers must be found, and that fighters must be far faster even to catch up visible invading formations. A Fury taking off with a Heyford flying at 15,000 ft overhead required thirty-three miles, climbing at optimum speed to reach the theoretical point of interception, but that was unlikely with merely visual guidance. The Watson-Watt experiments were therefore of vital value.

The best the Air Ministry could do in re-equipping squadrons could be gleaned from the current Press release that No. 7 would have Heyfords in place of Virginias; No. 8, Vincent instead of Fairey IIIF; No. 19, Gauntlet replacing the Bulldog IIa; No. 29, Demon replacing the Bulldog IIa; No. 604, Demon replacing Hart and Wapiti; No. 208 Army Co-operation, Audax instead of Atlas; No. 1 Coast Defence, Osprey replacing Fairey IIIF; No. 202 in Malta, Supermarine Scapa flying-boat replacing Fairey IIIF; No. 101, Overstrand instead of Sidestrand; No. 210, Singapore to replace Southampton; No. 3 and No. 5 FTS, Hawker Audax and Hart replacing Atlas, and Fury replacing Bulldog. At the Air Navigational School at Andover, the Avro Prefect would replace the Avro Tutor. The new cruiser *Ajax* would be issued with Osprey Fleet spotters to replace Fairey IIIFs of No. 443 Squadron.

Faireys, who maintained their manufacturing position against Hawker's RAF production largely through equivalent standardization by the Belgian Air Force of Foxes and Fireflies, had pinned their faith for future orders on the

Fantome cannon fighter – but hopes were dashed when the machine fatally crashed while Lieut-Cdr S. H. G. Trower, Staniland's assistant, was demonstrating to General Gillieaux, Chief of the Belgian Air Force, for the pilot probably imposed an accidental high speed stall, and the machine dived into the ground, fire breaking out on impact.

9

At Westland there was trouble. Serious delay had occurred in fulfilling spares contracts for the RAF's Wapitis in India, and final production of the enclosed-cockpit version of the Wallace was behind schedule because the hoods were incomplete. W. G. 'Bill' Gibson, whose untiring if unconventional, almost single-handed management of production had brought such success to Westland with cheap construction of Wapitis, was made scapegoat, and, in a brief stormy interview with the general manager Stuart Keep, threw in his hand. Though Keep took direct control confusion resulted, for the whole programme, while very clear to the rushing, confident Gibson, was not defined in detail on paper. Within a few months the AID reported to the Air Ministry, who endorsed their view, that Westland was in chaos and a new works manager must be appointed within twenty one days or face loss of future orders. Because of my experience as manager of civil aircraft production, Sir Ernest Petter asked if I would take over works management – but having enjoyed the freedom of test flying and the delights of winging far and wide over England, I declined that firing line.

Meanwhile it was announced at the twenty-fifth AGM of Petters Ltd on 16th July that its subsidiary, the Westland Works had been divorced from the oil engine business by the formation of Westland Aircraft Ltd with share capital of £250,000, and Sir Ernest as chairman, who stated that orders on hand were about £300,000 and cash working capital was £75,000. Capt Peter Acland, former key man of Vickers aircraft department, and recently in charge of Petters' London office, was appointed managing director; Teddy Petter was technical director, and tall, stern-looking, though in fact jovial, Air Vice-Marshal N. D. K. MacEwen, a great hunting man, was brought in as a non-administrative director to give worthy impress – but Keep, though still a director of Petters Ltd, had no seat on the new Board and remained general manager.

R. A. Bruce, who had managed Westland throughout its two decades, had retired six months earlier. Acland had already taken over his former office. This worldly, tall and handsome ex-cavalryman imposed new *bonhomie* of a somewhat spurious value on the fairly close relationships of other members of the staff whom he addressed by Christian name rather than surname, and became Peter, not the formal Capt Acland, to his senior staff. Nevertheless, his circuit of the shops every morning, after attending to incoming correspondence, was certainly appreciated by the workmen as indicating closer contact with the management – but it is doubtful whether work proceeded one iota faster, particularly as it soon became apparent that Keep and he were an incompatible pair. Within a few months he and young Petter forced Keep to retire.

Successful candidate for the post of Works Superintendent was fifty-three-year-old John Fearn, a doughty little Midlander with pendulous lower lip and thinning sandy hair, who had held divisional managerial posts at BSA but knew nothing of aircraft. Teddy Petter and I had the task of giving him elementary instruction on the complexities of aeroplane workmanship, but he opined: 'Believe me, my boy, there's nothing so highly indoostrialized as the pedal barcycle!' Undoubtedly he knew the intricacies of such bulk manufacture, but initially underrated the problems of aircraft production, particularly procurement, though he immediately began to build up numerically the scanty Works administrative staff. He could be benign or adopt a front of fierceness and authority, but soon sufficiently endeared himself to be referred to behind his back as 'Daddy' Fearn. Responsibility for shops' management became D. W. Nivison's, a grizzle-haired, moustached North-countryman, who as Gibson's production assistant had been notable for devising ingenious methods of detail fabrication. 'An idea a day is what gives me my pay,' he told me with a laugh. Like Gibson, he had the habit of striding through the Works with a great show of energy and intentness which kept the men on their toes.

By now, work on the Westland tail-less Pterodactyl had come to an abrupt end. After its designer Capt Hill left to become a professor at London's University College, Petter had no wish to continue its development or even to overcome the torsional flexibility of the wing which prevented more than a 200 mph dive with the stick fully forward. Herbert Mettam, key man of Hill's staff, had resigned, and joined A. V. Roe Ltd. Wimperis agreed that the prototype Mk V should be sent to Farnborough for general investigation. Within seconds of take-off to the east, over Yeovil, the coolant warning light flashed danger. Height was 100 ft above the houses. I instantly turned back down-wind, and the engine seized solid as the machine skimmed the hedge and landed. Inspection revealed that the front wheel of the articulated undercarriage had gone full travel and knocked the radiator cock closed. At Petter's instigation the machine was sent by lorry to Farnborough and never flown again.

Westland was somewhat the Cinderella compared with other aircraft companies, so its new guise created little stir, but when de Havillands announced that Hubert Broad had left that company there was enormous surprise and comment throughout the aeronautical world, for he had been with them fourteen years and was regarded not only as the star but the symbol of the DH business. Broad was as surprised as the rest. Without warning he had been called to Frank Hearle who, greatly upset, told him without explanation that the Board had unanimously decided to have a new test pilot. Even Geoffrey de Havilland had shirked the tragedy of telling Broad and therefore delegated it to his very understanding and tactful brother-in-law. Years later, Hearle and Walker separately told me that though Broad was often difficult he had been unsurpassable in what was virtually public relations work with clubs and pilots all over the world, but was so supremely skilful a demonstrator that he uncritically corrected handling deficiencies and thus failed to be sufficiently analytical to help the designer. In fact, with machines such as the great sequence of Moths, there had been little enough to criticize, and in any case D.H. himself had made many supporting flights and been satisfied. But with departures from the conventional rectangular-winged biplanes there were matters, such as stalling characteristics and resulting difficulties of landing, which had not been sufficiently examined and explained. This may have led to several disasters with the tapered-wing multi-engined machines, and now there were even worse tip-stalling difficulties with the taper-wing Hornet Moth which could be unexpectedly dangerous for amateur pilots. Fifty sets of wings were already in advanced construction. As Broad had accepted the tapered wings for the production version he had to take the blame – a state of affairs not unknown to test pilots, before or since, when 'carrying the can' for design errors. For the next production batch, re-ciphered DH 87B, more conventional mainplanes were being designed.

Making the best of matters, Broad, cashing in on his world-wide reputation as a test pilot, decided to continue as a free lance, but in any case had other business interests as a sideline. Nor had he lost prestige with the de Havilland School apprentices, for they remained determined that he would still pilot in the King's Cup Race the TK 2 monoplane they had designed and built under the guidance of Marcus Langley their technical instructor.

Though the prosperity of de Havillands was obvious, the removal of George Dowty's business of Aircraft Components Ltd from his little factory in Cheltenham to the splendid manor house and 56-acre estate of nearby Arle Court marked another important success. Here Dowty was joined by thirty-eight-year-old Rowland Bound as chief engineer with twenty years' experience of aircraft design, originally at Avros and latterly with Percival. Dowty and his team were already making world-wide sales for his internally sprung wheels and every form of undercarriage, fixed or retractable, complete in every respect.

A Dowty cantilever undercarriage was featured for the Avro 636 two-seat fighter trainer currently flying in glossy black and silver. Something of a cross between Avro and Armstong Whitworth design, it was a conventional single-bay biplane with outward-raked N-struts and staggered top wing of greater chord and span than the lower, and with a Panther XI attained 230 mph. The rear cockpit could be faired over and as a single-seat fighter it had two Vickers guns projecting through a marked hump at the blunt fore end of the cockpit decking. There seemed a possibility of export sales to smaller air forces, such as the Argentine, but eventually only a contract for four fighter trainers for the Irish Army Air Corps materialized. Biplanes were out. Every country was now visualizing monoplanes.

A more profitable possibility as a simple trainer was the latest Miles Hawk with trousered undercarriage and 130 mph Gipsy Major. Presently it would be developed as the Miles M 14 to Air Ministry specification, and eventually as the Magister would be ordered in quantity to condition pilots in monoplane techniques, though the difference in handling compared with biplanes was imperceptible, but it gave a better view and flatter glide.

Attempting to enter the private market was the new Heston Phoenix strut-braced high-winger. Despite wood and fabric construction, it had a modern elegance aided by its clean appearance with undercarriage retracting into a stub wing cantilevered from the lower portion of the fuselage and blending into the outrigger struts. Tested on its maiden flight by Edmund Hordern, a director of the Heston Aircraft Co and of the Wilton School of Flying near Boscombe Down, the Phoenix required no control alteration, and with a speed of 145 mph was faster than expected despite the generous size of the five-seater cabin. That it could compete in price with equivalent three-seaters of similar power seemed doubtful. However *The Aeroplane* reported that James Melrose's uncle, Noel Pemberton Billing, 'is talking about delivery of some Heston Phoenixes which he intends shall be used by Mr Melrose on the Adelaide-Melbourne service because he judges the Phoenix is the world's most efficient low-powered passenger-carrying aeroplane in that it carries four or five comfortably at reasonable cruising speed on a 200 hp Gipsy.'

Not only were light aeroplanes an everyday sight across the skies of England but whenever thermal up-current conditions were evident at week-ends sailplanes essayed cross-country flights – yet anything winged had its perils. On 30th July the greatest exponent of British sailplaning, twenty-seven-year-old George Eric Collins, Chief Flying Instructor of the London Gliding Club, was killed while giving an aerobatic display for Cobham's Air Day at Ramsey, Huntingdonshire. He had been towed to 3,000 feet and on releasing made two loops, then rashly dived into an inverted loop known as a 'bunt' and the left wing broke, the ensuing attitude preventing him from escaping by parachute from the cramped little cockpit. Long familiarity with the exuberant aerobatics of wire-braced biplanes had given the impression to many amateurs that aeroplanes and gliders were enormously strong; but sailplanes had only marginal strength under negative 'g', for light structure weight was paramount, and even commercial aeroplanes had relatively low load factors which were adequate for normal flight but might lead to disaster in the dynamically turbulent air of an Indian monsoon – though luckily the techniques of blind flying and radio fixes did not yet enable such excursions.

Increasing interest in gliding was emphasized by the BGA's competitions at the Yorkshire moorland ridge of Sutton Bank where the twenty-six gliders included not only German sailplanes such as the Rhönbussard, Grunau Baby, and Falke, but equivalent British designs – the gently murmuring Kirby Kite designed and built by Fred Slingsby; slender *Hjordis* designed for greater penetration by G. M. Buxton and owned by Philip Wills; the Baynes Scud II with diamond-section fuselage and distinctively tapered wing. There was also the miniature *Pegasus* in its 13-ft trailer as a late arrival overnight on Wednesday because tests on a Wallace took priority. There had been only time for a few short flights during the preceding week-end – but the first launch at Sutton Bank dispelled any doubt, for *Pegasus* manoeuvred and soared so easily that I gained my 'C' certificate. That evening I sailed around the bowl of Sutton

Bank, the air smooth as glass, the sky luminous, every splendid vista enfolded in quietness, *Pegasus* floating serenely wherever I headed her and only dusk and the first stars caused me to land.

Philip Wills, bespectacled, tall and thin, gained the British sailplane height record of 5,600 ft with *Hjordis*, the only sailplane with blind-flying instruments. Although owner of a Moth, he regarded gliding as sport of the select few, for it was proving costly; nor did he believe in attempts at flying with minimum power, such as semi-sailplanes with little engines like the Scud III, or the funny little tandem-winged Pou du Ciel which had become the rage in France for amateur construction and in Britain was sponsored by the Air League to encourage penniless youth into the skies as though riding a motor-cycle. Already one had been built by Stephen Appleby and successfully crashed on its first flight.

Though wide interest in amateur flying was evident, and the use of airlines increasing, the crucial turning point in British aviation was the outstanding success of the twin-engined Bristol 142 *Britain First* sponsored by Lord Rothermere. Only two months had been taken to complete the flight tests, fit equipment and furniture, and substitute two-pitch, three-bladed DH Hamilton-Standards for the wooden propellers. Uwins had taken this new concept in his stride, and in June flew it to Martlesham for handling trials. As Chris Barnes indicated in *Bristol Aircraft since 1910*: 'All who flew it were so favourably impressed with its performance and handling that the Air Ministry asked to retain it for full evaluation as a potential bomber. Rothermere, with a blare of trumpets, immediately presented it to the Air Council on whose behalf Sir Philip Cunliffe-Lister cabled grateful acceptance.'

Britain First was promptly taken over and given the RAF's tricolour targets on wing and fuselage, though these days rudders were left silvered lest painted stripes altered the static balance and caused flutter. Although damaged at Martlesham early in July when the cowling came adrift, necessitating a return to Filton for repairs, barely a month passed before Bristols received an urgent Air Ministry contract for 150 with nose altered to a bomber's station with fixed Browning gun, the wings raised 16 in to afford internal bomb storage, and a semi-retractable turret fitted abaft the trailing edge. Despite the elaborate design and new manufacturing techniques required, the Bristol management scheduled tooling, jigging, and the complex ordering system of materials and sub-contracts to meet a target date of June the following year.

Splendid though Bristol's achievement was in producing this light bomber, RAF strategists could see that America had secured a head lead with an aerodynamically superb four-engined 20 ton all-metal bomber of 103 ft 9 in span designed and built by Boeing Aircraft Company. Known as Type B-299, it was the prototype for the B-17 Fortress. With four 700 hp Pratt & Whitney Hornets, top speed was 256 mph, and six-tons of disposable load could be carried, but with more powerful engines might become nine. Heavier defence was the American mode, so there was one hand-operated 0.5 gun in the nose, a similar gun in a dorsal cupola, one in each side blister and one in a ventral blister. In true world advertising fashion it was flown from Seattle to the Army Air Corps test station at Dayton, Ohio, with the aid of a strong but unreported tail wind enabling it to cover 2,300 miles in nine hours, misleadingly quoted as a cruising speed of 255 mph – but disaster lay a few months ahead. While being demonstrated to the US Air Corps Procurement Committee at the beginning of November Leslie Tower, the Boeing chief pilot, took off with controls locked, a wing slightly dropped, and the great machine sank stalling to the ground, broke in two, and caught fire, killing Tower and co-pilot and one of the crew, and badly burning the others and some of the rescuers. Only a few hours later the Boeing assistant test pilot, M. T. Arnold crashed when an engine of their twin-engined airliner failed during a night-flying demonstration to United Air Lines, and he and the test crew were killed.

Probably the four-engined Boeing, later known as the Fortress, influenced Göring to agree to Generalleutnant Wever's insistence on the need for a long-range strategic bomber, for which a target date of a year had been set for the first flights. When in 1970 I asked the pre-war DTD Air Commodore Verney whether Intelligence had reported this German trend in strategic thinking he said 'No', but that it was known in DOR that the Germans had ordered metal-skinned fighter prototypes with retracting undercarriages from Messerschmitt AG and Ernst Heinkel. There was bitter antagonism between those two designers, fanned by Göring's deputy, Erhard Milch, who had long discredited Messerschmitt to the extent of declaring the advanced Bf 108 low-wing tourer unsafe, and got Wimmer to say that this outstanding designer was no use to them. As a result Messerschmitt had been on the point of accepting a Professorship at Danzig Technical University, but just in time his tender was accepted for the Bf 109 fighter jointly designed by him and Walter Rethel formerly of Arado Flugzeugwerke – and now in August it was ready for taxi-ing trials, but when Generalleutenant Udet saw it he said: 'It will never make a fighter!'

The Heinkel He 112 designed by Walter Günter under Heinkel's direction seemed even cleaner but was noticeably larger than the compact Messerschmitt, and like its forerunner the He 70, had an elliptical cantilever wing and oval duralumin monocoque fuselage. Like the Messerschmitt, it was temporarily powered with a Rolls-Royce Kestrel and fixed-pitch propeller because the intended 640 hp Junkers Jumo 210 was not yet ready. At the beginning of September both fighters made their initial flights. Soon it was found that the Messerschmitt with a speed of 290 mph was 17 mph faster than the Heinkel favourite.

When some years later I flew a captured Messerschmitt I recorded: 'Unlike British aeroplanes with normal hesitant carburettors, pick-up of the direct-injection engine was instantaneous and perfect. The Messerschmitt flew confidently, and even with the scarcely perceptible control movements of steady cruising flight felt lively, but further investigation disclosed minor snags with control and stability which in certain circumstances would reduce its potential as a fighter. The odds were that if a pilot could be tempted into a half-roll and dive at moderately low height he would not be able to pull out in time, and at lower speeds, or in very tight turns and violent manoeuvres, the wingtip slats opened with a jerk which upset gun sighting and must have let many a potential victim escape. . . .'

While the new German fighters were being tested the first of the British equivalents, the Hawker monoplane, later known as the Hurricane, required another two months for completion. Britain was also behind with her flying-boats, for the splendid hulls of the big four-engined Short shoulder-wing monoplanes were still under construction, and the first machine would not be ready for another ten months. Britain had not yet recovered the loss of design lead from cancellation of the six-engined Supermarine, so the USA not only had the large Sikorsky S-42 four-engined strut-braced high-winger of 114-ft span and 38,000 lb gross experimentally flying for Pan American on the South American route, but now the Martin Corporation caught up with a still bigger and more powerful boat of 130 ft span, with four two-row 800 hp Pratt & Whitney Wasps and achieved two world records with a payload of 23,914 lb at an all-up weight of 51,000 lb – but it had sponsons and seemed less seaworthy than the Sikorsky with conventional outboard floats like the Short design though the hull was not up to British practice.

10

With high summer came intensified flying activities, highlighted by the International Rally held at Lympne on 25th August under blue skies and hot sunshine which attracted scores of British light aircraft, thirteen from Germany, two from Czechoslovakia, three from Belgium, and an uncounted number from France of which the Caudron Renault-powered Simoun racer won the Wakefield Cup for a race which forgot earlier tragedies through collisions and had twelve left-hand turns and three right-hand turns in order that racing machines would always be in view of spectators. There were all the usual thrills of the fair, with aerobatics and formation flying, and demonstrations of the various small civil aircraft competing for all too few orders by too many small companies endeavouring to establish production. Star of the afternoon was a Focke-Wulf two-seat biplane, powered with a small Siemens radial, piloted by blond, bespectacled Herr Paul Förster who warmed up with a few loops and then proceeded with what seemed an endless sequence of rolls three times round and round the aerodrome, smoothed into two outside loops, a stalled turn which became a dive into a bunt, then a series of stalled turns inverted, finally pulling up inverted and rolling into level flight. Nobody had ever seen so dazzling a show.

The Rally made an effective prelude to the King's Cup Jubilee Race on 6th September over a course starting and finishing at Hatfield, but sweeping northward to Edinburgh, then a control point at Renfrew near Glasgow, down the coast to Galloway, a quick hop across the sea to land at Belfast, back to

England and a control at Manchester, another at Cardiff, and so to Hatfield. Finalists next day flew a tight triangle seven times from Hatfield to Broxbourne, Henlow, and back. The thirty starters had Edgar Percival as scratch flying his ivory white Mew Gull, entered by the Duke of York. 'Of course there were the usual grumbles about the handicaps,' reported *Flight*. 'Wally Hope decided to fly a Leopard Moth instead of his Comper Swift which he handed to Flt Lieut R. Duncanson who was without a mount because his Hendy Heck had been sold to Mrs Jill Wyndham who proposed to fly to India. But there never has been a King's Cup Race in which Mr Hope has been fairly treated by the handicappers! Even when he won a race by over an hour he was not fairly treated, or at least thought so before the race and other people thought so after the finish!'

At 7.30 am all machines were brought from the sheds except the new twin-engined DH 90 biplane as its Certificate of Airworthiness had not arrived. At 8 o'clock Mr Reynolds started the first competitor, the others following at thirty-second intervals. All were monoplanes except a Genet Avro Avian as nostalgic reminder of old-time flying, yet five years earlier the only monoplane had been Percival's Hendy 302. Within fifteen minutes the last of the entrants had gone and the officials were having breakfast, waiting for news to trickle in. Luis Fontes, well known racing driver, piled up his Miles Hawk Speed Six near Durham. R. O. Shuttleworth, wealthy collector of old cars, traction engines and aeroplanes, after taking off with his Swift's luggage locker open, returned to Biggleswade; Jack Armour with the recently-designed BA3 Cupid side-by-side two-seater retired at Newcastle; A. H. Cook with the Miles Hawk Trainer forced landed near Ayr with a burned-out exhaust but managed to patch it and take off again; Bradbrooke of *The Aeroplane*, flying a Miles Hawk de Luxe retired at Blackpool with engine misfiring. Ruth Fontes, entered as Miss R. Slow, passed the wrong side of Port Patrick turning point with her Miles Speed Six, so was eliminated; hefty Flt Lieut E. C. T. Edwards, flying a Gull, retired at Newtownards with his propeller spinner adrift and the port magneto coupling broken; the engine of twenty-one-year-old Alex Henshaw's Gull stopped while flying only ten feet above the water, and he stood up while stalling the machine onto the waves and was catapulted out, but was saved by the Isle of Man steamer, led by the circling olive-green Gull of Jim Melrose, who had thrown away his chance of winning by searching for him.

Edgar Percival led most of the time, and was the first to come in sight at the Hatfield finish, but to everyone's amazement, shot past on the north side and before he could discover his error and turn back, his rival manufacturer, Fred Miles, with his clipped-wing Hawk came streaking low down and passed the finishing line a minute ahead of the Mew Gull, having achieved an average of 154 mph, which was only 5 mph slower than Percival's machine, which the latter quite fairly said, had only been cruising. Four Gulls were next, with a startling moment for the last, piloted by Ken Waller, when a cyclist pedalled across his approach path. Miss Fontes came next, unaware of her disqualification, followed by cheery Tommy Rose in the Miles Falcon, Edwards in the Miles Hawk Trainer, and then a great cheer from the DH students as Hubert Broad flashed in with the yellow TK 2. In ones and twos came the rest, tails high, throttles wide open. Last was the only open-cockpit machine, the Avian, gallantly flown by A. H. Tweddle.

The finals next day, in glorious weather, had twenty starters. Tweddle was first away, and soon there was a medley of machines taking off and others flashing back. Not until well in the second lap was Percival flagged away in his Mew Gull. Spectators lost all sense of which lap was being flown or who was leading; but in the last lap broadly grinning Tommy Rose came first with Lady Wakefield's entry, the Miles Falcon, closely followed by Edwards flying one of the Hawk Trainers and Cathcart-Jones hot on his heels with the other, to cheer upon cheer from two hundred employees of Phillips & Powis who had arrived in seven coaches. Two seconds later came the neat but gaudy TK 2 flown by Broad, with Lipton and Harris in the Falcon close behind. Sixth, to his disappointment, was Percival with the Mew Gull, but, as always, complete with genial camera smile and jaunty black felt hat as though he had merely been on a brief pleasure flight instead of intently dashing around in the fastest machine ever to race in England, achieving an average of 208.91 mph. After him came bespectacled Charles Gardner in Gull No. 10, the first of the six eligible for the Siddeley Trophy which he won for the Redhill Flying Club.

For once there was no double-spread advertisements for de Havillands. Instead it was the privilege of Phillips & Powis Aircraft Ltd jubilantly stating: 'The Falcon Six has no equal', adding that the winning machine was in every way the same that could be bought for £1,325, flaps £60 extra.

II

At Coventry, Alan Campbell-Orde was busy with trials of the C 26/31 twin-engined AW 23 bomber-transport, but it was as cumbersome as it looked, and the Bristol Type 130 equivalent seemed more attractive and easier to load. Currently Handley Page decided that because his competing HP 51, initially flown by Cordes on 8th May, needed improvement, Lachmann should re-design it as the HP 54 with an extended tailplane and longer fabric-covered fuselage which he had previously vetoed, and at the same time replace the open nose and tail-gun positions with power-driven turrets and fit a mid-fuselage gunner's cupola abaft the wing. Meanwhile he was bent on securing a contract to cover the job, aided by Grp Capt Richard B. Maycock, who had joined in place of Tom Harry England as Services liaison officer, and significantly recently commanded the bomber base of Upper Heyford where Handley Page met him. A man of sterling quality and balanced outlook, Maycock not only had commanded the MAEE in the previous decade, but until a year ago was British Air Attaché in South America. In conjunction, Handley Page and he successfully argued their way to a contract for 100 of these machines effectively 'straight off the drawing board' to specification B 9/35. George Volkert, who had recently returned from the USA after his tour of major American aircraft companies, was allotted the task of planning production in association with James Hamilton, the works manager.

Vickers were equally successful with a production contract for their G 4/31 monoplane, despite a pile-up when landing on 23rd July due to failure of the port leg extending mechanism, resulting in extensive damage to the wing. At a cost of £30,270, private venture again had won, for on 5th July, Sir Robert McLean, after repeated strong representation to the Air Ministry, had written to Air Vice-Marshal Sir Hugh Dowding: 'I suggest it might be better to reduce these orders for the G 4/31 biplane, and in their place go into production of the monoplane as soon as tooling-up can be completed. Meanwhile, and until you can decide whether we shall be allowed to switch from biplane to monoplane, I do not wish to proceed with work on the biplane because, in my view, it is not a modern machine.' Two months later the company was instructed to cancel the biplane programme and proceed with construction of seventy-nine monoplanes to re-written specification 22/35, and production type-name of Wellesley. In the meantime the prototype was extensively reconstructed with enclosed cockpits, hydraulically operated undercarriage instead of manual, a broad chord rudder, and bomb containers pylon-mounted below the wings.

Undoubtedly the speed superiority of the 200 mph prototype was a telling factor in securing the order although the rate of climb was somewhat slower than other competitors; but it was also a personal triumph for Barnes Wallis, whom the Air Ministry rated highly for the ingenuity with which his R 100 design philosophy was translated into the geodetics of the Wellesley. Nevertheless its production facility and operational robustness had still to be proved, and that Vickers should embark on so daring an innovation emphasized the isolation of Mitchell and his Supermarine team from that of the Pierson-Wallis outlook – for with the Schneider monoplanes and metal-hulled flying-boats a proved technique of metal skinning was already available and could have been successfully adopted by Vickers at less cost than the geodetic system. Already there was a wealth of technical information on metal monocoque structures. Dependence on fabric covering, despite its retention for the new Hawker monoplane fighter, was already obsolete and justified only by increasing pressure from the RAF to obtain new types quickly.

It was a far cry from this British attempt at aerodynamic improvement to the 352.46 mph claimed as a landplane record by millionaire film producer Howard Hughes when he flew his special racing Hughes low-winger at Santa Ana, California, on 13th September – but because he flew a closed circuit of three kilometres instead of four times up and down a three-kilometre straight line the FAI disqualified it for not complying with regulations. Enormous though the speed seemed, it was less remarkable than the world's land speed record which Sir Malcolm Campbell achieved on 3rd September with 301.337 mph, but Hughes indicated the shape of things to come. Powered with a Pratt & Whitney double Wasp Junior delivering 1,000 hp for racing, the stressed-skin wing was only 25-ft span and the area 140 sq ft, yet sufficiently deep for the stalky cantilever undercarriage to retract cleanly into the leading edge. Scaled-up, this seemed the fighter of the future, but Hughes was not interested in making machines for war.

Preparation for war was becoming evident all over Europe. Germany held air defence exercises on a large scale near Brunswick 'to test new anti-aircraft guns and equipment, and give the personnel experience of war conditions.' The frontier was the Elbe, and the East Force attacked the West Force from the air, with Staaken aerodrome near Berlin as one of the objectives. The sky was filled with the thunder of Heinkel fighters seeking to intercept squadrons of Junkers tri-motor Ju 52 bombers which supplemented the top gunner's position with an extendable gun 'dustbin' for ventral defence. A few days later the Luftwaffe co-operated in North Sea and Baltic Naval manoeuvres. As though emphasizing their air-readiness, the Germans appointed Colonel Ralph Albert Franz Wenniger as their first Air Attaché at the German Embassy, London. Nor was Russia less active. The Moscow correspondent of the *Daily Telegraph* on 15th September reported: 'More than 1,000 men took part in a massed parachute raid far behind "enemy lines" during the Red Army manoeuvres yesterday. Armed with portable machine-guns they jumped from a fleet of big four-engined bombers in broad daylight, but in real war many would have been exterminated while floating to earth in full view of anti-aircraft batteries. None the less it was of great interest that specially trained men can be dropped with parachutes almost simultaneously onto a confined space of strategic interest behind enemy lines.' Germany – whose Swastika was now the national flag – was not slow to take notice.

Such a method of attack had not been apparent to the British Army during annual manoeuvres in mid September. Directed by the CIGS, Eastland and Westland were battling in an area some fifty miles by thirty, and started well by having most of their tents blown down by a gale that raged throughout the first night. Nevertheless: 'The Audax Army Co-operation Squadrons on both sides did splendid work. They knew just what the Army wanted, and in spite of gales, low cloud, and incessant rain storms, supplied rapid and accurate information throughout the manoeuvres. Even at night by use of parachute flares they were able to report on movement of troops and give essential information. There were also five Autogiros for photographic work and transport of Staff Officers – two for each Force, and the fifth, a neutral machine, was used by the directing staff.'

Italy was currently showing her teeth at Abyssinia who six months after admission to the League of Nations on 27th September 1934, had broken off direct negotiations with Italy after allegations that the latter's colonial citizens had been constantly raided by the Abyssinians, who in turn were demanding reparations for attacks by Italians. Because Britain was endeavouring to make a front with Italy against Germany neither Ramsay MacDonald nor Sir John Simon had dared mention Abyssinia at the Stresa Conference in April, but now in September, Simon's replacement, Sir Samuel Hoare, warned that Britain stood for 'steady, collective resistance to all acts of unprovoked aggression.' Short of direct attack on British territory by an Italian force it was not likely that

Britain would go to war in the Abyssinian cause. At best there would be 'sanctions'. But as C. G. Grey said: 'We English are treading delicate ground. The British Empire was built up by sheer piracy; we have salved our consciences with the idea that we were either introducing the benefits of Christianity or civilization to backward people – but we have done very well for ourselves. If Italy feels that the negro tribes, so long oppressed by the Abyssinians, need protection, who are we to object? Our negro tribes in the Sudan, Uganda, and Kenya for generations have been raided by the Abyssinian. Surely our duty is to put their oppressors in their proper place?'

That was the line which Mussolini took. In October he declared war on Abyssinia. First report to reach Europe was the air bombing of Adowa by aircraft led by his son-in-law Count Ciano. The League duly pronounced Italy the aggressor and called for sanctions – a course already proved barren when Japan began the conquest of Manchuria, for it merely resulted in Japan withdrawing from the League in 1933 and launching a full attack on China, whereupon the League did nothing.

12

Henri Mignet had flown to England in his unconventional tandem monoplane in August under contract to the *Daily Express* for a demonstration tour of several southern aerodromes. There seemed possible accommodation advantages through the weight disposition that a tandem afforded, so Davenport and I flew to a small field in the Isle of Wight to see it. Mignet, voluble and enthusiastic, accompanied by his charming wife who acted as interpreter, offered to let me fly the machine. 'No you don't,' said Davenport anxiously. 'I've never seen such a dangerous contraption – and anyhow "Lamps" wouldn't insure you.' Lamplugh had long been virtually dictator of civil flying, for his insurance approval was essential.

Within a week, the first fatal accident happened at Algiers. According to *Les Ailes* a Pou after circuiting the aerodrome at 200 ft began to switch-back and dived vertically into the ground. Inquiry confirmed that nothing had broken in the air. A facile explanation was that the stick had slipped from the pilot's hand. A month later came a fatal accident at Marseilles to the instructor of the local flying club when his Pou made 'a vertical dive when landing.' An announcement from General Denain, French Minister for Air, urged the utmost caution with these machines and stated that his Government had bought a sample for thorough inspection and test by experts. Coincidently in Britain, all airworthiness restrictions on ultra-light aeroplanes had recently been removed, but insurance against third party risk remained mandatory. Within two months there was another inexplicable fatal dive in France.

Accidents with all kinds of aeroplanes continued to occur indiscriminately. So skilled a pilot as Tom Campbell Black, who in March had married the comedienne Florence Desmond famous for her stage mimicry of Amy

Mollison, while attempting a record flight to Cape Town with the fifth and last Comet, *Boomerang*, had engine failure some seventy-five miles north of Khartoum, and found the power of the second engine insufficient to maintain height. The country was too rough for a forced landing, so he and his co-pilot John MacArthur, opened the cabin roof, struggled clear in the blast of wind, and parachuted down to face a long walk back.

That even more famous pilot, Sir Charles Kingsford Smith, arrived at Croydon on 8th October with his 550 hp Lockheed Altair which had been shipped to London, off-loaded to a barge and lifted ashore at All Hallows, whence he took off for the airport. Accompanied by J. T. Pethyridge he left for Australia on 25th October, but was back next day for a fresh start because ice formation over Greece caused minor damage, necessitating a forced landing at Brindisi. Before dawn on 6th November they began again, and landed at Baghdad at 10 p.m. Next day they reached Allahabad in thirteen hours, and left an hour later for Singapore. They were sighted in the dark above the Bay of Bengal by Charles Melrose who was *en route* for Australia in his Percival Gull, having left Croydon in formation with Harry Broadbent in another Gull heading for Port Darwin which he reached in 6 days 21 hours 19 minutes. But on landing at Singapore, Melrose learned there was concern for Kingsford Smith, and on 8th November he joined a massed search by two locally-based flying-boats and two flights of Vildebeests with the aircraft-carrier *Hermes* standing by. They had no success. The search continued until 14th November, and then Air Commodore Sydney Smith ordered it to be abandoned; but Melrose continued with his Gull, and presently was joined by Capt P. G. Taylor on learning of reports from the manager of a tin mine in southern Siam that an aeroplane flew low over the jungle on 8th November. No trace of the Altair and its crew was ever found. Not only Australia and Great Britain but the whole world mourned the passing of a great pioneering pilot. While all this was happening, David Llewellyn, flying with Campbell Black's pupil Mrs Jill Wyndham in her Hendy Heck, made what they described as an easy tour from Hanworth to Cape Town in 6 days 8 hours 27 minutes.

Private flying was finding improved facilities everywhere. The annual Air Ministry report covering the previous year's *Progress of Civil Aviation*, issued in October, listed ninety licensed aerodromes, landing grounds, and seaplane stations in Great Britain, of which twenty-one were municipal. There were 2,980 'A' licences of which 165 were held by women, and 498 'B' (Commercial) licences of which five were held by women, backed by 134 Navigators' licences, 140 W/T approvals and 190 R/T approvals, and there were 1,317 Ground Engineers' licences issued. At the beginning of the year 1,174 aeroplanes were registed, of which eighty were in regular air transport, 190 for taxi, joy-riding and general air work, 153 used by Schools and Reserve Training, ninety-five Club owned, ninety-nine for demonstration, seventy-nine held by agents for sale, 478 privately owned. Of the total, 22 per cent were DH Moths of all kinds – which made one wonder whether de Havillands had been wise in stopping production of so popular a type.

The Report stated that the UK airlines on a total internal and overseas route mileage length of 17,267 miles had totalled 4,560,000 miles and carried 135,100 passengers – compared with the USA which solely on its 50,800 miles of internal air routes had flown 48,786,551 miles carrying 561,370 passengers and added 390,600 route miles and 12,540 passengers from its South American service.

Of European countries, Germany led with 23,442 route miles and 165,846 passengers, and though France covered 21,295 route miles, only 50,019 passengers had been carried, while Holland with 11,816 route miles had 57,339 passengers. Globally the world's air routes covered 223,100 miles, an increase of 22,800 miles compared with 1933. Why America had such a commanding lead was revealed by its direct subsidy of £4,315,659 compared with £561,000 for the United Kingdom and its Empire route.

It was finance which worried many of the smaller British airlines. To strengthen their position there were mergers among the strongest survivors, while others were clearly going to fail. The major current amalgamation was Hillman's Airways Ltd, backed by d'Erlanger interests, with United Airways Ltd (which embraced Highland Airways and Northern & Scottish Airways) and Spartan Airlines Ltd, both backed by Whitehall Securities Ltd. At an extraordinary general meeting in London on 10th October Sir Charles Harris, chairman of Hillman's, said attempts had been made to discredit the company on the score that without the late Edward Hillman it lacked a vigorous guiding hand, and he admitted that in the last eight months it had lost £28,000, with working capital at dangerously low level. Gerard d'Erlanger had taken a large block of shares owned by the Hillman family and obtained a £15,000 guarantee against unissued shares, enabling the new company to start control of funds amounting to £100,000 for maintenance and expansion. On approval by shareholders of all three companies, the business was renamed British Airways, and it was planned to commence operations from Heston Airport on 1st January of next year.

The kind of operational problems which could arise even with a well organized company such as Imperial Airways was illustrated by a sequence of accidents in October. First one of their HP 42s, *Hanno*, was damaged at Kampala when landing with a burst tyre, causing the machine to swing violently and tip on to its nose with considerable damage. A few days later the Avro 652 *Ava* landed at Croydon with its undercarriage accidentally retracted, and, though it ran on the projecting portion of the wheels, the propellers were damaged. Then on 10th October the Short *Syrinx*, while being taxied by Capt A. S. Wilcockson to the sheds at Evère, uncontrollably swung round in a 65 mph gust and turned on its back, injuring one of the four passengers. Only a fortnight later the Boulton Paul *Britomart* undershot when landing at Evère Airport and was badly damaged, two of the seven passengers being slightly injured.

On 17th October Imperial Airways announced that a net profit of £133,769 for the year ending 31st March, and a balance carried forward of £64,503.

Except for the annual subsidy there would have been a loss of £366,000. However a financial commentator cheerfully wrote: 'The reserve of some £400,000 shows sound and conservative policy, for the company has four-fifths the original cost of their machines set aside for replacements. Through providing 25 per cent as obsolescence, the machines can be replaced after four years' work; but the eight aeroplanes of the *Hannibal* type are nowhere near the end of their useful life, although they have been used for nearly four years and have been flying more miles per year than most airliners.' Such gentlemen, solely engrossed with £sd, could not be expected to appreciate that the aerodynamically clean, multi-engined, all-metal American airliners had vastly changed the outlook towards greater speeds, and that though their first cost was greater they required less maintenance and could be left in the open instead of using large sheds. Nevertheless the order for the Armstrong Whitworth Ensign showed that Major Woods Humphery, the forceful manager of Imperial Airways, was fully aware of the need. The next step seemed the possibility of a smaller four-engined low-wing airliner seating twenty-two passengers instead of forty but cruising at the top speed of the Armstrong Whitworth. Ever since the beginning of the year de Havillands had been trying to convince the Air Ministry of the merit of such a design constructed in wood similarly to the world famous Comet, but had been informed on 8th March: 'It will not be possible for the Department to place a non-competitive order with any firm for a prototype, and your particular proposal cannot, therefore, be entertained.' Without an order de Havillands dared not proceed, for the Comet venture had been a costly loss because the nominal price paid by the three entrants amounted to little more than 25 per cent of the total direct cost. However, Woods Humphery and his technicians were now endeavouring to get the Air Ministry to order at least two of these machines. While negotiations dragged on, Hagg proceeded with the engineering.

Major H.G. Brackley, superintendent of Imperial Airways pilots, believed that marine craft, with their capacious hulls and ability to alight on any moderately sheltered area of water, were a more rational choice for the sea-encompassed Empire air routes rather than landplanes which, in his estimation, would require larger and larger aerodromes and eventually runways to take their weight. As a result Short's contract was increased to twenty-nine.

Endorsing the design parameters of these C-class flying-boats was the four-engined enlargement of the Short Scion, named Scion Senior S22, which Lankester Parker flew in twin-float seaplane guise on 22nd October. Recently married fifty-two-year-old Oswald Short was in confident mood, for on the strength of his test pilot's enthusiastic report he decided to lay-down a batch of four to give early delivery of the many orders he expected to receive for this efficient 55 ft span ten-seater – but there was intense competition from the DH equivalents.

Almost unintentionally the Scion Senior had become the aerodynamic scale model for the big Empire flying-boats, with broadly similar cantilever shoulder-wing, and nacelles integrated with the leading edge carrying Pobjoy engines half the diameter of the Bristol Pegasus. Stocky, wall-eyed Arthur Gouge, the chief designer, made the flying-boat twice the linear dimensions of the Senior. Numerous tank experiments led to the practicability of a narrower hull than dictated by theoretical considerations, resulting in vertical sides and elimination of tumble-home chine, although this had to be incorporated for greater beam in the almost identical S 21 flying-boat designed under Air Ministry contract as the lower pick-a-back unit of the composite system devised by Major Mayo. The S 20 upper unit was much smaller, with 73-ft span compared with the 114 ft of the flying-boat, using four Napier Rapier engines similarly installed, and the fuselage was a slim, beautifully shaped metal monocoque mounted on floats.

With such flying-boats and seaplanes in mind, a lecture by Brackley to the Royal Aero Club revealed not only a strong nautical flavour but a tendency towards marine methods of navigation and standards of training and discipline for officers and crews. He explained the technique of handling flying-boats; use of a tail-release line when leaving moorings, with a single slip line at the bow; paper balls spread from a motor-boat to show the water surface when alighting; refuelling in a heavy swell by anchoring the fuel barge ahead of the flying-boat; navigation methods and difficulties; night landings with an L paraffin-flare path visible on clear nights from fifty miles but difficult under misty conditions. He concluded by saying that pilots could not hope to reach the top of their profession if they found mathematics distasteful; a pilot must study and be a master of navigation. 'It is not a young man's job, as we used to think, and the physical condition of some Imperial Airways old-timers suggests their worth for many years yet.'

When some of those old-time pilots joined in the discussion Capt Perry said he did not think automatic pilots were essential on short routes as human pilots could make the passengers more comfortable, and Capt Wilcockson added that the trouble with blind flying was the necessity of learning it with every machine separately because of differences in time lag. But Major Wimperis diverted matters to the future by saying that the Italians were talking of speeds of 200 mph and heights of 60,000 ft in the stratosphere 'which would mean quite a new form of prime mover', though he did not define what that was, deeming Flt Lieut Whittle's turbine-jet proposals were top secret.

Conservative though the views of the Imperial pilots were, a zone system of traffic control had been instituted a few months earlier, and it was indicative of future requirements that the German-developed Lorenz blind-landing system was being installed at Croydon, using a main beacon transmitting on 9 metres at the far end of the best landing run. A signal beacon on 7.9 metres was located 3 km from the aerodrome and another on the approach boundary, both throwing a vertical fan of signals which were received in the aircraft as impulses in a neon tube contained in an instrument weighing 25 kg compared with 35 kg for the radio set. A pilot heard them as a string of dots and then turned to starboard until the dots merged into dashes, at which point he turned on to the approach course denoted by a steady note, maintaining level flight at 2,000 ft until the neon flickered on his dashboard as he passed the first beacon. For the next few seconds he must watch a meter indicating rapidly mounting values of radio intensity, and when the neon went out the reading had to be maintained until touch-down, for it emanated from the main beacon's inverted mushroom shape, down which the aeroplane must make a parabolic glide. Initially the tendency was to overshoot, put the nose down, and over-correct, resulting in a switch-back which could bring the machine uncomfortably near the ground, but though the system required great concentration, pilots managed quite successfully after a few hours' practice.

Courage and determination, even more than her navigational skill, were the keynotes of Jean Batten's latest exploit when she flew from Lympne on 11th November with her Percival Gull *en route* to South America by way of Casablanca, Morocco, Thies in West Africa and 1,700 miles across the South Atlantic to an abrupt landing in a swamp at Natal. She had equalled such men as Bert Hinkler and Jim Mollison, and though many French, Italian, and German pilots had flown that ocean since the late Capt Sacadura Cabral of the Portuguese Navy made the first crossing with a Fairey IIID seaplane in 1922, and currently the Germans ran a weekly service of flying-boats, in no way did that belittle her achievement. She knew the same anxiety at the change of note of the single engine; the compass hard to believe; constant worry whether wind direction had changed and altered the drift; rain, low cloud, mist, and obliterated vision; the unremitting vibration, even without the added discomfort of a bumpy atmosphere, which had effects on the bowel and bladder causing even greater problems for women. But why should she make such a flight? A challenge to her femininity? Rivalry with Amy Mollison? None of these: it was as indeterminate as man's prompting to climb Everest.

Five days before Miss Batten so resolutely set forth, George Bulman tested the silver-painted Hawker monoplane on which Camm had started design late in 1933 as a Goshawk-powered four-gun fighter with fixed undercarriage, modifying it the following August to specification F 5/34 as an eight-gun fighter with retracting undercarriage. Construction had been initiated as a private venture, for the Air Ministry contract to specification F 36/34 was not issued until February 1935. Sopwith, Camm, and a group of technicians came to Brooklands to see the flight of what seemed at that time an almost intimidatingly large fighter compared with the dainty Fury standing nearby. But it was a wise interim solution which had avoided design and manufacturing problems by retaining the easily manufactured, and therefore inexpensive, square-ended round tube structured fuselage of previous designs, fabric-covered like the wings which, though cantilever, were of conventional twin metal spar practice with wooden ribs. Essential to the design was the almost untried Rolls-Royce PV 12 (Merlin 'C') of 1,029 hp which weighed only 1,180 lb. No two-speed variable-pitch propeller had yet been built for such power, so a conventional Watts-designed two-blade wooden propeller was fitted. Take-off was uneventful, and Bulman climbed easily above the deep rim of Brooklands, heading into the west. He kept the undercarriage extended as a precaution, for it had somewhat uncertain hand-pumped hydraulic jacks. Presently he came gliding back, made one of his invariably smooth landings, taxied to the narrow strip of tarmac by the sheds, slid back the hood, and said when Camm clambered on the wing: 'Another winner, I think!'

Philip Lucas, his assistant pilot, forty years later recalled: 'I remember our surprise at the low approach speed and slow landing. I think Bulman also was quite astonished, for we had not previously experienced the combined effect of a thick wing section with the ground effect of a low-wing monoplane. This characteristic coupled with the very wide undercarriage was the main reason why squadron pilots found it easy to convert to the Hurricane.'

November also saw the total destruction of the Imperial Airways flying-boat *Sylvanus* after catching fire on the 9th while being refuelled in Brindisi Harbour. That was a poor augury for George Woods Humphery when he sailed on the 13th with Lieut-Col Shelmerdine and others to discuss with Pan American Airways and the US Government the possibilities and difficulties of the much desired transatlantic air mail services.

In a somewhat greater spirit of optimism the General Aircraft ST 18 twelve-seater made its first flight on 16th November from Hanworth.

13

There was mounting criticism of the National Government's failure to resist the successive aggressions of Japan, Germany, and Italy. At the Labour Party Conference at Brighton in October, Ernest Bevin so pungently emphasized that sanctions might lead to war that he caused a great split between the National Executive and those opposing both sanctions and rearmament, and George Lansbury, Leader of the Party, resigned because he disagreed with the official resolution supporting sanctions, carried by 2,168,000 votes to 102,000. Major Clement Attlee was then elected Leader. Such opposition to rearmament resulted in Baldwin 'going to the country.'

Support for the Government was overwhelming with 432 seats against the Opposition's 180. Baldwin, that astute politician in the guise of a reflective, pipe-smoking countryman, was firmly in the saddle. Ramsay MacDonald continued as Lord President, Neville Chamberlain was Chancellor of the Exchequer, Foreign Affairs were left with Sir Samuel Hoare, and Sir Philip Cunliffe-Lister on whom 'HM The King was pleased to approve that the dignity of a Viscountcy of the United Kingdom be confirmed upon him,' was reappointed Secretary of State for Air as Viscount Swinton. Sir Philip Sassoon, of incisive humour and frank exposition, similarly retained his post as Under-Secretary for Air. There had been expectation that Winston Churchill would be made War Minister, but Baldwin was far too cautious.

The Cabinet was facing many problems beyond those of producing airframes and engines in unprecedented peace-time numbers. Thus petrol supply was crucial to the RAF and causing some disquiet; the Navy and Army were no less dependent on oil. In 1934 Great Britain had imported more than $2^3/_4$ billion gallons of petroleum.

In a lecture to the Royal United Services' Institution in November, Lieut-Col W. A. Bristow recommended that large underground storage depots should be constructed to overcome the danger of supplies being cut off, but in any case more oil should be produced in this country from coal. He said that some 40 million tons of raw coal were annually consumed for domestic purposes alone, but if 50 million tons were carbonized it would yield 35 million tons of smokeless domestic fuel and 900 million gallons of crude coal oil which could be converted by hydrogenation into 150 million gallons of coal-petrol and $1^1/_2$ billion therms of gas, or could be distilled, yielding fuel oil, diesel, tar acids, pitch, and residues suitable for hydrogenation.

Certainly the days had long gone when the Navy steamed into battle with clouds of black smoke belching from their funnels, but there was a reminder on 20th November when Admiral of the Fleet Earl Jellicoe died at the age of seventy-five. Whether commended or criticized over the Jutland engagement of 1916, it had at least converted him to the vital role of fleet-spotting aircraft, but not until October 1917 had the Admiralty ordered all light cruisers and battle cruisers to carry fighting aeroplanes, the *Furious* to be fitted with a 300-ft aft landing deck, and the *Argus* to be used exclusively as a torpedo-plane carrier.

A similar old-time air surrounded the HP 42 Hannibals, butting majestically but economically into head winds – yet they were still favourites of Imperial Airways pilots, though it was five years since the first of the class had flown. Even older was the gallant old *Graf Zeppelin* which currently made its 501st routine transatlantic flight. In six years it had carried over 10,000 passengers in safety and flown more than 1 million kilometres, taking 100 hours or so from Friedrichshafen to Rio de Janeiro, depending on tail or head wind. Construction of the next Zeppelin was already well advanced.

The length of non-stop flights which airships commercially achieved was something of a feat of endurance for passengers, but Pan American with their impressive four-engined Martin flying-boat *China Clipper*, commanded by Capt Edwin Musick, was demonstrating that fast Pacific flights were practicable. Leaving San Francisco on the evening of 22nd November, he reached Manila on the afternoon of 29th November, having flown the 7,000 nautical miles in five stages in a flying time of just under sixty hours – a distance as great as London to Cape Town by the Imperial Airways route which took nine days and thirty stages, compared with six days, allowing for the International date-line, of the Pacific flight. Reuter said the USA's entire Pacific Fleet was in harbour and fired a salute and Capt Musick was presented with a key to the city. On 6th December the homeward journey commenced. All of five years had gone in preparing for this opening of a regular service.

Men such as Professor Geoffrey Hill were already proclaiming that there could be *Travel in the Stratosphere*, as described by him on 4th December to the Royal Society of Arts. After summarizing the stratospheric phenomena of atmospheric pressure and temperature extremes, and the difficulties to be overcome for man and motor, he suggested that pressure cabins could either be kept full of air at sea-level pressure which would need great strength in the structure, or could be filled with oxygen at reduced pressure which was easier but brought more risk of fire. To make the long climb and descent worthwhile, the distance flown would have to be at least 1,000 miles between landings, and necessitate variable wing area and controllable-pitch propellers with multi blades. His ideal was a cleaned-up transport version of the Comet with a cruising speed of 275 mph – which he feared would disappoint many of his audience thinking of 2,000 and 3,000 mph. 'Unfortunately there is a limiting speed of sound in air, roughly 700 mph, where air ceases to part ahead to make way for the moving object and no longer follows the streamlined shape but builds up a "shock" bow-wave.' In his charming, teazing manner, Hill concluded with the comment that undoubtedly his audience had come to the meeting hoping to hear what was going to happen in the future – whereas, if he knew, it would be happening in the present! Surprisingly, Tizard, despite his knowledge of secret work on gas turbines, commented that: 'It was rather odd to strive to do something, such as stratospheric flying, which when done was of no practical use – but at any rate we must get the height record from the Italians.'

There had been considerable secret investigation by the RAE and NPL of that possibility. Bristols at Air Ministry invitation had tendered in September of the previous year for the construction of two high-altitude research two-seat low-wingers Type 138A, powered with a two-stage supercharged Pegasus. For attempts at the record they could be converted to single-seaters with protective cockpit canopy, the pilot wearing a pressurized suit and helmet. Airframe weight had to be a minimum and for simplicity was wood, and the big span of 66 ft was dictated by the basic requirement of light wing and span loading. In January the design had been approved by the Air Ministry, and at the time of Hill's lecture the components of this spruce-framed, ply-covered machine were ready for sub-assembly.

Though stratospheric flying was regarded by Tizard, of all people, as a doubtful feature of the future, Wimperis had backed it in comparable manner to his early vision of the value of rotary wings. The practicability of the Avro-built C 30A Autogiro had been proved beyond doubt. British private owners, several flying clubs, and the RAF, had bought a score, and there were orders from Holland, France, Germany, Switzerland, Poland, Italy, Australia. Several constructional licences had been issued to various manufacturers abroad, as well as to A. V. Roe & Co Ltd, G. and J. Weir Ltd, Westland, de Havilland, Parnall, Comper and Airwork.

Westland had completed the eyeable side-by-side two-seat cabin CL 20 jointly designed by Petter and Lepère, but because there were some doubts during rotor running-up tests it was decided to dismantle the machine and send it by road to Hanworth where H. A. Marsh, now very experienced at testing, made the first flight, but found a deficiency of lift, though otherwise controllable. This was puzzling. Neither Cierva nor Lepère could find a reason, for the Pobjoy gave only 10 hp less than the power of the original much heavier direct control C 30. Experiments dragged on, but the machine had a ceiling of only a few hundred feet when carrying a passenger. In rash expectation that this very interesting little machine would presently prove satisfactory, Petter sought agreement from his father to lay down a batch of six welded fuselages and order machined parts.

The Weir W 2 single-seater designed by Dr J. A. J. Bennett, had also been disappointing, but re-designed as the W 3 with a slightly more powerful Weir engined designed by C. G. Pullin, it had a new type of two-bladed 'autodynamic' rotor to give a jump take-off like the experimental C 30. The rival Kay single-seat Giroplane, despite its outstanding advantage of collective pitch control, seemed to have faded into obscurity through lack of money for development. However, Raoul Hafner's much neater looking AR III Mk 2 single-seater powered with a 90 hp Pobjoy Niagara had obvious performance superiority over the similarly powered but 200 lb heavier Westland CL 20 and the 75 hp Kay, for its rotor disk loading was much less.

14

At the beginning of December, Vickers Ltd announced changes in their Board and that of their subsidiaries Vickers-Armstrongs Ltd and the English Steel Corporation Ltd. Commander Sir Charles Craven, chairman of Vickers Ltd became chairman of all three, and Mr F. C. Yapp, former director of Vickers (Aviation) Ltd, and Maj-Gen Sir John Davidson were appointed to the main Board of Vickers Ltd following the death of financial expert Sir Mark Webster-Jenkinson, who until May 1933 had been a director of the aviation business. Since his demise that company had in effect been autonomously operated by Sir Robert McLean who virtually overwhelmed Sqn Cdr James Bird of the Supermarine subsidiary, and H. H. Davall the secretary. The stage was set for a clash between the all-powerful Charles Craven and the strong personality of McLean. The action of the latter in throwing away what seemed a valuable order for the GP biplane against the hazardous chance that the Air Ministry might accept the monoplane had annoyed Craven who resented the independence of the aircraft company and the freedom with which McLean authoritatively made policy decisions direct with senior members of the Air Ministry or even the Secretary of State for Air. Matters had been exacerbated when the Supermarine 'development' multi-gun fighter was initiated as a private venture by McLean, who, it was alleged, had told Air Ministry 'experts' that they must not interfere with Mitchell's conception of what a fighter should be. Despite growls from Craven, the ascetic-looking McLean pursued his way. Meanwhile Craven lost a valuable adviser through the death of Sir John Valentine Carden, Bart, a director of Vickers-Armstrongs Ltd, who was killed when travelling in a Sabena-owned Savoia Marchetti which crashed on 10th December near Biggin Hill while making an approach to Croydon Airport through low cloud.

Carden, keen owner of a Klemm, not only was a car and engine designer, but, with his friend Vivian Lloyd, was responsible for design of the light tank built by Vickers-Armstrongs, which had become standard equipment in the Royal Tank Corps. As his obituary stated: 'Sir John had many ideas on improvement of aeroplanes, and thought much further ahead than the average professional designer of aircraft. Consequently it is reasonable to assume that if he had lived his great foresight, active mind, and intense energy would have been of great service to the technical progress and consequent expansion of British aviation.'

There were problems, too, of personality and policy disturbing the British Government. Sir Samuel Hoare on 10th December resigned his post of Secretary of State for Foreign Affairs as the result of Parliamentary fury on learning that, on his way through Paris for a holiday in Switzerland, he had agreed with Pierre Laval, the French Premier, regarded as cynical, hard and coldly logical despite a superficial temperamentalism, to propose that Italy should cease its war against Abyssinia in return for large tracts of that country. In the House Sir Samuel, precise of mind and courageous, boldly justified his policy in the belief that it would prevent war between England and Italy, or even a European war, and declared that sooner or later a return would have to be made to his solution. Nobody dared disclose that both Italy and Britain were interested in obtaining Abyssinian oil, so there were grounds for conciliation.

It was the end of all real belief in collective security. Baldwin was apologetic, but let Hoare go. Lord Cecil spoke of 'respecting the territorial integrity of Abyssinia' although it was composed of minorities conquered fifty years ago. Sir Francis Lindley said, with great common sense: 'The Paris proposals do not accord either with League principles or abstract justice – but without war no settlement can be obtained which is either the one or the other, so the practical question is whether the proposals offered Abyssinia a chance of a better settlement than she would have obtained if they were rejected.' For the time being Sir Samuel Hoare must take comfort from his family motto *hora venit* and await his opportunity. He was replaced by Anthony Eden.

Australia also was considering replacements – but by twin-engined American aircraft instead of British. On 13th December *Lepana*, one of the DH 86s operated by Holyman's Airways *en route* from Launceston to Melbourne, crash landed on Hunter Island, thirty miles from the Tasmanian coast, after a flying wire broke. This was the fourth DH 86 wrecked in Australia – the first when Capt Prendergast was delivering one to Qantas, and two others of Holyman's, one of which completely disappeared, and wreckage of the other was found in the sea. The Australian Minister of Defence suspended the Certificate of Airworthiness of all DH 86s in Australia pending investigation of structural design by their own technical people. A few days later the *Sydney Morning Herald* reported: 'Capt L. J. Brain, Flight Superintendent of Qantas, returned to-day from a conference in Melbourne regarding DH 86 planes and said he could make no disclosure on the outcome but would fly one of the type to Singapore on Wednesday. From that, he added, it could be inferred that the machines had come out of the inquiry quite satisfactorily and that he was satisfied with them.'

On 15th December the distinguished scientist Sir Richard Tetley Glazebrook died, aged eighty-one. In 1899 he became the first Director of the National Physical Laboratory at Teddington, and for nearly a quarter of a century, from 1909 to 1933, made important contributions to British aeronautical research, first as chairman of the Advisory Committee for Aeronautics, and then as chairman of the Aeronautical Research Committee which succeeded it in 1920. As one of his staff said: 'He was a particularly good chairman of committees. He would listen to everybody talking and tell them what the conclusion was and get them to agree to it!' But his qualities were far greater: he had a tremendous memory and a penetrative mind that rapidly assimilated and assessed the value of every scientific development.

In a sense his passing marked the end of the first epoch of British scientific research in aeronautics and the beginning of the industry's immersement in overriding problems of big production and big business during which those early devoted employees of every factory became submerged among the new entry of management and men from industries outside the aircraft business but skilled in their particular specialities. As S. R. Worley, chairman of Handley Page Ltd, said at the annual staff dinner just before Christmas: 'We are no longer working for ourselves, but for the nation. Many of our new employees have come from the nearly-finished *Queen Mary* on the Clyde, and as for the rest, James Hamilton, our works manager, has assured them that he could not wish for a better lot of people – even if there is not a Scotsman among them!' One of the new difficulties was that many an old hand, though greatly skilled in producing a prototype or a small series, had not the ability to explain how he would set about larger production, nor could he demonstrate with charts and calculations the guideline of expenditure of money against production progress and the involved timing of ordering and delivery dates for material and parts which now had become the familiar routine of those trained in mass production of cars. Sir Philip Sassoon had talked of automobile manufacturers turning over to aircraft business in order to fill the aeronautical expansion programme. Already there were rumours that the Austin Motor Co and Rootes Brothers Ltd were prepared to do that.

So to the end of the year and the Imperial Airways Calcutta flying-boat *City of Khartoum*, on the day-long haul from Brindisi to Alexandria via Crete for refuelling, crashing just short of the breakwater at its final destination after all three engines stopped at 600 ft during final approach to the flare path. Normally it would have taken four hours for the 360 mile stage from the island, but the airliner was more than an hour overdue when its W/T operator informed Alexandria that he was winding in his aerial prior to alighting. By then it was dark, and when the machine failed to appear, naval vessels began to search, but it was five hours before *HMS Brilliant* reported picking up the pilot, and half an hour later found the wreckage containing the bodies of all nine passengers and the remaining three crew. Long investigation followed, for it seemed likely that the machine was inadequately refuelled at Crete, but the pilot insisted he had constantly observed the petrol gauge, and that a minute before the accident the starboard tank showed empty but the port registered 25 gallons which was sufficient for another 20 minutes flying. Many weeks later the Consul-General at Alexandria, as coroner, said it was impossible to deduce the cause, but he rejected a theory that alteration to the fuel jets was the primary factor as it could equally have been an air-lock which was an 'act of God'! He recommended to Imperial Airways that alterations affecting fuel consumption should be brought clearly to the pilot's notice, and he criticized the delay in communications and rescue measures.

Communications and rescue measures – was that what the world needed? Surely Britain was not lagging? The United Kingdom had the longest regular air service in the world, reaching to Australia, and this year a direct telephone service had been initiated to Japan. Early next year it was planned that the British public would have even more dynamic contact with events than afforded by radio, for there would be the new medium of high definition television using the Shoenberg electronic system manufactured by EMI. Was not the motto of the BBC: 'Nation shall Speak Peace unto Nation'? There were also those unfathomed possibilities of Radar, for as Sir Robert Watson-Watt later said: 'Man had discovered a method of seeing without eyes,' and that might prove of incalculable value in guiding aircraft and ships through darkness, storm, and fog. Already the compass bearing, height, and distance of an aeroplane could be accurately established from the ground.

But what of political rescue methods? With Eden in office at the Ministry of Foreign Affairs it was evident that to conciliate public feeling the League of Nations must remain at least a nominal force, though long proved an incompetent channel through which to defeat aggression. Mussolini was accelerating the invasion of Abyssinia, and Hitler ever more blatant with his resurgent, rapidly re-arming Germany. Newly industrialized Japan, unhampered by League affiliation, was ready to continue war with China and posing a threat to Anglo-American strength in the Pacific. Beyond the eastern horizon of Europe loomed the enigmatic figure of Stalin and the red threat of Communism, though that had not prevented a recent trade agreement with Russia.

Despite the vocal minority of conciliators in Britain, it was evident that re-armament must be accelerated under guise of defence, and even for unilateral action if necessary. Nevertheless the government's air-raid precautions pamphlets which had been popped through the nation's letter boxes, stirred only a passing ripple of interest and a general belief that the warnings were a move to ensure that a National Government was kept in power. It had been of much greater interest when that long standing landmark of southern London, the Crystal Palace of Queen Victoria's day, was destroyed by fire, for whatever the game of politics, when it came to their traditional background all Britains were conservative at heart. They did not want a changing world – only peace and the prosperity which this recently extending re-armament programme was endangering.

CHAPTER II

1936: *New Types for Old*

'No revival of economic stability is possible without international political confidence, and this is largely dependent upon a sense of security, which is also the foundation of peace.'

Brigadier-General P. R. C. Groves (1936)

I

The New Year dawned with a bitter east wind that matched a course set for war; yet the British people seemed reluctant to believe that rearmament had a deadly purpose; indeed the extent of preparations was unimaginable amid the new prosperity of rising incomes, cheaper goods, easier transport, football pools with ten million participants, and twenty million going weekly to the grandiose cinemas of Gaumont and Odeon. Why not leave it to others to settle the problems of money, depressed areas, and international squabbles? Leave it to good old Baldwin! Another war? Unthinkable! Had not Colonel Lindbergh and his wife come to live in England because it was so safe? And that doughty cripple, Franklin Roosevelt, virtually dictator of the richest country in the world, had announced that in the event of war between any nations he would cut off American oil supplies to the belligerents – probably more effective than anything the League could do.

Customarily the year's opening promotions and postings presaged the augmentation of policies. Thus fierce-looking Air Vice-Marshal Frederick 'Ginger' Bowhill and calmly confident Air Vice-Marshal Charles Burnett were promoted to Air Marshals and gained KBEs. Air Vice-Marshal Barratt was appointed Commandant of the RAF Staff College at Andover where he had been chief instructor six years earlier, and his recent office of Director of Staff Duties was transferred to Air Commodore William Sholto Douglas.

Tall, quiet, Lieut-Col Francis Shelmerdine as DGCA was awarded a KB for his success in the delicate task of reconciling British and Canadian air transport interests with those of the USA. In this he had been assisted by Ivor McClure, a social character of devastating wit, who originated the AA's Air department which had proved more influential than the Royal Aero Club in providing information, maps, and world-wide touring permits for private owners, and even for record-breakers. This service now continued in the hands of his efficient and round-spectacled former assistant O. J. Tapper. However it was not McClure but John Galpin who was appointed Deputy Director of Civil Aviation. Air Vice-Marshal Sir Hugh C. T. Dowding, who as AMRD had so ably striven to give the RAF its new look in weapons, monoplane fighters, and bombers, was appointed C-in-C Fighter Command. His sobriquet 'Stuffy' was aptly descriptive. Humourless, withdrawn, he masked his quick perception and kept his distance at meetings with designers and directors of the aircraft industry, yet his staff found him to be a magnificent leader, balanced in his judgments and fearless in upholding them. He was succeeded by Air Vice-Marshal Wilfrid Rhodes Freeman, with effect from 1st April. No one could have offered greater contrast, for Freeman was a taller, younger man of charm and handsome appearance to whom the industry quickly responded with mutual friendliness – though, as one of his colleagues commented, 'behind his ready smile and quick mind was a cynicism and moodiness which could baffle his staff, some of whom thought him unreliable because of changes of mind for reasons he did not explain.' Often this was because he accepted arguments made by designers whose views might considerably differ from those of DOR.

Of rising Air Commodores, not only A. S. Barratt but E. L. Gossage, J. E. A. Baldwin, and R. E. C. Peirse, had been promoted to Air Vice-Marshal. Promotions in other ranks revealed that the RAF Command was aware of the merit of many other outstanding junior personalities.

All over the world there were portents of war. Italian bombers were attacking every column of marching Abyssinian troops they could see; Japan had launched a new aircraft-carrier, the *Soryu* of 10,000 tons capable of 30 knots; the Grand National Assembly of Turkey had authorized £3½ million for more aircraft to add to their Supermarine Southamptons, Curtiss Hawk fighters, and Breguet reconnaissance bombers. In the USA, Brig-Gen W. G. Mitchell was still battling for a more effective Army Air Corps, and bitterly said: 'Not a single American aircraft unit is suitable for war service. Of the £600 million spent by the USA on aircraft since the War, none has gone to real progress in aviation. It has been thrown into the hands of financial manipulators.' Nevertheless Roy Fedden and Major Bulman, separately visiting the USA, were much impressed with the enormous advance which American aeroplane and aero engine designers had made. However Bulman confirmed that: 'The USA Army Air Corps is handicapped by bureaucratic machinery of their procurement system limiting contracts to work which must be completed within the current financial year,' and perhaps with Brig-Gen Mitchell in mind added: 'It takes much greater strength of character for an individual to drive through the fog of officialdom there than it does here.'

The Americans not only had established the new look in metal-skinned airliners and bombers but also specialized in 'rugged' aircraft for 'bush' conditions, such as Canada. However, that country was attempting to fight back with its own designs. Bob Noorduyn, a Dutchman whose mother was a Churchill from Somerset, had recently initiated a company to build a remarkably efficient 51 ft 6 in span, strut-braced high-winger with welded tube-framed fuselage for easy repairs. Noorduyn had been one of Sopwith's earliest draughtsmen, then joined Armstrong Whitworth Ltd in 1914, becoming responsible for 'Big Ack' production before joining BAT as Koolhoven's design assistant. When Sir Samuel Waring shut down that enterprise he became general manager of Fokker's factory in the USA, then vice-president of the Bellanca Aircraft Corp, subsequently joining Pitcairn Aircraft as designer of their cabin Autogiro, but when these found few sales he migrated to Canada. Noorduyn, in collaboration with his associate Don Martin, had canvassed the requirements of many small aviation operators and it quickly became clear that they needed a machine simple enough to be handled and maintained by one man on lakes far from civilization, and have room and power to carry mining equipment with large doors for access, and though primarily a first-class seaplane must also stand up to the terrific racking of skis. The result was the Noresman. Powered with a 420 hp Canadian-built Wright radial, and aided by Handley Page slotted flaps, it could take off in twenty two-seconds with a payload of 1,250 lb at an all-up weight of 6,050 lb. Whether so specialized a business could make good seemed doubtful, despite an immediate order from Dominion Skyways Ltd – but if Noorduyn could keep his overheads low he saw a chance of emulating the success in England of Percival and Miles in competition with the established de Havilland business.

Profits in England were certainly low. At the annual general meeting of Fairey Aviation Co Ltd shareholders found that the profit was £39,112 10*s* 5*d* compared with £47,534 2*s* 4*d* for the previous year. Whether that included the Belgian factory was not made clear, for this was a profitable sideline taxed only in that country. Towering head and shoulders above his audience, Dick Fairey assured them that orders in hand and in prospect exceeded anything they hitherto had; three types were going into production, of which delivery of one had started, and they had hopes of orders for yet another type. Hinting at the rumour of shadow factories, he emphasized: 'Only our highly specialized industry can fulfil the large and urgent requirements now being made. The effort to obtain the necessary aircraft of proper quality from other sources, or by organizing national factories, would fail because it could not possibly acquire the knowledge and technique within the necessary time, if at all – but I am conscious that anything I may say relative to the prospects of the company or the industry is liable to be taken down, stripped of its context, and even distorted in order to be used as evidence by those opposed to private manufacture of aircraft.' Certainly his purchase of the Heaton Chapel factory at Stockport had been vindicated for it was twice the size of the Hayes Works and some 750 people already were employed there, managed by Major Tom Barlow. Meanwhile both the Hayes plant and the Hamble Works were being enlarged.

A few days later, on 16th January, a man who was almost a legend died, aged seventy-six, following an operation in a London hospital. He was Rudyard Kipling, patriot poet and novelist, joy of the youngsters but also prophet of flying, for early in the century he wrote *With the Night Mail* vividly forecasting a night flight between London and New York with a machine supported by an anti-gravity device instead of wings. He also publicly championed France and showed an inveterate antagonism to Germany. *The Times* commented: 'It will take time to clear away the clouds of partisanship that this enthusiasm has drawn over his memory. He may be found to have been mistaken; he will never be found to have been anything but sincere, courageous, a single-hearted lover of his England.'

But even the memory of him became subdued in national mourning on 20th January when King George V died at Sandringham after catching a chill. With his death an era vanished in a world declining into international suspicion and preparation for war. As his eldest son said: 'My father waged a private war with the twentieth century, for he believed in the old-fashioned attributes of common sense, tolerance, decency, and truth.' A columnist commented: 'King George V will go down to history as the King of England who above all others shared the lives and joys and sorrows of his people. There are those who imagine that the King of England is a figurehead who does what his Ministers tell him, but that is not true of the present Royal House. Queen Victoria was an immense force behind the scenes at home and abroad. King Edward VII was a diplomat of unexcelled ability. King George, above all things, was the Father of the English people.' He was also Commander-in-Chief of the Royal Air Force.

Five days after the death of the King, Sir Alliott Roe's great business partner, John Lord, managing director of Saunders-Roe Ltd, died at Cowes following a stroke. This great, ebullient, little Lancashire man had been the backbone of the Avro business ever since commercial construction began in 1910 at the Everard Works owned by A. V.'s brother, H. V. Roe. As C. G. Grey said: 'From that time, John Lord became a prominent figure in the aircraft industry. His shrewd common sense, capacity for hard work, commercial knowledge, and friendly attitude towards all from the highest government officials to workmen in his own shops, were of the highest value to the Avro business. And he never failed to present in their most favourable light the products of that able team, Messrs Parrott, Chadwick, and Dobson, who during and since the War have been responsible for the design and production of the successful series of Avro aeroplanes.' And with John Lord went a great fund of funny stories.

A reshuffle of Saunders-Roe affairs followed. Forty-year-old Sqn Ldr C. J. W. Darwin, DSO, who since 1928 had been London manager of the Bristol Aeroplane Co Ltd, was appointed managing director with Harry Broadsmith as technical director and the irreplaceable Henry Knowler as chief designer. Their new main assembly shop on the east bank of the Medina dominated the Cowes landscape. Alongside was the old stone house where Sam Saunders once lived but it was now the drawing office. Said a visitor: 'You do not get a fair idea of the immensity of the shed until you see two large flying-boats being built nose to nose and looking quite unobtrusive. A large number can be simultaneously erected and can then be pushed out of the shed, run down the slipway, and onto the water at the rate of one a fortnight. Across the back of the main bay are offices for the Works staff and planning department. The manager's office commands the whole floor. All this is additional to the other Saro shops which have recently been re-equipped with the latest machine tools.'

On the 28th the funeral of King George took place at Windsor, after lying in State at Westminster Hall, London, guarded by officers of the Household Brigade and Gentlemen-at-Arms. Meanwhile the youthful-looking, popular but unconventional Prince of Wales was proclaimed King Edward VIII. On the first day of his reign, 21st January, the new King, accompanied by the Duke of York, flew to Hendon from the RAF aerodrome at Bircham Newton near Sandringham to attend a meeting of the Privy Council, thus becoming the first British monarch to use the air for transport. A few months later he ordered Smith's Lawn in Windsor Castle grounds to be converted to a landing strip for his private and official aeroplanes.

Not far away, near the village of Langley, on the north side of the Great Western Railway London to Bristol line, Hawkers had bought an extensive arable area as an aerodrome for prototype construction and testing instead of at small and crowded Brooklands. Similarly the company's associated firm, Sir W. G. Armstrong Whitworth Aircraft Ltd, transferred activities from their cramped Whitley aerodrome to the new Coventry Municipal aerodrome at Baginton, three miles from the town, where an initial factory building, 600 ft by 150 ft, had been erected. There was every possibility of extending the area, for it was in the middle of 1,100 acres owned by the Coventry Corporation.

Westland's narrow little grass aerodrome at Yeovil was also being lengthened to give a 1,200-yard east–west length to cope with increased landing runs of future aircraft. A large new fitting shop and store had also been built, and in it on the day of the King's funeral, Sir Ernest Petter held a brief Commemorative Service attended by all his employees.

So extensive was practical interest in aerodrome requirements that a three-day conference was held by the Aerodrome Owners Association under the aegis of the Society of British Aircraft Construction at British Industries House, London. The official lunch was cancelled in the belief that it showed disrespect to the late King – so everyone lunched unofficially and even indistinguishably from the expected one, but rallied to drink the health of the new King after a respectful silence for the old. Thereafter the large audience was addressed by the darkly handsome chairman of the Association, Councillor Richard Ashley Hall, a thirty-five-year-old paint manufacturer who was the moving force in private and commercial aviation at Bristol. *The Aeroplane* reported: 'The chairman and various high officials of the Air Ministry took their coats off; everybody smoked, and an atmosphere of informality communicated itself to the proceedings. People of considerable academic attainment told the truth about technical things in a way we have never heard equalled at any scientific meeting. Responsible but comparatively minor officials of aircraft constructing companies said what they thought of other designers and constructors and the Air Ministry and its methods in a way we should have thought would assure their dismissal on a week's notice! The conference will probably stand out in memory of all aeronautical technical people as the most completely joyous incident in their dull academic lives.'

Discussions hardly left time to inspect the splendid exhibition of specialist equipment, and on the last evening everybody rushed to Croydon by motor coach to watch night flying, but were disappointed to find that although the weather seemed perfect, the programmed Airspeed Envoy prototype, recently fitted with split flaps and re-engined with 250 hp Wolseley Scorpios, was not allowed to make its intended demonstration flights despite bright moonlight and a mere breeze.

'They daren't risk it,' someone maliciously whispered. 'Those accounts of the recapitalized company show a loss of £4,350, and this machine is their only asset!'

That was partly true, but though the Envoy was a non-starter this cold January night, the situation was saved by the Lufthansa night mail Junkers which took off across the floodlit undulations of Croydon, swung round into a climbing turn, its downward-shining landing light picking out the huddled spectators on the Airport roof, then roared away towards the North Sea, regardless of weather warnings of worsening conditions in Germany.

While Airspeed was struggling to establish sales for the Envoy, its somewhat heavier, more powerful, and rather more practical equivalent, the Avro 652 was proving very successful with Imperial Airways because of its exceptional cruising speed of 165 mph. More importantly, the first of the RAF Coastal Defence production versions with centre-bay internal stowage for two 100 lb and four 20 lb bombs had been flown on 31st December by Geoffrey Tyson, and was being favourably reported upon by the A & AEE. The next three were almost ready, and 170 were following, with promise of a further order in a few months. Since early prototype flights, the tailplane span had twice been extended, and the rudder reverted from a horn balance to the inset shielded balance of the Imperial Airways version, also the 310 hp Siddeley Cheetah IX engines were moved a foot further forward than those of the civil aircraft. Despite the fixed nose gun and Lewis in the hand-operated cupola of the aft cockpit, the Anson was not intended to fight it out, but run for home taking advantage of what was considered a high top speed of 188 mph. Undoubtedly it was one of the easiest aeroplanes to fly, with control response appropriate to its modest inertia, but the undercarriage winding handle at the side of the pilot's seat still required such a huge number of turns that it was exhausting work.

The rival DH 89M adaptation of the Rapide meanwhile had required a curved fin extension sweeping the leading edge into the fuselage, and was obviously a frailer and more complicated construction than the Anson, but primarily was ruled out because it was some 30 mph slower and had inferior climb – though a limited number were ordered as communication aircraft for No. 24 Squadron at Hendon.

De Havillands in any case were busy. The civil version of the DH 89 was in use by twenty-eight operating companies and the DH 86 Express four-engined airliner by Imperial Airways, Qantas, Railway Air Services, British Airways, Jersey Airways, British Continental Airways, Holyman's Airways, Misr Airwork, and Union Airways of New Zealand. Sixteen more were on order. Hornet Moth production had been held up until a more conventional mainplane with only the slightest taper and almost square ends was designed, issued to the shops, tested, and ultimately approved by the Air Registration Board. Owners were invited to trade in their original Hornet wings, and most of them did so. Thereafter the machine was known as the DH 87B, and one was purchased by Lord Londonderry. Many went to clubs, and the largest fleet, ten in number, were for the chain of flying clubs operated by the Straight Corporation founded by young Whitney Straight. A surprise at Hatfield was the production line of sub-contracted Gloster Gauntlets, despite Geoffrey de Havilland's avowed concentration on civil types. In the experimental shop there was evidence of the Air Ministry's decision to order Hagg's big wooden airliner, like a scaled-up Comet, powered with four of the as yet untried 525 hp twelve-cylinder inverted vee-type engines designed by Halford. Compared with the DH Express this was a design of tremendous aerodynamic refinement and beauty, thanks to Hagg's artistic eye. The long tapering fuselage of circular cross-section was planked with cedar ply outer and inner skins and balsa between in the manner devised by Hill for his first Pterodactyl. The 105 ft tapered wings were being built in similar style to the Comet with diagonal spruce planking in two layers around a big box spar and supplementary rear spar carrying flaps and ailerons. There was also the new five-seat DH 90 Dragonfly tapered-wing biplane of 43 ft top wing span and 38 ft lower and only 288 sq ft area, the unique feature being a cantilever lower wing root taking the entire loads of both wings from an inverted V-strut and outer-braced cellule. Fuselage construction was closely based on the Comet. At £2,650 this twin-engined *Grande Tourisme* was £700 less than the standard Dragon. What with this and expanding sales of the other types, together with an output of three Gipsy engines a day and 5,000 already in use all over the world, Alan Butler at the AGM had cause to be cautiously optimistic. A Series II Gipsy Six was being bench-tested with hopeful results, fitted with the DH version of the Hamilton two-pitch propeller manufactured in a special section of the Works as a trial run before establishing series production of propellers suitable for the new era of powerful fighters and bombers. To assist Frank Hearle, now general manager of the company, he was joined by Flt Lieut John J. Parkes, who learned to fly at the DH School in 1926 and for six years had been at Airwork, first as an instructor and latterly as technical manager. Within a few months the indefatigable Francis St Barbe and that ingenious designer Arthur Hagg was appointed to the Board.

De Havilland production of the Tiger Moth was almost exclusively for the RAF and Air Forces of foreign governments; none were available for the majority of the flying clubs, of which thirty-three were subsidized by the Government. Of these, twenty-seven submitted returns for the past year showing they totalled 7,738 members compared with 7,490 in 1934 and 4,800 in 1933, and made over 90,000 instructional flights in almost 40,000 flying hours using 117 machines averaging 335 hours each, resulting in 559 new 'A' licences and 123 'B' licences.

On many a bright day the skies seemed to hum with little aircraft making cross-country flights or heading abroad, and professionals were still intent on records. Two such pilots set forth on 6th February bound for South Africa. One was the ever popular Tommy Rose with the Miles Falcon demonstrator owned by Phillips & Powis Aircraft Ltd. Starting from Lympne just after midnight, a series of long hops averaging some 1,200 miles brought him to Cape Town a few minutes after 6 pm on 9th Feburary, having achieved the 7,134 miles in three days seventeen hours thirty-seven minutes thus beating Amy Mollison's time by thirteen hours nineteen minutes, and despite only two hours' sleep, this tough rugger-playing test pilot seemed not in the least exhausted and greeted everyone with his usual broad grin. The other memorable effort was by David Llewellyn, recently appointed pilot to Light Aeroplanes Ltd operated by 'Bones' Brady, and remembered for his Hendy Heck record flight to the Cape the previous year. His new mount was the ultra-light, wire-braced high-wing, two-seat Aeronca, introduced from the USA by Brady's firm which was in the throes of purchase, together with Lang Propellers Ltd and Aircraft Accessories Ltd, by the newly-formed Aeronautical Corporation of Great Britain Ltd. Directors were Humphrey Roe, brother of 'A. V.', and founder of the Avro business; Brig-Gen Guy Livingston the pre-war manager of Grahame-White and recent representative of Vickers Ltd in South America; John V. Prestwich son of the founder of the JAP Motor Co who originally entered aviation by supplying engines for Alliot Roe's triplane and Harry Ferguson's Irish-built monoplane; S. D. Davis, recently from Hawkers and earlier of Vickers design departments who was engaged as chief engineer; and W. G. 'Bill' Gibson, that vigorous ex-works manager from Westland.

The first of these Aeroncas, powered with a two-cylinder Aeronca of 40 bhp, was sold to a private owner on South Africa's Gold Reef, so Llewellyn had undertaken the most daring ultra-light cross-country flight so far attempted. Flying by easy stages in the course of twenty-three days from Hanworth to Rand Airport, Johannesburg, he had no trouble, but it took six hours to cross one stretch of the Mediterranean; and when faced with the desert he managed over 600 miles non-stop by augmenting his fuel from a hand-pumped tank on the baggage rack behind him.

Simply built, using a triangular-section fuselage of welded steel tubes faired with ply formers and stringers and wire-braced to the simplest of untapered wooden wings with plank spars, it was an amazing little machine with ample room in the cabin to seat two in comfort, and remarkably easy to fly. Despite such low power, take-off in a light wind was under 150 yards, but climb was little more than 300 ft/min and a cautious flight path was necessary to avoid trees and buildings. Full out at 2,500 rpm somewhat smoothed the noticeable engine roughness, and resulted in a shade under 80 mph. Approaching at 50 mph was a simple exercise ending with an absurdly easy landing despite the uncanny sense of scraping along the ground because the undercarriage was so low, but the differential brakes proved delightfully effective after the more usual brakeless light aeroplane. With dual ignition engine made by J. A. Prestwich Ltd, it was expected to sell at £385, and because it cruised a little faster than the earliest Moths the sales prospect seemed not unreasonable – until one tried flying in a brisk wind and realized how much easier ground handling and landing was with a machine of much heavier wing loading.

2

Further emphasizing the end of one era and the beginning of the new were two diametrically different machines flown by George Bulman that month – the Hurricane prototype K 5083 to Martlesham on 7th February, and a week later when he flew the last biplane ever to be designed by Camm and his men – the attractive little Hector Army Co-op two-seater, powered with the twenty-four-cylinder H-type air-cooled Napier Dagger. No piston engine could be smoother. To add yet another type of powerplant to the RAF's supply and maintenance system seemed questionable, for though the speed was 17 mph faster than the Kestrel-powered Audax and climb to 10,000 ft two minutes better, the real reason was that the Air Ministry wished to keep the disappointed but brilliant and ambitious Halford and his employers, Napiers, in business in case war came.

The company had a trading loss of £19,960 in 1935 and £50,000 had been taken from general reserves and £15,000 from the contingency reserve to write off development expenditure on Halford's engines. What pulled the company through was almost £107,000 from their mainstay revenue of investments, to which was added recovered income tax and a balance from 1934, enabling a dividend of 8 per cent to be paid. Undoubtedly the directors were heartened by an order of 300 engines for production Hectors which, in the great rearmament move of the previous year, had been programmed for construction by Hawker Aircraft's associate A. V. Roe & Co Ltd who were now too busy on the Anson to undertake it; but since reorganization of Westland was now deemed by the Air Ministry to be satisfactory, an ITP was placed with them for 178 aircraft and the contract ratified in April.

By Air Ministry edict, delivery of the prototype Hurricane had been made after only ten flights totalling eight hours. As Philip Lucas explained: 'Encouraging as the first few flights had been, we soon ran into the inevitable teething troubles one expects with a prototype, and because of the urgency to fly it to Martlesham, such limited flying as was possible had to be restricted to brief evaluation of stability and control and testing modifications to the sliding hood, retractable undercarriage and engine installation, all of which were giving considerable trouble. The sliding hood could not be opened in flight above a rather slow climb speed, and the first came off after the third flight. The undercarriage was almost impossible to retract and lock using the hand-operated hydraulic pump, for there was no such thing as an engine-driven pump. Even the engine started to develop serious mechanical faults, and because of the need to nurse it, very few performance measurements were possible beyond snatch readings in level flight to get some idea of the suitability of the wooden propeller. We could do no diving to check stabilty and control at speeds greater than level flight because the engine was limited to only 5 per cent above maximum permissable, nor was the aircraft spun or aerobatted. Although initial difficulties were corrected at least to a usable standard, we were far from happy at letting Service pilots fly our precious prototype before we had time to find out much about it ourselves.'

Nevertheless Martlesham reported on the remarkable ease of handling and good control at all speeds down to stall. A top speed of 315 mph at 16,00 ft was established, thus handsomely beating the Air Staff's requirement of 275 mph. 'The only thing which marred the otherwise very satisfactory trials was continued unrealiability of the engine,' said Lucas. 'There were at least three engine changes during the first two weeks due to a variety of defects, the most serious being internal glycol leaks causing rapid loss of coolant, coupled with distortion and ultimate cracking of cylinder heads because of much higher operating temperatures possible with this type of coolant. Soon it was apparent that the engines required a great deal more development before it became sufficiently reliable for Service operation. All this delayed Martlesham tests and the machine's return to Brooklands for development flying and performance measurements. Meanwhile Rolls-Royce decided that the troubles could only be overcome by intensive flight development with re-designed cylinder heads. We learned that Merlin I engines would not be available for production Hurricanes and that the modified Merlin II would not be ready until autumn of 1937, some three months after the first production Hurricane was due off the line. Worse still, we were told that only a bare minimum of engines would be available to keep the prototype flying.'

Default by Rolls-Royce would disastrously hit the ambitious new programme of fighters and bombers. Early doubt of the speed of rearmament infiltrated Parliament, resulting in renewed proposals for a Ministry of Defence and a Royal Commission to enquire into organization, equipment, and control of the Fighting Forces – but after due debate these motions were withdrawn. Concurrently pressure increased for more sub-contracting. General Aircraft Ltd was a lucky recipient as Hollis Williams recorded: 'Early in 1936 it was obvious that costs were out of hand and our financial affairs getting difficult. We were building the twin Pratt & Whitney ten-seater Croydon, and it was ready for test at the beginning of March. Though that remarkable man Eric Gordon England was convinced of a civil market, cash was running out and something had to be done. At that point we managed to land a contract for eighty-nine Furies. That was the only time a firm outside the "ring" got a direct contract from the Air Ministry for complete aircraft – and this of course made the Society of British Aircraft Constructors tighten its belt and decide this must not happen again!'

Though the Fury was not so fast as the Gladiator, the RAF heavily relied on it to fill the gap until modern machines were ready. Seventy-two Furies particpated in the small-scale Defence Exercise against an equal number of Heyfords on 18–19th February, in which bombers attacked Biggin Hill, Hornchurch and Dagenham by day, and Staines Reservoir, Ponders End railway station and the Ford Works at Dagenham by night. Layers of cloud favoured bombers on the first afternoon, but at night the sky cleared and twenty of the thirty raids were intercepted. Next day there was fog, and the bombers were forced down all over the country, two Heyfords crashing with loss of six lives, but in general the bombers again proved that most could get through to their targets.

Russia was demonstrating attack in more dramatic fashion. Their publicity film of the Red Army's recent manoeuvres showed not only 700 parachutists dropping with their machine-guns and light field-guns but also a fleet of 'giant aeroplanes carrying tanks, armoured cars, lorries, and field-guns slung beneath. When the aeroplanes landed the crews jumped out, manned their equipment, and in a few moments the whole force was in movement striking rapidly at the enemy from the rear.'

At the opening session of Russia's Central Executive Committee in the Kremlin, Vyascheslav Molotov, President of the Soviet of Peoples' Commissars, stated that more money must be spent in 1936 to strengthen the Red Fighting Forces, and inferred that the two prospective enemies were Germany and Japan. Meanwhile Eden, as Foreign Minister, seemed to have been assured by Maxim Litvinov that 'the British oil fields in Iran and Iraq are quite safe against Russian bombers in the event of war.'

Germany was well aware that Russian Communism was the chief threat, and projected rearmament was designed to equate with Soviet strength, but the new Zeppelin LZ 129, named *Hindenburg*, launched at Friedrichshafen on 4th March could not have been intended for warlike purposes. Though this was the greatest airship yet made, Dr Eckener, who took command for the first flight, found so little fault with its handling that he stayed up for three hours and did not bring the monster in until misty dusk, with the ground illuminated by ninety-six floodlights and a crew of 500 standing by to manhandle it into the hangar. A month later it made its opening transatlantic voyage to Rio.

Unpublicized but of more significance to Britain was the refined and elegant Supermarine eight-gun fighter K 5054, soon known as the Spitfire, flown at Eastleigh Airport by 'Mutt' Summers on 6th March. Painted a highly polished blue-grey smoothly hiding every joint and rivet, it seemed the most exotic and awe-inspiring fighter ever produced – yet had a remarkable similarity to the smaller span Messerschmitt, although the latter had straight taper wings and the Spitfire's were elliptical, based on the Heinkel. Reggie Mitchell, clearly a very sick man, was there to watch. As Joe Smith said: 'During flight trials of a prototype one did not indulge in silly chatter in light-hearted manner, for he was always worried for the safety of his pilots, and although he witnessed the first flights of so many aircraft he never grew accustomed to it. He was in continual tension lest a pilot be injured or killed and felt that he carried a personal responsibility in the matter.'

Though suffering bouts of pain, Mitchell continued supervising design, spending almost all his time in the design office discussing details with the draughtsmen. His latest and last design was an impressive four-engined bomber with bombs stowed internally across the entire span between port and starboard outer engines. The specification was almost certainly inspired by the Intelligence discovery that the four-engined Dornier Do 19 and Junkers Ju 89 'Ural' heavy bombers would be ready towards the end of the year.

Several months after the first flight of the Spitfire, I flew it at Martlesham – the first non-Vickers industrial pilot to do so. Though my notes were of the usual technical jargon I recorded elsewhere such impressions as 'the rolling gait of the narrow undercarriage as I taxied out; the dropping wing and emphatic swing as the over-course fixed-pitch wooden propeller laboriously gripped the air, dragging the machine into a run faster and longer than anything I had experienced. Longitudinal control felt excessively sensitive, and because the undercarriage retraction pump was on the right I could not stop my left hand on the control column moving in sympathy, so the flight path violently oscillated until the thuds of the legs locking into the wings reassured me, and the machine settled into steady flight path. Suddenly a Gladiator appeared 1,000 ft above, its fixed cantilever undercarriage extended like an eagle's claws, offering the opportunity of a dog-fight. I drew the stick back in the manner I had become accustomed, forgetting the lightness of control. A vice clamped my temples, my face muscles sagged, and all was blackness. My pull on the stick relaxed instantly, but returning vision found the Spitfire almost vertical and the Gladiator a full 2,000 ft below.'

Every pilot who flew that Spitfire instantly recognized it as a winner. The prototype was 34 mph faster than the fabric-covered Hurricane, and both were faster than the Messerschmitt Bf 109 in its early version with Kestrel and wooden propeller, but now was being tested with a Jumo engine. News that a production order had been given for 310 Spitfires to production specification F 16/36 soon reached German Intelligence, resulting in pressure to speed competitive trials between the Messerschmitt and the shapely Heinkel He 112.

Never before had Supermarine received so large an order. The problems of such advanced and complex a metal monocoque structure were well known from experience with the Schneider racers and production of metal flying-boat hulls, but this time expensive jigging and tooling would be required, and the Works' capacity was insufficient, so extensive sub-contracting had to be planned – and urgently, for Germany on 7th March repudiated the Locarno Treaty, and on the 8th the Rhineland Zone was reoccupied by German troops, resulting in the League Council on 19th March declaring that Germany had infringed the Versailles and Lacarno Treaties.

On 10th March the silver-painted low-wing Fairey Battle prototype, K 4303, made its debut in the hands of Chris Staniland after some months delay due to the Merlin engine snags found by Hawkers with the Hurricane. A two-seat fighter-bomber equivalent of the war-time Bristol Fighter was still deemed necessary. Because of the urgency of rearmament, the Air Ministry had already placed a pre-flight contract for 155 on the understanding that there would be a guaranteed speed of not less than 195 mph at 15,000 ft, a figure which the prototype exceeded by 68 mph, though it became less for production machines. The rival Armstrong Whitworth Tiger-powered AW 29 was still far from completion, and could not compare in elegance with the slim Battle, but it offered superior rearward defence because of the Whitley-type gun turret, whereas the Fairey had a patented spring-loaded gun mounting shielded only by tilting the rear canopy to form a wide windscreen for the open cockpit.

I flew the Battle not long after, and found it very easy to handle though heavy; it had an innocuous stall, and was simpler to land than a Moth because of the cushioning low wing – though my purpose was to try the DH-Hamilton two-pitch propeller and experience the benefit of fine pitch for take-off. Fairey hoped to replace it with one of Graham Forsyth's patented design, based on his early hydraulically-operated constant-speed propeller tested on the R-R Condor of a Horsley in 1926; but this project was adding greatly to the load on his department, for he was developing a sixteen-cylinder version of the twelve-cylinder Prince, intended to compete with the R-R Merlin, and also designing a still more powerful engine consisting of two side-by-side vertically opposed twelve-cylinder units on a common crankcase with separate contra-rotating crankshafts driving two co-axial propellers. For maximum fuel economy half the engine could be throttled or cut out altogether. Despite the ingenuity, and Fairey's virulent sales talk, the Air Ministry was reluctant to have yet another type. What was needed were extended production facilities for Rolls-Royce and Bristol engines, so Lord Swinton was convening a meeting with major automobile firms with a view to setting up a shadow engine industry.

Meanwhile to finance his prospects, Fairey proposed issuing 1½ million ordinary shares of 10*s.* each. Existing shares immediately shot from 10*s.* to 40*s.*, but it was also revealed that Fairey was entitled to £3,350 per annum for travelling and entertainment, and that caused enormous criticism, for top salaries were still rarely more than £1,000 a year – yet on such a figure one could hardly maintain the world's largest sailing yachts in the manner of Sopwith, Fairey, and Sigrist.

A week after the Battle flew, the silver-painted, twin-engined, heavy-looking Armstrong Whitworth AW 38 Whitley prototype K 4586 made its first flight at Baginton piloted by Campbell-Orde. Though somewhat lacking in lateral stability, all seemed promising. Already the shops had a line of bull-nosed box-like Whitley fuselages and wings with great box-spars, for eighty of these bombers had been ordered seven months before the prototype flew, and many of the 795 hp Tiger engines were being assembled, though it was proposed to equip later machines with the more powerful Mk VII engines.

To co-ordinate all manufacturing programmes and investigate hold-ups, the Air Ministry appointed Lieut-Col Henry Disney as Director of Production in the AMSO's department with effect from 31st March. This was viewed with some misgiving both by contractors and within the Air Ministry, but Disney, an ex-RFC pilot who in 1917 was deputy assistant-Director of Aircraft Equipment and subsequently aeronautical adviser to the Ministry of Labour, had extensive industrial experience, particularly with major telephone and radio companies where he held a number of directorships. 'After all that,' blissfully recounted C. G. Grey, 'he remains as carefree and joyous as when he became a Lieut-Colonel, and will be very effective in speeding production among importee or dilutee firms whose managements are not accustomed to aviators and their ways, unlike the professional aircraft firms. We may be assured that he will secure co-ordination and synchronization through that talent for organization which has made him so successful in business.' Nevertheless one of his most senior technical colleagues said: 'One felt an increasing uneasiness from his obvious intention of overcoming the huge difficulties on his own, aided by men he would introduce, and he almost deliberately ignored offers of help or advice from those closely knit with the aeronautical industry ever since the 1914–18 War.' As a safeguard, Lord Weir was co-opted by the Air Ministry as adviser on production affairs.

3

The annual Air Estimates on 17th March marked the fifth annual occasion on which Sir Philip Sassoon had introduced them. They were by far the largest since the War, with a gross total of £43½ million. He emphasized that by the end of the 1936 the first-line strength of the RAF would have been doubled in two years. 'The industry will need reinforcements to cope with possible war demands, and arrangements are being made whereby additional firms, upon which we shall have to rely in the event of war, will create large extensions to build aircraft for the present programme. When this is done they will form a valuable part of the war potential. Austin Motors and Rootes Ltd have already agreed to give the Air Ministry the benefit of their great production experience.' He said that the manpower problem was particulary complicated, for the RAF required 25,000 additional personnel in the next two years, though 14,500 ground crew and 2,000 of the 2,500 pilots needed had already been accepted from civil life. Initially it was intended to enrol 800 further Reserve pilots from civil life in each of the years 1936, 1937, and 1938; also fifty new RAF stations would be needed of which twenty-nine sites had been acquired. In the next three years the Air Ministry expected delivery of more machines than in the entire seventeen years since the War. Sassoon said that the British aircraft industry was now about equal in strength to the USA's, the number of employees having risen in the last three months of 1935 by 6,500 and was still increasing. Perhaps ominously he added that 'The departmental machine for price control has been strengthened, particularly with technical costing staff to investigate manufacturing costs and firms' overheads. Work is proceeding on a basis of "Instructions to Proceed" (ITP), by which, if a fair price cannot be agreed, the final figure will be determined after examination of the contractor's books by Air Ministry accountanats. The aim is to ensure no profiteering, and to see that industrial enterprise has its fair reward.' On matters of civil aviation he said that vote had been increased by 28 per cent to cover improved ground organization and equipment for the Empire air transport scheme which would extend the Imperial Airways routes from 21,243 miles to 41,405 excluding extensive internal routes operated by the Dominions.

The usual bickering between Labour, Liberals, and Conservatives followed in the guise of debate, during which Churchill warned that future naval engagements would be completely swayed by aircraft, and aircraft would also play a major role on Trade routes. Clarification of Admiralty and Air Ministry responsibility for operational control was therefore necessary as the Navy had neither flying-boats nor bombers. A motion by Labour 'That in view of the peril to civilization latent in air warfare this House calls for immediate and sustained effort to secure the abolition of military and naval air forces and international control of civil aviation' was in customary manner defeated, and the Estimates accepted. 'Taking it all round,' said C. G. Grey, 'the Debate showed far more intelligence among MPs than in previous years. This is probably because so many MPs to-day are either air pilots or use air transport. In fact one might say that the House is rapidly becoming as much air-minded as it is hot air-minded.'

The following day Dick Fairey, with the impressiveness of a big bowler-hatted eagle, held a special Press display of the Battle prototype at his 'Hunterized' grassy Great West aerodrome, the Air Ministry having readily given permission because politically it was desirable to show Britain in the van of progress. Chris Staniland made his usual impeccable show, cameras clicked and whirred, drinks were handed round, and everyone seemed delighted with a pleasant afternoon and the prowess of what Sir Philip Sassoon described as 'the new Fairey air-raider.' Protagonists of 'dash-and-run' bombing were convinced enough to assert: 'No single-seat fighter could successfully engage a bomber of the speed and manoeuvrability of the Battle.' They had yet to see the Spitfire.

Ten days earlier Germany had made great show of strength with dive-bombers and fighters in connection with re-militarization of the Rhineland; they circled Cologne Cathedral so that everybody could see them at midday, two Staffeln remaining at Cologne aerodrome, one each at Düsseldorf and Mannheim, and two at Frankfurt. There was a tremendous outcry from the French, who preferred to dictate to the League of Nations what Germany should do. Hitler made a nine-point proposal to alleviate the situation:- four months' standstill on the Rhine in which neither Germany, Belgium or France should reinforce their troops or move them nearer the frontier; representatives of Britain, Italy, and a third Power to supervise the standstill; negotiations, immediately after the French elections in May, for twenty-five year non-aggression pact between those countries, with Great Britain and Italy as guarantors, and the Netherlands if it wished; Germany and France to agree to avoid any publicity likely to impair good relations; any agreement between nations to be ratified by plebiscites; Germany willing to make non-aggresssion pacts with her Easter neighbours; an international Court of Arbitration to supervise observance of pacts; Germany willing to return to the League of Nations if German colonies be amicable discussed and the Treaty of Versailles eliminated from the Covenant of the League of Nations; a Conference to be held on limitation of arms.

There was no response. The British and French Governments were more interested in arrangements for non-intervention in the Civil War in Spain where for years there had been violence and disorder of Communists against Fascists. When revolution was suppressed in 1934 Franco had been hailed as the Saviour of the Republic and made Army Chief of Staff. Now he came hurrying from Las Palmas, for the interest of his country, whether under King or Republic, was paramount.

In Britain there was some strenuous infighting over extension of the 1930 Air Navigation Bill to empower subsidy agreements up to £1½ million per annum until 31st December, 1943 – an increase of £500,000. Darkly G. S. Johnson, Labour, warned: 'The vultures are gathering round. Before the public knows where it is, millions will be lost by simple minded investors unless the Government steps in with a National Investment Board and prevents the rackets which are beginning in air affairs. In the USA, air mail appropriations have been expended merely for the benefit of a few favoured corporations which could use the funds as the basis of wild Stock Exchange promotions, resulting in profits of tens-of-millions-of-dollars to promotors who invested little or no capital.' The fiery Communist, Mr Gallacher darkly added: 'Until we get rid of the National Government who gives subsidies to their friends there will be no development.' But the Bill went through.

For the proposed transatlantic service, the Estimates allocated £75,000 towards the cost of flying-boat and landplane bases. *The Aeroplane* reported: 'As the first of the Imperial C-class flying-boats is nearly ready for tests, and a second will have long-range tanks for preliminary trials on the North Atlantic service, such bases will soon be needed. Also the DH 91 landplane should be ready for Atlantic experiments some time this year. The Estimates provide £20,000 for payments to Imperial Airways for experimental Atlantic flights, and £18,000 for an annual subsidy to Imperial Airways for the service between New York and Bermuda, one-fifth of which is to be paid by the Bermudan Government. The Bermuda–New York service is a promising proposition, and as Lufthausa is about ready to fly the North Atlantic by the Azores-Bermuda route, British developments ought not to be delayed. At any rate we should not depend too much on an amicable hook-up between Imperial Airways and Pan American Airways, which although it may give our people something of a pull over other nations, does not give us a monopoly of the North Atlantic.'

Scaremongering though Labour was on investment, the British public were ready to buy shares in aircraft projects, believing that rearmament was opening rosy prospects even for smaller manufacturers. Certainly the newer entrants were striving hard with new designs, such as the twin-engined Double Eagle six-seater built by the British Aircraft Manufacturing Co to the designs of George Handasyde – whose surly manners countered his efforts to entrench himself in peace-time aviation despite miscellaneous successful revisions of other people's designs. In any case the chances of the handsome BK 1 Eagle, based on the Klemm L 32 were not great, and those of the Double Eagle still less. Twin-engined, it was a strut-braced mid-winger with centre-section cranked upward from each engine mounting to shoulder height on the round-section wooden monocoque fuselage, and the undercarriage retracted into the engine nacelles. Split flaps were fitted, but gave nothing like the lift of the rival Short Scion's arc-sectioned flap which extended rearward and down while retaining a smooth top contour with the main surface, increasing the wing chord by 30 per cent and lift by the same amount. A near equivalent of the Double Eagle, known as the Peregrine, designed by Miles was already in an advanced stage of construction, and as a low-winger with retractable undercarriage seemed more practical.

Next door to the British Aircraft Manufacturing premises at Hanworth, to the confusion of many, was the simple hangar of the British Aircraft Company founded by the late Lowe-Wylde, and now operated by the famous Austrian sailplane pilot, Robert Kronfeld, who was trying to secure British naturalization. Lord Sempill, despite the disaster of NFS, was a director and as a publicity boost flew the latest Super Drone ultra-light to Berlin, some 570 miles. A special 16-gallon tank gave twelve hours' flying time if the 23 hp Douglas Sprite engine was throttled to 2,600 rpm. Weather was bad, with low cloud, but at 6.45 am on 2nd April he set out from Croydon, was nearly forced back, crossed the Channel at 200 ft, passed Calais at 8.40 am, then battled across Holland with visibility only a few hundred yards, eventually reaching Temepelhof on Berlin's outskirts in a raging rain squall, having taken slightly under eleven hours with two gallons unused. Three days later he flew back, heading into a strong north-west wind, literally skimming the waves, and landed near Canterbury almost out of fuel. Total petrol cost for the two-way flight was 26*s*., and the risk enormous.

Two Drones had been acquired by Charles Scott's Flying Display squadron, which was a reconstruction of Cobham's National Aviation Display business which had closed after seven years' intensive campaigning. Scott was chairman, with Cobham's manager, D. L. Eskell, managing the new venture, and that famous Avro joy-rider, cheery Flt Lieut P. Phillips, was one of the four directors. For the season's 'Air Day', the fleet included the Airspeed Ferries, three Avro Cadet three-seaters, a Tiger Moth, two Avro 504s with Mongoose and Lynx respectively, a C 30 Autogiro, a Comper Swift, a Pou, and Joan Meakin's sailplane.

The Reich Air Day on 21st April had no resemblance to Scott's efforts, but was the anniversary of the death of Manfred von Richthofen – who became the subject of Göbbels' propaganda in every German newspaper; flowers were laid on his grave, and sixteen Berlin street re-named after him and fifteen other famous German pilots of World War I. Göring had just been promoted to General-Oberst, and on reviewing detachments of the Luftwaffe at Dachau, where he presented flags bestowed by Hitler, told them: 'May these flags show to our descendants that Germany has arisen out of its deep distress, and that they may lead our Fatherland to honour and renown.' More significantly, all the Armed Forces that day were celebrating the 200th anniversary of the death of Prince Eugene of Savoy, 'under whom,' as Göbbels announced, 'fought all the German tribes for the future of our common Germanies and against the powers of the Orient in the South-East and those of French Imperialism in the West.'

In the *Morning Post*, Air Marshal Sir John Higgins, former chairman of the Armstrong Whitworth and Avro Group, wrote: 'A large proportion of Conservatives believe that France, by making an alliance with Soviet Russia, broke the Locarno Treaty and put Germany in such an intolerable position that it is difficult to blame her for taking the steps she had done. They hold that this alliance absolves Great Britain from the obligation to support France under the terms of Locarno. But is it not traditional policy to prevent the Low Countries being occupied by a possible enemy? Has not Mr Baldwin said that Britain's frontier is the Rhine? A more foolish aphorism has seldom been made than this dictum. Taken literally it means our destinies are entirely bound with those of France, and that we can have no Foreign policy apart from that country's. I do not believe that supporting France, and therefore Soviet Russia, against Germany is the way to do it. The teaching of history is against a French alliance. If we are forced to make alliances with anyone, let us choose aright. To give unqualified support to France, is merely to give her inducement to attack Germany. A conference between the General Staff of Great Britain and that of France is not, as Mr Eden appears to consider, a mere formality, but is a highly provocative and dangerous action which, I beleive, is entirely disapproved by the majority of people in this country.'

They were much more interested in Amy Mollison, who was in the public eye again. Wealthy young François Dupré, a director of the George V Hotel in Paris, was backing her with new captial in a joint venture registered as Air Cruises Ltd. A pale-blue Percival Gull had been purchased for another assault on the London-Cape Town record, but her first attempt on 3rd April ended in disaster at Colomb-Bechar on the north edge of the Sahara where she landed to refuel, and on starting her take-off, swung violently to port, causing the undercarriage to collapse. On 29th April, elegantly dressed, she was back at Gravesend with the repaired Gull, determined to succeed. Unfortunately Jim Mollison also had returned, somewhat down at heel after his amours in the USA but ardent with reconciliation; so with farewell kisses on 4th May she set off again, and in five long laps reached Cape Town in three days six hours twenty-five minutes, beating Tommy Rose by just over eleven hours. On the 10th she left Cape Town Airport for the return trip, which she achieved in four days sixteen hours seventeen minutes, beating Tommy Rose by one and a half days. Even *The Times* was gravely enthusiastic: 'This is no mere flash in the pan but an achievement in keeping with her distinguished aeronautical career.'

Tributes flowed in, the fan mail restarted. The *Express* bought her story. 'The publicity orgy, and relentless penny squeezing from reflected glory that ensued upon Mrs Mollison's arrival was inevitable when record-breaking must seek financial means from newspapers,' cautioned C. G. Grey. 'The gallant pilot may yet be robbed of her due meed of recognition, sacrificed to yet another Fleet Street holiday. Mass hysteria of idolatory, so easy to enflame and so profitable to the idol's so-called discoverers, is familiar currency – but the worshippers can become just as hysterical in abuse of the heroine, jaded from an escapade which would try the nervous stamina of a hippopotamus, if she fails to treat them to the ready-made gush of a cinema actress.' Within days Jim was admitting more infidelities, and Amy was on the verge of a nervous breakdown.

With the African records so much in the news, H. L. Brook was prompted to try with a 36 hp Czechoslovak-designed Praga Air-Baby loaned by F. Hills & Sons Ltd, furniture-makers of Manchester, who had the British manufacturing rights. Like the Aeronca it was a 36 ft span side-by-side two-seater, with two-cylinder horizontally-opposed Praga B engine which Jowett Cars Ltd were licensed to build. Painted a vivid yellow, this aerodynamically clean cantilever monoplane, with cabin below the leading edge, had been part of Scott's Flying Display, but in Brook's hands left Lympne on 6th May and reached the Cape sixteen days four and a half hours later, thus bettering the twenty-three days of Llewellyn. Meanwhile Hills had started production, renamed the Hillson Praga, and advertised it at £385.

Two-cylinder 'poppers' whether Aeronca or Hillson, had no great appeal, particularly as the slow rate of climb could be eliminated by a strong down-draught, but were not dangerous compared with the Pou. There had been half-a-dozen inexplicable fatal accidents to these homebuilts and now Flt Lieut A. M. Cowell, official test pilot for the Air League, was killed while diving one at Penhurst. The Air League, with 'Lamps' Lamplugh's endorsement as the *éminence grise* of insurance, advised the Air Ministry to ban all flying on the type pending investigation. Urged by Air Commodore Chamier, the DSR reluctantly agreed that the Royal Aircraft Establishment should carry out full-scale wind-tunnel tests. Similar investigation was undertaken by the French in their big wind-tunnel at Chalais-Meudon. After several months, it was established that though the little machine was suitably stable in normal flight it trimmed into an irrecoverable dive beyond a critical attitude, and thereafter pulling back the control column to increase the incidence of the front wing had no effect due to aerodynamic interference with the close-coupled rear wing. That was the end of the Pou, though hundreds had been built, and but for this fatal characteristic it was the easiest machine in the world to fly. At least it had shown that everywhere there were enthusiasts ready to build a constructionally simple, cheap little aeroplane.

The private owner's market remained the perquisite of machines like the Leopard Moth or the Miles low-winger whose success was underlined when Rolls-Royce Ltd financially participated in Phillips & Powis Aircraft Ltd, and stated: 'The main reason for the association will be study and development of problems in connection with installation of engines in prototype aircraft, as the engine manufacturer now has to take much responsibility in this matter. Messrs Phillips & Powis will continue with their present successful private-owner, trading, and light commercial aircraft, but will endeavour to cater for a share of Air Ministry requirements – mainly in training types. Rolls-Royce, through this association, will consider distant requirements for commercial aircraft, and endeavour to retain even closer technical liaison with the industry than in the past.' Undoubtedly Fred Miles with his charming wife 'Blossom' – publicized as 'designer' – was going from strength to strength. More derivatives of the original Hawk were either being built or devised on the drawing board. The ply covering instead of being laboriously tacked with gimp pins, obtained gluing pressure by using a conventional wire stapler, and because Miles earlier built machines with his own hands there were many other simplicities aiding economic construction.

On the heels of the Rolls-Royce announcement came news that Reg Parrott, that valiant man on whom Sir Alliott Roe relied from his earliest days at Manchester, had joined Phillips & Powis Aircraft Ltd as general manager. One of his first occasions was to show Colonel Charles Lindbergh the latest Miles M 11A prototype two-seat side-by-side low-winger which had been ordered by the Whitney Straight Corporation. Pleased with the handling of this Falcon derivative, Lindbergh ordered a special tandem two-seater with sliding cabin top, the M 12, known as the Mohawk and fitted with an American-built 250 hp Menasco Buccaneer, resulting in a top speed of nearly 200 mph.

4

The Season's flying meetings opened with the Royal Aeuronautical Society annual Garden Party on Sunday afternoon, 10th May, and the 2,000 guests were received by the president, Lieut-Col Moore-Brabazon, MP. *Flight* duly reported: 'The Party at Fairey's magnificent Great West aerodrome should have delighted the pacifically-minded; whereas last year's resounded to the roar of fighters, Sunday's meeting was an all civil affair. There were only two military types, and neither took the air.' These were quiet but evident publicity for the Swordfish and Battle. 'Of the term "Garden Party" nobody could complain. There were all the ingredients – trim grass, charming frocks (and their wearers), the RAF Central Band, and tea by Gunter's. Literally everybody who is anybody in the aviation world was there. Just as representative were the visiting aircraft, privately owned and otherwise: Imperial Airways and British Continental DH 86s; the beautifully finished Heinkel He 70 used as a flying test-bed by Rolls-Royce; a twin-engined high-wing monoplane built in Austria to the design of the Archduke Anton of Hapsburg; and even an old but sturdy little Westland Widgeon, guarded by a wire terrier which, from his place in the cockpit, said rude things to anyone who approached.' That dog owned me.

Next day saw the first flight of the 66 ft span Bristol high-altitude low-winger, Type 138A, structurally designed almost single-handed by that pre-war Bristol pioneer draughtsman Clifford Tinson, who commented to me: 'This pleased me quite a lot because it was the only Bristol prototype, as far as I am aware, that came out a few pounds under estimated design weight.' The wings had three spars with mahogany flanges and plywood webs, skinned overall with 0.8 mm plywood, but the fuselage had somewhat thicker ply glued and screwed to mahogany longerons and stiffeners. Lest there be trouble from grease freezing in bearings, greasing was eliminated and all controls had ball-bearings dipped in thin gun oil which was drained off, then fitted with sealed covers. Uwins found the machine handled sufficiently well for its purpose, though only using a standard Pegasus IV and a three-bladed propeller to give initial handling experience to the chosen pilot, Sqn Ldr F. R. D. Swain at the Royal Aircraft Establishment after which its special 500 hp Pegasus RE 6S with four-bladed fixed-pitch wooden 'climb' propeller would be installed for the record attempt. In the shops the second machine was being built as a two-seater to be powered with a special 500 hp Rolls-Royce Kestrel S fitted by the RAE.

Aircraft factories had a kind of split personality these days. Experimental work went quietly on, but the complex of production was a world apart. The new target set by the CAS, Sir Cyril Newall, was now 3,800 aircraft within two years, yet the industry was unaware of the overall extent and in any case was finding difficulty in recruitment to meet even current contracts. Engines, as in the past War, would be the crux. One or other type of aircraft ordered off the board might well be discarded if not a success operationally – but it was essential that engines not only were completely reliable but available in huge quantity with adequate replacement and spares backing. To that end Sir Herbert Austin, Douglas Burton of Daimlers, the Rootes brothers of Humbers, Spencer Wilks of Rovers, and Lord Nuffield of Wolseley Motors were invited for discussion with the Air Ministry. Rolls-Royce had informed the Government that they would make their own arrangements for expanding production, however great. Bristol engines were therefore the requirement, and it was proposed that each participating firm would construct separate factories, financed by the Government, with the aim of producing some 4,000 Mercury and Pegasus engines within two or three years in addition to Bristol output. That abrupt and suspicious man, Nuffield, declined to participate, but in the course of several meetings the others agreed to the questionable plan of each making a particular sub-division of the engine's components instead of complete engines. One well directed bomb might therefore prevent assembly of completed engines elsewhere. Napiers, with smaller contracts, could handle all likely production – but if success attended the 2,000 hp water-cooled, scaled-up twenty-four cylinder, Dagger-like engine for which Halford had been lucky enough to secure a prototype contract, then ways and means of production would have to be considered.

Unrecognized, there was a sign of even more important things to come. In March a small firm named Power Jets Ltd, with nominal capital of £20,000, but actually only £2,000 loaned by investment bankers O. T. Falk and

Partners, got under way with Sir Maurice Bonham-Carter as chairman, W. L. Whyte as a director representing the bank, and Flt Lieut Frank Whittle, who had just gained a First Tripos at Cambridge, as honorary chief engineer and technical consultant, though his patents on jet propulsion had lapsed. Magnanimously the Air Ministry agreed that although Whittle was a serving officer he could accept the post provided he did not devote more than six hours a week, and free Crown usage was demanded. Falk and Partners then placed a contract with the great electrical firm of British Thomson Houston, builders of steam turbines, for an experimental Whittle unit comprising a small-diameter turbine turning a centrifugal compressor at 17,750 rpm, and using a single combustion chamber burning an unprecedented 200 gallons per hour. In little more than six months combustion tests began amid deafening noise and dense clouds of smoke.

Compared with this project, the first flight of the Fairey Seafox seemed a step backward. Like the TSR and Swordfish, it was a thoroughly conventional biplane, but slightly smaller and considerably lighter. The two-bay 40 ft, fabric-covered wings were two-spar metal structures, but the fuselage was an Alclad metal monocoque stiffened with Z-section frames and was completely sealed for emergency flotation as a light reconnaissance two-seater for the FAA. There was no provision for bomb sights, though it carried two 100-lb anti-submarine bombs, and sole armament was a single .303 Lewis on a rocking pillar normally stowed in the rear decking. Only the observer's cockpit was enclosed, for naval pilots preferred being in the open because of hypothetically easier escape if ditched. For minimum touchdown speed Lobelle had fitted camber-changing flaps to upper and lower wings, but the heavy cutaway at the centre-section avoided difficulties of downwash and certainly made folding easier as the flaps automatically drooped to give clearance against fuselage and hood. Pilots liked the pleasant handling as well as the vibrationless running of its 395 hp Napier Rapier VI – a characteristic of similar appeal when I flew the prototype Hawker Hector K 3719.

Looking much more up-to-date was the little 750 cc Douglas-powered single-seat Tipsy low-winger which Chris Staniland demonstrated at Fairey's aerodrome. Designed by pioneer E. O. Tips, who had become managing director of Avions Fairey in Belgium, a dozen were to be built by Aero Engines Ltd at Kingswood, Bristol, at an intended selling price of £265. Starting the little engines was as frustrating as any outboard motor, for like them they had a pulling cord wound round a bobbin – but in any case, did people want single-seaters? Low-powered two-seaters had a slightly better chance. In the Hanworth to the Isle of Man race on 30th May, on a day of low cloud and murky atmosphere, it was a 36 hp Hillson Praga which won, and an Aeronca was second.

Though such meetings and races had become commonplace every week-end the opening on 6th June of Gatwick Airport alongside the Southern Railway was a big event. 'Everybody who ever had anything to do with aviation was among the guests, distinguished or otherwise, and thousands came to renew or acquire a long range acquaintance,' ran one report. 'As in all the best functions of this sort the organization became slightly swamped. Some hundred aeroplanes, varying from 23 hp to 1,500 hp and of every variety and shape, were arranged along the runway and edge of the landing area. The Secretary of State for Air, Lord Swinton, arrived from York in the Air Council's Dragon Rapide and opened the Airport.'

Several years earlier it had been a small club aerodrome suitable for light planes, but the original area was now obliterated by the twin hangars of British Airways and a big two-tier rotunda and passenger waiting hall from which star-like covered ways projected to embarkation points. Stretching away westward was the relatively long grass airfield. A Lorenz blind-approach system was to be installed and Chance lights for night operations, but though not yet fully equipped except for radio, already several hundred passengers had passed through the airport and it had big reserve capacity for further exploitation.

After lunch there was the inevitable flying programme during which scores of aeroplanes were demonstrated and occasionally aerobatted, three Gauntlets of No. 19 Squadron easily stealing the show with formation aerobatics while tied together with ribbons; but the most heart-stopping was the so-called 'birdman', Clem Sohn, who emerged from a Dragon and came steeply gliding down on outstretched wings, swooped around, got into a spin, closed his wings and pulled his parachute rip chord, but nothing happened until a mere 200 feet above the ground, when two parachutes opened and he landed normally. As one reporter recorded: 'He is not likely to come much closer to sudden death than that, and the crowd certainly got the thrill that brought most of them there.' Predictably a few months later, something went fatally wrong, possibly because his arm was dislocated in a tangle of wing and webbing, and he fatally fell with parachute unopened.

Civic interest in airports was active. Even such small places as Weston-super-Mare had built an aerodrome from which car-agent Norman Edgar's Western Airways had operated a Fox Moth service since 1932 and now initiated a DH Dragon service to Cardiff and Birmingham. At Portsmouth the City Council was considering a contentious scheme for an Imperial flying-boat base at Langstone Harbour, but it entailed dredging two costly alighting strips a mile long and 200 yards wide to give sufficient depth and facilitate night operations by reducing the difficulty of mounting guidance lights. The scheme would take years to complete but the first of the splendid Short Empire flying-boats, *Canopus*, was within a fortnight of its first flight, with *Caledonia* close on its heels – so Southampton Water must still be used.

Short Brothers (Rochester and Bedford) Ltd currently found that commitments necessitated a £100,000 increase beyond the registered capital of £150,000, and £400,000 ordinary shares were issued to the public at 5*s*. That seemed insignificant compared with the doubled capitial required by Hawker Siddeley Aircraft Co Ltd, who issued one million £1 cumulative preference shares and four million ordinary shares of 5*s*. which were eagerly snapped up. Within a few weeks Sopwith's office *maestro* Frank Spriggs was placed in charge of the entire organization of the Armstrong Whitworth Siddeley Group at Coventry, A. V. Roe at Manchester, and Air Service Training at Hamble.

In a different way Blackburn Aeroplane & Motor Co Ltd were overcoming their depressed finances, and a declaration of solvency was filed on 2nd June with an authorized captial of £225,000 fully paid up. In addition to Robert Blackburn, the directors were named as Frank Bumpus the chief engineer, Reginald R. Rhodes, (director of the holding company Blackburn Consolidated), Edwin Hudson as director of the associated but no longer operative North Sea Aerial & General Transport, and James L. N. Bennett-Baggs formerly of Armstrong Whitworth.

5

Mid June saw the launch of three prototypes – two on the 15th when the twin-engined Vickers B 9/32 Wellington and the Westland A 39/34 Lysander made their initial flights, piloted respectively by 'Mutt' Summers and the author. The third was the Vickers Venom.

To show their confidence in the geodetic bomber prototype K 4049, both Wallis as the man answerable for its novel structural integrity, and Trevor Westbrook who had been responsible for its intricate construction, accompanied Summers on the first flight. Nose and tail gun locations had been faired with stringers and fabric, so the pair stood close to the pilot peering through his windscreen. The machine climbed easily away from the track-encircled grass of Brooklands, and Summers found no difficulty in handling, although there were directional anomalies with a hunting rudder, and the elevators were somewhat overbalanced but masked by slight longitudinal instability despite tailplane location relative to the wing-chord axis in similar position to that of the stable Wellesley. The appearance of being a mid-winger was due to the addition of a bomb bay beneath, resulting in a deeper fuselage. Because of light structural weight afforded by the geodetic trellis work, equivalent to taking a monocoque skin and cutting rows of diamond-shaped plates from its entire surface and enveloping with fabric, the potential load capacity was far above the original specification bomb weight of 1,000 lb with a range of 720 miles, and presently resulted in a bomb capacity of 4,500 lb and range of 3,200 miles at 180 mph cruise.

When the silver-painted Lysander prototype K 6127 was flown at the quiet, long-deserted grass aerodrome on Boscombe Down near Amesbury, only twelve months had elapsed since the issue of the contract. As might be expected from Petter's devotion to France, the machine had a French appearance, yet embodied Davenport's pioneering experience of strut-braced high-wing monoplanes and Westland's extensive data on wing slots. Petter was no draughtsman, but as his later personal assistant Denis Edkins said: 'He always understood and acted upon the premise that in a complex system, such as an aeroplane, the detail design is all important once the basic choices have been made. He therefore spent a great deal of time going round the drawing boards, for he was a creative designer both in the overall sense and in detail, so there were many good ideas embodied in the Lysander.' His pioneering use of light alloy extrusions was notable, but even this would have been impossible without the close co-operation of bustling Devereux, and it was as the result of discussion with him that the hairpin undercarriage was devised.

The growth of a new design of aeroplane from its parts to sub-assemblies, main assemblies, and ultimate completion becomes part of the sequence of one's life; by the day of flight trial every aspect is familiar including the theoretical background of trim, stability, engine manipulation, and performance. There is every certainty that the machine will fly – but exactly how, and with what unsuspected snags, is the question.

A few days earlier at Yeovil I had made a few slow speed runs with the machine in its rust-red undercoat, but now I made two 'straights' to get a better idea of the feel at low flying speed, then taxied back for a final ground inspection. Half an hour later, heading into a gentle west wind, the throttle was steadily opened; already the tip slats were fully out, and within the first few yards the root slats pulled forward and partly lowered the interconnected flaps. In less than 150 yards the machine was climbing away, and as speed gathered I changed to coarse pitch. The view was magnificent. At 5,000 ft I began trying the controls. This took some time. Response was good, but ailerons too heavy. There was a measure of longitudinal instability with controls free. Engine throttled, I tried the stall: tip and root slots extended fully, and there was perfect control with the stick right back. I mentally added 20 per cent to the stalling speed, swept wide around Salisbury, and made my approach to Boscombe Down. What was immediately clear was insufficient tailplane adjustment, for I was pulling back on the stick to maintain slow enough speed, and when flaring out, using a little engine, the stick was already at its limit. As the machine began to drop for what would otherwise be an untidy landing I gave a burst of engine to increase slipstream over the elevators, and she settled gently, but the tailwheel was still just off the ground. Youthful Teddy Petter looked relieved at this first trial of failure or success, and certainly my report was favourable but warned of difficulties with tail trim. The wind-tunnel, as so often, had failed to predict so big a change of downwash, and this necessitated several re-designs of the trimming gear to secure an exceptionally big negative tailplane angle for landing. That too had its danger.

Two days later, 'Mutt' Summers flew the Aquila-powered Vickers Venom fighter designed to the now obsolete F 5/34 specification. Essentially it was an up-dated version of the F 20/27 Jockey which had spun into the ground in 1932, but with stiffer fuselage and a dorsal fin flaring into the cockpit enclosure, and the Jockey fixed undercarriage was replaced by one cleverly retracting into the relatively thin RAF 34 wing, already obstructed with eight guns.

Despite precautionary fitment of a tail parachute for the first time in test flying, the Venom was not spun by the Vickers pilots as the Air Ministry considered this should be done by the RAE – but in fact no extended spins were ever made, despite the readiness of Summers' new assistant at Eastleigh,

Flt Lieut Jeffrey Quill, to do so. With Summers involved with the Wellington bomber at Brooklands, Quill, as an ex-fighter pilot, was busy at Eastleigh with both Spitfire and Venom, and found the latter extremely manoeuvrable and with a mere 625 hp it achieved almost 320 mph. The Air Ministry became interested in its possibilities as a low-priced, low-powered, light-weight fighter, compared with the costly Spitfire, and proposed that as both the Vickers and Supermarine factories were fully engaged, the Venom should be built by some other company, though the idea was presently abandoned.

Whether the eight-gun fighter was the full solution remained a matter both for practical experiment and discussion. 'Consequent upon the main problem of time required for intercepting and attacking a bomber, it was by no means certain that the single-seat single-engined fighter would succeed,' recorded Air Marshal Sir Ralph Sorley ten years later. 'By accepting slightly reduced performance it appeared possible to carry a four-gun power-operated turret and increased fuel capacity in order to attack bomber formations from a standing patrol. Even though the speed margin was less than with a single-seat fighter it gave the advantage of attacking from any angle rather than the essential position astern of a single-seat forward-firing fighter. This led to specification F 9/35, and tenders were accepted from Boulton Paul and Hawkers for what became the Defiant and Hotspur. Both had the turret mounted behind the pilot with all-round field of fire in the upper hemisphere on the assumption that the target would be usually above the fighter. Special four-gun turrets had to be conceived. Using the basic scheme of a Frenchman, Boulton Paul redesigned and developed it as the BP 4 turret. Bristol also tendered with a variant of their Army Co-operation Type 148 for which Fraser-Nash, who had joined Parnalls, introduced one of the first remote-control systems for a battery of four guns separated from the gunner in a flush-fitting inverted "dustbin" turret but because of preoccupation with the Blenheim this monoplane was not built, and the turret was used for the Hawker.'

Not only the Hurricane, but the Henley, a two-seat dive-bomber to specification P 4/34, had been occupying Camm and his little known assistant chief designer Roy Chaplin, whom a senior member of the Hawker staff described as 'a good sound engineer and administrator who was very loyal to Camm though the latter never gave him his due.' Chaplin had been with Hawkers ten years and had taken a large share in design. He was now engaged on the Hotspur F 9/35 fighter re-design of the Henley both of which utilized the outer wings, tailplane, and fin of the Hurricane.

Camm himself was regarded at Hawkers as: 'A curious mixture in character. He was a good family man with a strong code of conduct. Basically artistic, kind, generous, and with a keen sense of humour; but he could nevertheless be cruel and quite unreasonable at times, thoroughly obstructive to introduction of essential modifications unless he originated them, and with an embarrassing habit of tearing strips from his senior staff in front of visitors. But if anyone stood their ground when he was in one of these unpredictable moods, he would calm down and become amenable to constructive suggestions. We never thought of him as a genius, but in his day he was a first-class designer, notably scooping the pool in the RAF and export market with the Hart and Fury biplanes. Both were mainly the product of his own skill and ability, but when aeroplanes started to get more complex from the Hurricane onwards, he became more dependent upon the technical ability of his staff rather than his personal flair, though he had only a small and not all that brilliant staff to lean upon when the transition took place.'

The fighter field still required backing with more powerful armament. As Ralph Sorley said: 'Once construction of the Hurricane and Spitfire got off the ground the possibility of using 20-mm Hispano guns came within reach. Because the thick wing of the Hurricane was designed essentially for stiffness it offered the installation solution for heavy armament. With four such guns in the wings a decisive result essential in two seconds of fire appeared obtainable because of the devastating effect which heavier shells had upon the target. Undoubtedly there were engineering problems to mount the gun within this wing, but it was decided that the Hurricane should be so developed and specification 10/35 was revived.'

This also led to proposals for a twin-engined single-seater mounting four cannon, and specification F 37/35 was circulated to the industry in the spring of 1936. By now Petter had encouraging support from the AMRD, Air Vice-Marshal Wilfrid Freeman, who was impressed by the ingenuity displayed in the Lysander design. Petter felt that here at last was an intellectual RAF man who understood his point of view and would smooth all difficulties. Davenport's assistant, John Digby, was making the initial drawings of an ambitious design with magnesium-skinned slender fuselage, high aspect ratio wings, and two Goshawk-derived Rolls-Royce twelve-cylinder 885 hp Peregrines which were of smaller cross-section area than the more powerful but trouble-involved Merlin and therefore offered less drag. The four cannon were concentrated in the nose where they could be readily armed and serviced. Wind-tunnel tests with the initial twin-finned model indicated a possible 375 mph. Bristol tendered for the same specification with a twin Aquila machine, Type 153A, of such similar configuration that both designs seemed collaborative.

In the USA, Dr George W. Lewis, director of research of the NACA, told the Institute of Aeronautical Sciences that whatever the power, speeds with current aerofoils would be limited to about 575 mph, because of what he termed 'compressibility burble'. Ciné films taken in their new 750 mph super-speed tunnel showed the airflow beginning to develop turbulence at a quarter chord from the leading edge at a speed of about 570 mph, and was so critical that any slight increase caused the flow to break down entirely, causing all lift to be lost. Nevertheless Germany had an unpublicized clue to the solution, for Professor A. Busemann was experimenting with the mitigating effect of sweep-back, which Geoffrey Hill had earlier predicted as the likely future shape of high-speed aeroplanes, including the delta which had been investigated in the Westland wind-tunnel but revealed control response problems for which there was no ready solution other than by expenditure which the company could not afford nor the Air Ministry provide. Thus one loses the race.

The requirements of operating high-speed aircraft dictated decentralization in reorganizing the Home Commands of the RAF in June. The ADGB would have four Commands, with Air Marshal Sir John Steel as C-in-C Bomber Command at Uxbridge controlling four Groups of bomber squadrons; Air Marshal Sir Hugh Dowding as C-in-C Fighter Command at Bentley Priory in delightful grounds at Uxbridge, controlling two Groups of regular fighter squadrons, an Army Co-operation Group, and a Group of Auxiliary fighter and Auxiliary Army Co-operation squadrons; Air Marshal Sir Arthur Longmore to be C-in-C Coastal Command at Lee-on-Solent, controlling two Groups of flying-boat and general reconnaissance squadrons together with administration and shore training of FAA squadrons; Air Marshal Sir Charlse Burnett to be C-in-C Training Command at Tern Hill, Shropshire, who, and with a few exceptions, would control all Home training units, the Groups under this Command comprising one flying and one ground training establishment, and there would be an additional armament Group.

Overseas, the British Forces in Iraq at the Hinaidi HQ near Baghdad were commanded by Air Vice-Marshal W. G. S. Mitchell administering an armoured car company, supply depot, a hospital, and three light bomber squadrons of Hawker Hardy, Vickers Vincent, and Westland Wapiti, a bomber transport squadron of Vickers Valentia and a flying-boat squadron of Short Singapores 'to support the Iraq Government and to defend British interests, including the oil supplies of Northern Iraq and the air route to the East.' In the Middle East the RAF HQ in Cairo was commanded by Air Vice-Marshal C. T. Maclean, supervising a training school, an Army Co-operation squadron of Hawker Audaxes, a bomber-transport squadron of Valentias and three bomber squadrons of Fairey Gordons – for 'with the development of air power, Egypt had become the most important strategic centre in the world.' In Palestine and Transjordan the RAF HQ in Jerusalem was commanded by Air Vice-Marshal R. E. C. Peirse who had an armoured car company and five light bomber Gordons, and 'although the Command has been free from major operations there have been many raids over the borders which have needed combined operations to put right, but dissensions between the Arabs and Jews are normally handled by infantry units under his orders.' The Aden Command was under Air Vice-Marshal E. L. Gossage, and included units of the RAF, Army and Aden Protectorate Levies with armoured cars and a bomber squadron of Harts, but a warning of bombing to recalcitrant tribes was usually sufficient. The RAF in India had HQ at Delhi in winter and Simla in summer, commanded by Air Marshal Sir Edgar Ludlow-Hewitt, and comprised an aircraft depot at Karachi, four Army Co-operation squadrons of Audax, one bomber-transport of Valentia and three bomber squadrons of Hart and Wapiti – for there had been constant trouble on the frontier, where 'in recent years co-operation with the Army has been continuous and effective.' In the Far East the RAF HQ at Singapore was commanded by Air Commodore Sydney W. Smith, with two torpedo-bomber squadrons of Vildebeests and a flying-boat squadron of Singapores, but he also administered units of the FAA on carriers, capital ships and cruisers – 'mobility, the most important aspect of air power, is practised, and the squadrons make regular long-distance flights to such places as far away as India'. Finally there was the Mediterranean Command with HQ at Valetta, Malta, commanded by Air Commodore P. C. Maltby, with a flying-boat squadron of Supermarine Scapa and administration of several FAA units, though primarily Malta was merely a base for the Fleet's shipborne Nimrods and Ospreys in the Mediterranean.

6

Competing with the Wellington was the Handley Page HP 52, later known as the Hampden. This was the third machine for which Lachmann had full responsibility, all showing similar design characteristics, but the B 9/32 bomber marked a considerable advance in his technology. At this stage Volkert took over again and Lachmann continued with research investigations dictated by Frederick Handley Page.

For the first flight of the HP 52 the gunners' positions, like those of the Wellington, were covered over. Nevertheless Lachmann's machine seemed to have greater defensive potential, for the 3 ft wide central fuselage portion had deep fairings above and below terminating abruptly in dorsal and ventral gunners' positions, and the nose had a smoothly rounded gunner's enclosure. On 21st June all was ready at Radlett, and Jim Cordes flew the prototype next day, making a series of brief handling flights which showed that though minor modifications were needed it was satisfactory enough to be flown to Hendon for the seventeenth RAF Display of 27th June. Painted dark green, it seemed dull among the brightly-finished prototypes in the New Type Park comprising the Hurricane, Spitfire, Venom, Battle, Bristol *Britain First* with RAF roundels, the Wellington prototype with a curious bulge beneath the rudder and the fabric of its airship-like fuselage ridged by its stringers, contrasting with the straight-lined prototype AW 38 which had the thin metal sides of its fuselage similarly slightly furrowed, and nearby was the Lysander, still unmodified. To the credit of Phillips & Powis Ltd, the presence of their Nighthawk side-by-side Gipsy-powered trainer showed they had broken into the sacrosanct 'ring' of favoured firms. For the first time there was not a single biplane – yet no less strikingly, every event featured the RAF's cherished biplanes except for three Ansons and three Autogiros. 'Hendon certainly did us proud on Saturday,' enthused C. G. Grey. 'We had good reason to be thankful for the large grey clouds as sunshades, for nothing gives a blinding headache as much as watching an air display in bright sunlight, even when, as at Hendon, the sun is at one's back. Certainly the RAF never put up a better show. One perennial spectator remarked that each year one sees fewer and fewer aeroplanes there and better and better flying.' But for the first time there was no Royal Box at Hendon. There were increasing rumours about the King and an American woman, dark and elegant, but not pretty, Mrs Ernest Simpson. She had been by his side when the heralds proclaimed his accession. Now he was on a Mediterranean cruise with the yacht *Nahlin*, and foreign papers featured pictures of the couple aboard, though the British Press, news-reels and radio, unanimously imposed a rigorous self-censorship. Yet the rumours were growing, particularly as the American Press was reporting all it could glean on the royal romance. Baldwin was keeping a stiff upper lip and so was the Archbishop of Canterbury.

For many, the most fascinating part of the RAF Display was the 'live' exhibition of historic aircraft beginning with a replica of the Wright Flyer under power but mounted on a tricycle trolley to enable taxi-ing, a graceful Antoinette which had to be towed because of refractory engine, an Anzani-powered Blériot XI which taxied under power but was not fully airworthy, a big, wire-singing, Renault-powered, khaki-painted 'Horace' Farman flown by Sqn Ldr D. V. Carnegie, a dainty Anzani-powered white Caudron flown by Sqn Ldr H. A. Hamersley of earlier Avro fame, a Sopwith Triplane with humming Clerget flown by Flt Lieut N. R. Buckle who had organized the event, a still aggressive Camel which its owner R. O. Shuttleworth taxied in a halo of blue smoke from its blipped engine, an SE 5a with war-time markings of No. 60 Squadron flown with verve by Flt Lieut J. Hawtrey, and finally a Bristol Fighter rumbled into the sky flown by its owner Flg. Off Nigel Tangye, RAFO.

Their display was so nostalgic that it almost spoilt the fly-past from the New Type Park, though all eyes concentrated on the impressive, pale-coloured, ice-smooth Spitfire, described as 'like a baby Schneider Racer which folds up its feet', its elliptical wings making striking contrast with the square wingtips of the spiteful-looking Vickers Venom which followed a few moments later. The Spitfire still had its fixed-pitch wooden propeller, but the radial-engined Venom had a two-speed DH Hamilton thanks to the persuasion of Roy Fedden who was a valuable ally to his friend Rex Pierson.

Though *The Aeroplane* described the Display as 'a vast advertisement for the Hawker-Gloster-Siddeley combine, barring the flying-boat parade', there was a startling moment when the pilot of a Heyford slow-rolled it after his bombing attack. The tied-together aerobatic drill of three Gauntlets seemed fantastic, but no less remarkable were the aerobatics of Flt Lieut Harry Broadhurst during which he broadcast detailed explanation of every manoeuvre. A later event suggested that the British Empire was beginning to crack, for the programmed 'tribesmen' despite their warlike cries had been forbidden to blacken their faces lest offence be given. That we also had an eye on Russia was revealed when fighters attacked two bright red Virginias which then belched clouds of red smoke and a deluge of sixteen parachuted dummy 'troops' emerged. A great set piece, this time of a power station, ended in the usual big bang and a huge column of smoke. 'Altogether it was a great Display,' wrote an enthusiastic reporter. 'The flying was worthy of the world's best Air Force. Machines and motors were worthy of the world's greatest aircraft industry. The shadow of the shadow trade did not obtrude. Behind all is knowledge that the renewed aircraft industry is importing fresh ideas and introducing new designs which prevent fossilizing under the continual drip of government regulations and technical interference. In fact everything in the aeronautical garden looks not only lovely but encouraging and very amusing.'

Many foreign diplomats and technicians attended the SBAC party two days later at Hatfield – not for amusement, but to study how technically dangerous Britain might be. Not all the latest machines were there. The mid-wing Blenheim bomber version of *Britain First* had made its first flight four days earlier but was too secret to show. As *Flight* reported: 'Several aircraft had nasty morning trips to arrive in time, and some were weather-bound. Mr J. Summers timed his entry with a Spitfire convincingly displayed just when people were coming out after lunch. The little Vickers fighter arrived late and flew later – and the geodetic bomber did not appear. Flt Lieut A. M. Blake showed off the solid qualities of the Blackburn Shark – a thankless task with a large machine which has neither flaps to flap nor retractile wheels to wave at the spectators. Mr C. K. Turner-Hughes showed that the Whitley was agile for its size. Flt Lieut C. S. Staniland did what he was allowed to do with a Fairey Battle despite the announcer's unfortunate statement that "the Battle is the only Fairey machine which will fly", though he meant that the Fairey biplanes would not be demonstrated. Major Cordes duly flew the Handley Page medium bomber, and it was clearly much faster than the Vickers. P. E. G. Sayer, with a Gladiator now fitted with cockpit coupé, performed outstanding aerobatics, and George Bulman with the Hurricane drew the crowd from the tea tent to watch his rolls and rocket loops during which he vanished into the cloud. Bristols staged an impressive trio, with Cyril Uwins flying the RAF's Rothermere, the civil version flown by Mr C. A. Washer, and the big Bombay, fixed-undercarriage transport-bomber flown by Mr A. P. H. Coleman. Last of the military machines was the Lysander which had not yet been aerobatted, but its ability at slot hanging and take-off and landing in the length of a football pitch was demonstrated.'

Interspersed with the military aircraft were demonstrations by the civil machines – two small Monospars, one flying at top speed the other stalling along; the Heston Phoenix with a slow-flying display which looked dangerous; the Percival Vega Gull which was sharply looped at the moment the announcer unwittingly said 'This machine can be folded up by one person.' De Havillands stand-in test pilot, ex-apprentice R. J. 'Bob' Waight, a licensed ground engineer who learned to fly in 1932, showed the short take-off and steep climb of the DH 86A fitted with constant pitch propellers, and Hugh Buckingham of the sales staff demonstrated the Dragonfly with one engine stopped. The British Aircraft Double Eagle made its first public appearance flown by Flt Lieut J. B. Wilson, a former expert RAF meteorological pilot. John Lankester Parker flew the new four-engined Short Scion Senior, emulating Buckingham by throttling both engines on one side, and C. E. Gardner showed off the twin-engined Pobjoy Scion. Finally Fred Miles threw the Whitney Straight around with great effect while the Nighthawk and Hawk trainers were being aerobatted.

Said a reporter: 'The guests were duly impressed. We talked to General

Milch, Chief of the German Luftwaffe under General Göring; and to General Lindquist, head of Finland's Air Force; and Capt José Cabral, Chief of the Portuguese Air Force. All were delighted with the show – and revealed sound judgment in picking which were the best machines. General Milch's estimates of performance showed extraordinary accuracy.' That could not have been difficult, for trials of the Heinkel and Messerschmitt had been completed, resulting in the selection of the latter by General Udet, head of the Technical Department, and ten pre-production machines had been ordered for squadron evaluation. A similar batch of Heinkels was eventually allocated to a Japanese order for thirty.

There was antagonism between Göring and Milch who was highly efficient and had the ear of the Führer. Göring had caused irritation by removing Udet from Milch's supervision, resulting in lack of co-operation between procurement and technical development, and this had worsened since the death in an aircraft crash on 10th June of General Wever, the first Chief of the Air Staff, who had been a pilot of great technical ability and a brilliant strategist as shown by his concept of the 'Ural' long-range strategic bomber to enable Germany to strike at any corner of Europe. Junkers and Dornier had almost finished their respective Ju 89 and Do 19 prototypes – machines more advanced than anything under development by Britain or France, though Air Commodore Verney as Director of Technical Development had in the spring of 1935 issued specification B 1/35 for a twin-engined heavy bomber of under 100 ft span to carry 2,000 lb of bombs for a range of 1,500 miles, resulting in the acceptance in October of tenders from Vickers, which became the Warwick, Bristol Type 155, transferred eventually to Armstrong Whitworth as the AW 41 Albemarle, and Handley Page for the HP 55 which was never built. Now that there was evidence of the 'Ural' bombers, equivalent aircraft were formulated and specification B 12/36 for a long-range four-engined bomber was issued in July, followed by P 13/36 in September for a medium-range bomber powered with two big, twenty-four cylinder, Vulture engines, nominally of 1,760 hp, which Rolls-Royce were urgently developing.

German outlook on bombers coincidently changed. Wever was succeeded by Lieut-Gen Albert Kesselring, an ex-Army Staff officer formerly in charge of administration of the Luftwaffe which he regarded as an Army tactical support force rather than a strategic striker. Göring instructed him to survey all prototypes under construction and development, and with his connivance both 'Ural' bombers were struck from the inventory without knowledge of Milch. Nevertheless it was agreed that the two prototypes should be completed and tested.

On the subject of Britain's defence, Sir Samuel Hoare, reinstated to become First Lord of the Admiralty, said: 'If extreme partisans of the new arm are right and the aeroplane can force impotence upon surface ships, Great Britain and the British Empire, as an oceanic power and dependent upon surface ships, would be faced with a new and tremendous danger. It is the duty of the Admiralty to develop in closest contact with the world of science, defensive means of dealing with the aeroplane whether by guns or armour or design. My advisers are determined to build a Fleet that can go anywhere, and to use the fullest use of air power to make the new Fleet once again the world's predominant and most mobile force for peace.

'Why is it necessary to build a great Fleet? Two lessons emerged from our experiences of the last six months – the first was that collective security meant the presence of the British Fleet in the Mediterranean. If pacifists had their way and abolished the Fleet, there would have been no collective security. The second lesson is that although in theory the combined strength of fifty nations seems a formidable military force, in practice it depends entirely on the willingness of those countries not only to go to war with an aggressor but to be prepared by the time the aggressor makes his attack. How then at the present state of world opinion are we justified in assuming we can depend upon collective help in distant regions of the world where the leading powers are not members of the League? Let us not foolishly play the game of our enemies and harp upon the loss of British prestige. We are determined to stop the drift to war. By being stronger in the future than we have been in recent years our influence in the cause of peace will be greater and not less. Thus shall we remain true to the traditions upon which the British Commonwealth of Nations is founded.'

A few weeks later, when introducing a Supplementary Defence Estimate of £13,262,000, Sir Thomas Inskip, Minister for Co-ordination of Defence, rashly said: 'If expansion had taken place earlier, our Air Force would have been equipped with machines which would have been out of date for any emergency that they may have to meet in the future.' Winston Churchill pounced: 'I am sure that such statements will be received with interest and surprise in the ranks of the RAF. The Minister's claim that had we begun expansion three years earlier we should have been encumbered with a mass of inferior machines is a new excuse for the Government's miscalculations of the relative strength of British and German air forces. If our aircraft factories had been set to work three years ago they would have been all the more capable of making the new types, and deliveries would have flowed in far greater volume at an earlier date. If the argument is carried to its logical conclusion we should have a more up to date Air Force if we waited another two years!' With a typical growl he added that while the British were currently on holiday or dreaming of the Coronation, the remorseless hammers of which Göring had spoken were descending day and night in Germany, and the most warlike people in Europe were being welded into a tremendous fighting machine. He asked the Government that the intermediate stages between peacetime and inevitable war be proclaimed a state of emergency in view of increasing dangers gathering around the British people. MPs maintained a stony silence but passed the Estimate.

The atmosphere of the first AGM of the reconstituted Bristol Aeroplane Co Ltd had an altogether more hopeful air because W. R. Verdon-Smith, the twenty-four-year-old barrister chairman of the company, announced profits of

£237,224 after provision for income tax and directors' fees, enabling a final dividend of 15 per cent, making a total distribution of $22^1/_2$ per cent for the year. That was the kind of return investors looked for, so after explanation of the urgent necessity of provision for extensive additions to aerodrome, factories, machinery and equipment, there was a ready response to a resolution increasing the ordinary capital with 1,200,000 shares of 10*s*. offered to shareholders at 25*s*. Clearly there was money to be made by the public from this rearmament business.

Verdon Smith reported: 'A large new engine factory was brought into operation at the beginning of this year, and on the aircraft side additional buildings came into use towards the end of last year. The aerodrome at Filton is insufficient for the latest types of aircraft, so arrangements are in hand to increase the area. We have also acquired additional training facilities for Reserve officers by purchasing an aerodrome at Yatesbury, Wiltshire, and training began there in January and is in full swing.

'As stated in Parliament, national requirements for air defence are of such magnitude that the Government is supplementing the country's existing manufacturing resources by creating a "shadow industry". Our company's aircraft and engines have been chosen for manufacture under this scheme. Additionally the company is fully engaged in producing large quantities of the latest series of our well known radial air-cooled Mercury and Pegasus engines, and it will be of interest to shareholders that the development of a new type is well advanced. This is the Bristol sleeve-valve air-cooled series upon which the experimental department has been working for some years past.'

That was one of Fedden's greatest triumphs. He had been closely interested in Harry Ricardo's government-financed research on sleeve valves as described to the Royal Aeronautical Society early in 1930. Crucial patents held by Burt MacCulum had luckily lapsed. Backed by David Pye, Deputy DSR (Engines), Fedden was granted funds for preliminary work entailing thousands of hours of single cylinder running and modification, aimed at a 500 hp engine. Meanwhile there were big problems of metallurgy, distortion, overheating, and fatigue to be overcome, let alone difficulties of accurate manufacture. Major Bulman recalled: 'From about 1934 Fedden was one of the first to realize the inevitability of war with Germany, driving himself and his team with terrific, almost fanatical dedication, and beyond question became the leading aero engine personality of the world during the 1930s, and even more famous as an individual in America than at home. To cope with this very great genius, and live with him, one had to be very clear headed, unshockable, understanding, and tough. Perhaps he was unconsciously biased by what he thought was good for Bristol, while I fundamentally was concerned with the good of the Service – but I always had the advantage that it was for me to agree or not to agree Air Ministry financial support for his projects!'

For what might be described as 'utility' aircraft compared with those with the greatest possible speed, radial engines reigned supreme and had the particular advantage of accessibility and ease of daily servicing. This was extremely important to Imperial Airways, and therefore the impressive Short Empire flying-boats were powered with Bristol Pegasus XC engines enclosed in quickly detachable, drag-reducing long-chord NACA cowls with adjustable cooling gills similar to those of the Lysander. The requisite 760-mile range required two cylindrical 326-gallon fuel tanks which were located between the engine nacelles within the huge open-girder wing spar. For the first time in British flying-boats the hull was so deep that a two-deck layout was feasible, affording ample headroom, with smoking lounge and promenade cabin in addition to mid and aft cabins, toilets, galley, and baggage compartment, but in time-honoured manner the interior was bare when the prototype, *Canopus*, was launched on 2nd July after some days of engine adjustments on the Medway slipway. Next day the machine was cleared for Lankester Parker to make fast water runs, but the light wind happened to be along the length of the river, and the big boat handled so well that he took off and made a fourteen-minute flight, finding little wrong but could only partially lower the variable-area flap. On the 6th he flew *Canopus* for an hour, and three days later made a full-load trial. Everything indicated that the Empire boat was going to be a tremendous success.

Years later Oswald Short ruefully said to me: 'British aircraft constructors were very late in following my example of building aircraft on stressed-skin principles, and had they taken more notice of the *Silver Streak* at Olympia in 1920 they would not have waited until 1936, the year of the Empire flying-boat, to turn their attention seriously to my method of construction. Rival designers about this time were writing about stressed-skin metal construction in a manner implying that they also were responsible for its success.' Of course they were in a measure. Certainly Oswald Short had long priority in patenting his method, but Reggie Mitchell had developed his own techniques at Supermarine, and Bristol metal construction was largely the result of laboratory experiments by Harold Pollard in conjunction with the design ideas of Barnwell, Frise, and technical studies by A. E. Russell, based like Lachmann's designs, on Wagner's researches in Germany.

The full panopoly of the industry's latest designs was on view at the A & AEE Martlesham Heath on 8th July when the King, in RAF uniform, accompanied by Air Chief Marshal the Duke of York, made a tour of four stations of his Air Force. In his Royal Rapide, piloted by Flt Lieut 'Mouse' Fielden, he flew first to Northolt to inspect the Fury and Gauntlet Fighter Squadrons, then visited No. 11 Flying Training School at Wittering, afterwards to Mildenhall to inspect Hind and Heyford Bomber Squadrons, and so to the A & AEE where he was received by Air Commodore R. H. Verney and the Station Commander, Grp Capt A. C. 'Cissie' Maund. On the tarmac, backed by the original World War I hangars, were the Spitfire, Hurricane, and Venom, all with gun muzzles aggressively protruding, the hitherto unrevealed Bristol Blenheim medium bomber, Westland Lysander, Fairey Battle, Vickers Wellesley, Vickers Wellington, Handley Page HP 52 Hampden nicknamed 'flying panhandle', and Armstrong Whitworth Whitley. The King went aboard the Wellington and received a dissertation on gun turrets, spending five minutes personally operating the nose turret. Thereafter Sqn Ldr E. G. 'Ted' Hilton, Officer Commanding the bomber test Flight, demonstrated the Blenheim, and Sqn Ldr D. F. Anderson, OC the fighter test Flight, flew the Spitfire. Reported *Flight*: 'The Spitfire roared past the Royal Standard at well over 300 mph, followed by the Blenheim, the speed of which was a revelation of what a modern monoplane bomber can do. There seemed little to choose between them. We certainly have a bomber which can outfly any fighter in service in the world to-day. The Spitfire landed quite slowly; one gathers it comes in at about 85 mph and sits down at about 70. The actual touchdown of the Belnheim was reasonably slow, but it seemed to go on running for ever. The long landing and take-off runs compared with our familiar lightly loaded aeroplanes was the most evident characteristic of all these new machines and caused much speculation, though it was comfortingly pointed out that in emergency they could be landed safely on their bellies with undercarriage retracted and put in a smaller field than a Gauntlet – but it is expensive!'

Not long after the Royal visit, 'Mouse' Fielden, a discreet and engagine, military-looking personality who was anything but mouselike, was appointed Captain of the King's Flight. There were also major changes in the RAF higher command. Air Marshal Sir Arthur Longmore became the first RAF officer to be appointed Commandant of the Imperial Defence College; Air Marshal P. B. Joubert de la Ferté, took Longmore's previous post of AOC Coastal Command; Air Commodore A. W. Tedder was appointed AOC Far East; Air Commodore Roderic Hill, brother of Professor Geoffrey Hill, was appointed AOC British Forces in Palestine and Transjordan.

7

Once again came that great annual aeronautical event, the King's Cup Race, with Hatfield as starting and finishing point on 10th and 11th July. It was preceded by news that twenty-two-year-old James Melrose, whom everyone regarded so highly, had been killed in Australia with his passenger on 5th July when his luxurious Heston Phoenix broke up in the air in somewhat similar fashion to earlier Puss Moths. His was with the second production machine, and disaster such as this was likely to prove fatal to future sales. Though the company had flown their new blue and silver demonstrator at the Royal Aeronautical Society Garden Party, it was not entered for the King's Cup.

The Race had three divisions, with eleven machines up to 150 hp in Class A, thirteen of more than 150 hp in Class B, and three multi-engined machines in Class C. Scratch machine was Percival's Mew Gull entered by the Duke of Kent. Amy Mollison was the only woman entrant and flew a BA Eagle in lieu of her back-staggered Beechcraft B17 which she had wrecked when forced-landing in a fog. Capt de Havilland was the oldest competitor and with his son Geoffrey Junior flew a DH 90 Dragonfly. Bob Waight, now officially chief test pilot, was flying the cleaned-up, glossy green TK 2. Favourite was Tommy

Rose, no longer with Phillips & Powis but operating his own taxi service using a Miles Hawk. His two rivals in long-distance flying, Charles Scott, and Arthur Clouston each had a Miles Falcon. Other rising notables were Charles Hughesdon with a Miles Hawk, Bill Humble with another, Alex Henshaw with a Leopard Moth, and Roland Falk with a Percival Gull. Wealthy Charles Gardner, who owned a small airfield near Croydon, accompanied by his University friend Giles Guthrie was flying the latter's Vega Gull. An old-timer was Sydney Sparkes, pre-war instructor of the Grahame-White School at Hendon. Tough broken-nosed Hugh Wilson, chief instructor of the Blackburn Flying Training School, was flying Wally Hope's Comper Swift, and Hope was flying a BA 4 Double Eagle. J. B. Wilson, currently test pilot of British Aircraft Manufacturing, flew a similar machine. The remainder, though not so well known, had considerable experience of racing, particularly the keen Flt Lieut D. W. F. Bonham-Carter of the A & AEE and Flt Lieut Hugh Edwards who had flown in more races than most.

Saturday's eliminating round required two laps each of a 612-mile course across the low eastern counties to Norwich, Nottingham, south to the Whitchurch control near Bristol, High Post near Salisbury, the control at Shoreham, then north across the Chilterns to Anstey near Coventry, and acutely back to Hatfield. Dawn was cheerless, with low cloud and rain. Nobody felt enthusiastic. In that poor visibility it was safest to send the fastest machine first to avoid dangers of overtaking, so the slightly lengthened Mew Gull was dispatched taking a longish run. The others were flagged away in order of speed. The crowd of spectator umbrellas and sodden raincoats streamed across the mud to the crowded club house. Protested C. G. Grey: 'Only those with cars or aeroplanes or burning enthusiasm would ever think of going to Hatfield except to buy de Havilland aeroplanes, and though they are the most widely used civil aircraft in the world, buyers are not yet so numerous as to cause a crowd at any one time. But I was surprised at the number of people who came to the race. I had no idea there were so many rain-minded people.'

From time to time came news of competitors. The Mew Gull continued leading with ease, and others were changing place. John Kirwan had broken the Heck's undercarriage at Bristol; Peter Reiss nosed-over and bent his propeller at Shoreham; low oil pressure caused Alex Henshaw to forced-land at Brockworth; Jack Matthew, flying Mrs Battye's Hawk Major, overran into a ditch at Shoreham; Wally Hope was flying on one engine. At Hatfield, shortly after noon the Mew Gull appeared, landed, scurried to refuel, and took off for the second lap. In succession the others sped in, and presently all were on the second lap. A few minutes after 5 pm the Mew Gull came streaking through a drenching rainstrom, having averaged the lap at 171.75 mph including stops. Not far behind was Bill Humble's Hawk averaging only 10 mph less, then Waight with the TK 2, its beautiful paint chipped with hail. In a relative *rallentando* came the rest, but it was well after 8 pm, long after they had landed, that Hughesdon arrived 'having rounded all the prescribed turning points even if not all at the time of asking.' Amy Mollison had retired in trouble with the retractable undercarriage, and so had Wally Hope because of an oil-tank leakage. Geoffrey de Havilland, when asked how he had fared with the appalling weather typically replied that he and his son had had a very pleasant trip.

Next day dawned in glory, but low cloud began to spread, and rain was falling that afternoon when the fourteen finalists took off for the prize-winning six circuits of the tight triangular 26-mile course which had Hatfield as its inverted apex. Bonham-Carter with his Miles Hawk went first, followed by Clouston's Falcon, the de Havillands' Dragonfly, and Richardson's Comper. By the time H. J. Wilson started, the leader reappeared and was rounding the pylon. One by one the flag dropped for the others, but it was forty minutes after the first machine that Percival took off. So many were the changes on each circuit that only the umpires with their orderly timing could follow the sequence and forecast who seemed to be winning. 'The finish itself was not the closest we have seen, but no one could complain of the handicapping,' reported Frank Bradbrooke. 'Gardner, the only one to exceed his estimated speed, won comfortably, and neither Rose nor Wilson as second and third were hard pressed. Percival had again to be content with a good show and fastest time on both days. That his latest model, the Vega Gull, had won will no doubt console him fairly effectively.' On the strength of it he took a victorious double-page advertisement the following week.

For all participants, particularly on that first day over the difficult British countryside, it had been a triumph of map reading and navigation – but in the House of Lords the second reading of the Air Navigation Bill was a big step towards freeing flying from restrictions of a different hue. Viscount Swinton said they all agreed in passionately desiring that the airways of the world should be the ways of peace. Nevertheless on Thursday, 16th July, when the airship *Hindenburg* was approaching the Irish coast from New York, her commander, Capt Ernst Lehmann, received a radio order from the German Air Ministry forbidding him to cross England that night. Said a reporter: 'That was due to a request from the British Government which has weakly given way to the agitations of anti-Germans headed by that ardent but misguided patriot Mr Churchill, who in default of a political following seemed to have placed himself at the head of this gang.' But it was not only Churchill; many people were increasingly concerned that this airship was taking photographs of towns, cities, and docks as bombing targets.

When Capt Harold Balfour asked whether the Under-Secretary was aware of the grave misgivings with regard to these airship flights, and whether he would take action to see that they were not allowed to continue indefinitely over certain areas, Sir Philip Sassoon replied that under the Anglo-German Air Agreement liberty of passage was guaranteed to each country, but there was a proviso requiring formal permission for the operation of a regular route and the *Hindenburg* and *Graf Zeppelin* must avoid passing over Britain except when obliged by weather or other urgent reason. 'There is no reason to suppose that the regulations are not properly observed by the *Hindenburg*, and it is not kept under special observation any more than the regularly operating craft of other countries which fly to Croydon.' He added that weather decided the course on the last occasion.

From Germany, Austria, Belgium, France, Holland, Hungary, Morocco, Poland, Rumania, Spain, Switzerland, 180 private owners and their passengers flew on 22nd July to join the splendidly organized three-day International Week-end *Aérienne* at Heston. Weather was atrocious. From Redhill, Penshurst, Hawkinge, and Lympne came messages that craft of assorted nationalities had landed at those places and could not proceed. The crews continued by train and car for a cocktail party that first evening. Next day they went back to fetch their machines for flying visits to Cambridge and Oxford. That evening there was a great party given by the Marquess and Marchioness of Londonderry at their splendid London residence, attended by almost the entire Air Council, many MPs, and everyone who could claim the remotest connection with the event. On Saturday visitors and hosts went to the Elizabethan house of Great Fosters, and after lunch to Eton 'conducted by a platoon of aviating Old Etonians.' Weary but determined, that evening they all dined at Grosvenor House as the grand finale presided over by Sir Phillip Sassoon who set the mode with a felicitous speech, to which a national of almost every country made reply, led by Herr F. Florian who spoke enthusiastically of the *kameradschaft* of flying and its influence in promoting peace.

Next day, the official party over, nearly a hundred aeroplanes flew through sunny skies to Lindsay Everard's private aerodrome at Ratcliffe, setting out for their Continental homes later that afternoon via Heston. 'One Rumanian did a spectacular shoot-up of the remaining machines on the ground. Herr Kropf did a peculiarly naughty take-off, trailing the right wing of his high-wing Focke-Wulf and his tail along the ground until half-way across the aerodrome, then came right side up and shot away like a rocket.' But one subdued pair, Señor Ivan Pruneda and Capt Morato, who had flown from Spain in a Hornet Moth, were undecided where to go, for though they were close friends they were of opposite political belief and thought one or other was likely to be shot on arrival.

On 17th July revolution had erupted in Spain. Francisco Franco lost no time in asking Mussolini for aircraft to ferry 15,000 Moroccan soldiers to Spain. The Duce at first refused, but the danger of Spain becoming a Bolshevist foothold caused him to change his mind, and on 29th July he sent twelve three-engined Savoia Marchettis to Franco. Hitler similarly offered help. Ober-Leutnant Rudolf Freiherr von Moreau was given command of the air lift, and the first of twenty Ju 52s was flown via Stuttgart to Morocco, and on 1st August six Heinkel He 51s and twenty-two 'ack-ack' guns followed via Seville to protect the Junkers. By that time Franco had managed to get 800 men across as the initial army for his operations on Spanish soil. Steadily his force built up.

Several units of the Spanish Air Force joined Franco, but the Communist Government also had aircraft, and on 20th July they dropped bombs in the Straits of Gibraltar in an attempt to prevent transport of troops from Morocco. In retaliation Franco's Royalist aircraft attempted to bomb two Spanish warships during the afternoon of 22nd July but were driven off by anti-aircraft fire. A second attempt that evening resulted in the British liner *Chitral* and *HMS Shamrock* being hit by splinters. The British Consul-General in Tangier was immediately requested by the Royal Navy's C-in-C at Gibraltar to make vigorous protest to Franco. On 14th August the first German bombing attack was with a Ju 52 commanded by von Moreau, who put the battleship *Jaime I* out of action. A week later the first air lift of food, two tons, was dropped for the beleagured Alcazar of Toledo. By the end of September 12,800 men had been ferried across with 400 tons of equipment. Franco began to advance from Algeciras toward Madrid, and Germany began to infiltrate more Luftwaffe Volunteers. The British Press made little of the news, and the general public cared less, for Spain seemed a far country which had no impact on their lives.

With fine impartiality the quick boys were selling any aeroplane they could lay hands on, either to the Communist Government or the Revolutionary Royalists. The Foreign Office announced that no responsibility would be taken for anyone going to Spain. However a fee of £150 a trip was sufficient to secure the services of free-lance pilots for ferrying.

Nevertheless most Englishmen were patriots, and there was growing feeling among younger men, particularly the better educated, that they must be ready to defend their own country. Recruitment for the Territorials had an upward trend; the week-end fliers of the RAF Volunteer Reserve were numerous and enthusiastic, and their flying, as everyone saw at the RAF Display, was outstanding; there were also the students of the Cambridge and Oxford University Air Squadrons who spent six weeks of summer leave attached to the RAF at Abingdon. Though taught *ab initio* on Avro Tutors, six seniors in each group had recently graduated to Hawker Harts.

The specification for Harts had been issued in May 1926 but they were still the mainstay of Hawker production. At the third AGM of Hawker Aircraft Ltd on 14th July, Tom Sopwith, that rarely seen magnate whose familiar initials belied familiarity on the part of his employees, said: 'Output during the last few months has increased. The Fury has been sub-contracted to General Aircraft Ltd and in due course royalties will accrue to your company. Our new monoplane fighter, the Hurricane, has completed trials with very satisfactory results, and has been ordered in quantity. The Gloster four-gun fighter, the Gladiator, will shortly be in production, and certain prototype machines will be submitted for test in due course.

'Our trading profit amounts to £196,101 0*s*. 11*d*. Against this, provision must be made for income tax, capital redemption, and preference dividends, and there is a bank overdraft of £253,000. On the other hand work in progress has increased by £215,000. As chairman both of Hawker Aircraft Ltd and the Hawker Siddeley Group, I would like to assure you that, because of the strong position of the Group, benefits should accrue to both parties as the result of our full merger.'

Parnall Aircraft Ltd also was involved in the expansion programme, and proposed to increase its flotation capital of £92,000 and £50,000 working capital from shares held by the Electric & General Industrial Trusts Ltd, to £450,000 by creating 600,000 ordinary 5*s* shares offered at 5*s*. 6*d*. 'Manufacturing capacity at Yate has been increased far beyond that originally contemplated,' explained the chairman. 'To meet Air Ministry demands, new Works costing £100,000 have been built at Tolworth, Surrey, and have been at full pressure for three months. Besides the Fraser-Nash gun-rings for the Hawker Demon, for which further contracts have been received, the company is to produce new versions for each type of new aircraft in which power-driven armament is required. The company's powered armament is the only one which meets the needs of modern military aircraft, and the Yate factory has large contracts from Sir W. G. Armstrong Whitworth Aircraft which will extend over two years.'

Hubert Scott-Paine, that earlier driving force of Supermarine, was also making good, for the RAF had ordered numbers of his latest 64 ft hydroplaning rescue boats powered with three 500 hp Napier Sea Lion engines, and his premises at Hythe, adjacent to the Imperial Airways base where the Empire flying-boats would dock, were being extended.

There was also consolidation by amalgamation between British Airways Ltd and British Continental Airways Ltd founded by Graham Mackinnon, a Lloyds Underwriter who was the son of Sir Percy Mackinnon, chairman of Lloyds. Major J. R. McCrindle remained chairman of the enlarged British Airways Ltd. P. W. Lynch Blosse, formerly chief pilot of Spartan Air Lines, was air superintendent. Most of the pilots had worked with Hillman's, and surprisingly one was R. H. McIntosh who for a few years had flown for Imperial Airways but became bored with the routine and launched out as a freelance. On 27th July British Airways started the first night mail service to Cologne in conjunction with the Swedish AB Aerotransport, using DH 86s.

General Aircraft had been greatly disappointed that British Airways, after their pilots had tried the Croydon ten-seater, decided that America's Lockheed 14 seemed more attractive because of its metal-skinned construction compared with the fabric covering of the Monospar. However Major C. R. Anson of Anson Airways agreed to purchase the prototype if it made a record flight to Australia where he proposed to operate it. With Colonel the Master of Sempill the ostensible pilot for publicity purposes, but the amiable 'Timber' Wood, the original chief pilot of Hillman's Airways, as safety pilot, together with two crew, the laden Croydon departed from Croydon on 30th July.

Hollis Williams told me: 'All went well until Karachi where Sempill made a heavy landing and the tailwheel broke. Crocombe, as chief designer, went out to supervise repairs, but there were difficulties with Sempill who went home in a huff, so Crocombe took his place and they continued to Australia – but by that time the purchaser repudiated the contract because of delay. It was therefore decided to attempt a record-breaking flight to England. They set off across the Timor Sea to Koepang, but kept receiving warnings from Air Traffic Control that there was too much north in their course. Wood made appropriate corrections, but several times more he was told they were too far north, and presently fuel was running low with no sign of their destination but only the vast

Indian Ocean. Wood began searching on different courses, but it seemed their compass was wrong. They spotted a coral atoll, and he decided to land wheels and flaps down, which he brilliantly accomplished, but again damaged the tailwheel because it sank into the coral. Some native boats were fishing, and our crew managed to induce the men to take them aboard, though the only word they had in common was Koepang. What little shipping there was had been alerted, and miraculously, after four days a cargo boat loomed up, stopped, and they were saved – for a check on the charts showed that those natives, steering instinctively, were accurately on course for Koepang. But that was the end of the ST-18, for there was not enough money nor good enough prospect to build a replacement.'

A string of air accidents marred late summer and early autumn. No airline was immune. On 10th August the Vickers Vellox operated by Imperial Airways crashed almost immediately after taking off from Croydon for Paris, striking some houses and bursting into flames. The pilot, Capt L. F. H. Orr, and crew of three lost their lives, but there were no passengers as the machine was carrying only freight and mail. Two days later a British Airways DH 86 on the new night mail from Hanover to Cologne and Gatwick crashed thirty-five miles SE of Cologne and was destroyed by fire, the radio operator being killed outright and the pilot Capt C. S. Gill died from injuries five days later. Wireless bearings were believed the source of trouble as the machine was thirty miles off track and the ground was 2,000 ft high in that vicinity.

On 22nd August there was disaster to Imperial Airways flying-boat *Scipio* even though piloted by so skilled a veteran as Capt A. S. Wilcockson, senior pilot of the company. Alighting at Mirabella Bay in the normal course of a flight from Alexandria to Brindisi, wind and sea were so abnormally rough that a wingtip touched, the big boat swung round and was wrecked with such force that it sank, and two passengers were killed and the other five injured. The Inspector of Accidents placed the blame squarely on Wilcockson and First Officer Long: 'The approach was under very disturbed air conditions, and the pilot made an error of judgment necessitating opening up the engines at the last moment, but the tailplane had inadvertently been set by the wireless operator to maximum incidence thus rendering the boat increasingly nose-heavy as it neared the water, and on opening up the engines this produced a downward pitching moment which rendered the pilot at such low altitude unable to level up before reaching the water.' One wondered whether the Inspector had the piloting skill even to attempt such conditions.

In mid September another British Airways DH 86 crashed a few moments after taking off from Gatwick on the regular night service to Hanover, having turned as though with engine trouble, stalled, and nosed steeply down from several hundred feet, smashing into a tree and catching fire. Three of the crew, including Capt Walter Fraser Anderson, chief pilot of British Airways, were killed, and another crew member injured. There was no doubt that Anderson was very experienced, for ever since leaving Bomber Command in 1930 he had been on airline operations.

On 25th September there was yet another disaster when the Imperial Airways twin-engined Boulton & Paul *Boadicea*, which was being flown by Capt A. C. Thomas, accompanied by a radio operator, disappeared over the Channel after having been reported crossing the coast. Aircraft and life-boats put out and searched throughout the afternoon and night, but nothing was found, and the body of the pilot was washed ashore a month later. How could people be induced to think that flying was safe, or that the sky was the future highway of the world?

Certainly Imperial Airways was setting the highest standards of professionalism. Since February 1935, they had added 107 pilots to the previous sixty-seven, of whom eighteen held Master Pilot's certificates. Nobody was accepted unless with more than 1,000 hours of which 500 must be twin-engined, and experience of night flying and competence in navigation was essential. To maintain the requisite standard, a subsidiary named Air Pilots Training Ltd was established near Croydon, and two Westland Wessex and one Short Calcutta were used for advanced tuition on a 12-month course for qualification as First Officer. 'Considering the enormous number of pilots who will leave the RAF yearly after the present expansion, Imperial Airways will have more than enough from which to make its choice,' was the happy forecast in expectation that Empire air travel would be fully established within a few years.

As a step towards transatlantic flying-boat services, Sir Philip Sassoon announced: 'A joint operating company will be incorporated in due course with three directors each from Imperial Airways and companies nominated by the Irish Free State and Canada. The chairman will be the managing director of Imperial Airways which will hold 51 per cent shares, and the other companies 24½ per cent each. Reciprocal landing rights will be granted in the USA, Canada, Ireland, and Great Britain to the aircraft of Imperial Airways and Pan American Airways. Preference will be given to the direct route from England by way of the Irish Free State, Newfoundland and Canada, but the Bermuda route may be used at first during winter. Rights of landing in transit are guaranteed to Imperial and Pan American Airways exclusively for fifteen years.'

8

There was trouble in the higher echelons of the Air Ministry, for its Permanent Secretary, Sir Christopher Bullock, was dismissed by the Prime Minister after a special Government committee had investigated discussions alleged to have taken place between Sir Christopher and representatives of Imperial Airways, though there was no question of corruption. After serving in the RAF first as an observer and then as pilot during the Great War, he had risen to eminence as private secretary to a succession of Air Ministers and finally as Permanent Secretary to the Air Ministry. As C. G. Grey said: 'He has probably done more for the Air Force than any one man except Lord Trenchard, Sir Samuel Hoare

e Airspeed Envoy AS 6 Mark III was of wooden construction, with a stressed skin plywood wing, retracting
dercarriage, metal flaps, and carried eight passengers.

alon was the Imperial Airways Avro 652 with one-piece wooden mainplane and fabric-covered welded steel tube
elage.

totype of the military version of the Avro 652A which had a hand-operated gun turret.

The Northrop all-metal low-winger was purchased from the USA by the Air Ministry and sent to Martlesham for RA
handling trials and inspection by the British aircraft industry.

Less elegant than the famous Supermarine Southamptons, the more capacious and powerful Stranraer showed the san
family design.

Britain First, shown
its maiden flight, crea
a furore because it wa
50 mph faster than th
latest fighter, the Glo
Gladiator biplane.

ɪigned for those private owners who still preferred biplanes, the DH 87 Hornet Moth had sociable side-by-side
ting and dual flying controls.

ɪd-tunnels were originally standardized at 4 ft by 4 ft, but the RAE built one big enough in which to test a complete
ter.

velopment of the neat Westland CL 20 Autogiro, powered with a Pobjoy engine, was abandoned because of
duction pressure for military aircraft, though a big market had been expected.

The production version of the impressive Rolls-Royce Kestrel-powered Fairey Hendon II was the first metal-structured, though fabric-covered, low-wing cantilever monoplane to go into squadron service with the RAF, but or fourteen were built.

The Vickers G 4/31 contender was a private venture which secured a big production order of which Wellesley K 7556 was the first machine.

Presented to the Air Council by Lord Northcliffe, *Britain First* became an RAF prototype for the subsequent bombe version.

ɴly one prototype of the Armstrong Whitworth AW 23 twin-engined transport-bomber was built, but the general ɴnfiguration was the basis for design of the subsequent Whitley bomber.

he Bristol Type 130 Bombay prototype was the successful rival of the AW 23, accommodating twenty-four fully-quipped troops or an equivalent cargo, and had gun turrets in nose and tail.

he HP 51 was Handley Page's contender for the bomber-transport specification by converting his previous nsuccessful biplane transport to a monoplane, and in revised configuration for production became the HP 54 Iarrow.

Derived from the Blackburn M 1/30A, the Shark III with Pegasus IX engine was a production version of rugged strength and simplicity extensively used by the Fleet Air Arm and Royal Canadian Air Force.

As practical exercises the de Havilland students built a succession of light aircraft, of which TK 2 was a single-seat, long-range racer, and in the modified form above had a maximum speed of 182 mph.

Built in the remarkably short time of nine months, the prototype Hurricane made its initial flight on 6th November 1935, fitted with a fixed pitch wooden propeller. A winner from the start, 600 were ordered in June 1936, and it entered service in December 1937, superseding the Gauntlets and Gladiators.

ıe Avro Anson production version had full length cabin windows, and became the first monoplane with retractable ıdercarriage to be used by the RAF.

reened from view in the de Havilland workshops in 1937 was the beautiful de Havilland DH 91, later named batross, designed as a transatlantic mailplane, powered with four 520 hp Gipsy 12 inverted vee engines.

ıegant and fast for its power, the DH 90 Dragonfly was a tapered wing, five-seat, luxury tourer and many of the ıty-six built were sold abroad.

King George VI, Chief of the Royal Air Force, inspects the smoothly finished Spitfire prototype at the famous A & AEE test base of Martlesham Heath, Suffolk.

Similar to the Northrop, the Fairey Battle was the key two-seater chosen for early equipment of the expanding RAF, and made its first flight on 10th March 1936, entering service a year later.

The first metal skinned twin-engined heavy bomber to go into RAF service was the Armstrong Whitworth Whitley with Armstrong Siddeley engines, but the Mk V with Rolls-Royce Merlins was the most numerous variant and with the Wellington became the early mainstay of Bomber Command.

e all-wood Miles Merlin was typical of the low-wing, three-seat cabin tourers of the late 1930's but had the novelty
pneumatic flaps inter-connected with the throttle.

e Short Scion Senior was a ten-seater available either as landplane or seaplane powered with four Pobjoy engines,
t it also represented the half-scale configuration of the Empire flying-boat.

aking over assets of the Comper Co, the Heston Aircraft Co produced the Phoenix – a high-winger of unusual but
easing configuration with seats in two pairs and a fifth behind them, but was one of the more expensive private
craft at £1,980.

Great prestige attached to records. The 66-ft span single-seat Bristol 138A represented successful British endeavour at capturing the world's altitude record by achieving nearly 54,000 ft in 1937.

Last of the sweetly handling variants of the Hart was the Hector Army Co-op machine powered by the Napier Dagger, but when the first production machine was delivered to the RAF in February 1937 it was already outdated.

The prototype Vickers B 9/32, twin-engined geodetic day bomber was more attractive than the Armstrong Whitworth Whitley, and led to specification 29/36 for a production version named Wellington but irreverently known as the Wimpey.

ɪe unique, fully slotted Lysander made initial hops with unfaired undercarriage, non adjustable tailplane, and fixed
ch wooden propeller before making its maiden flight as K 6127 in 1936.

e Hampden, designed to the same specification as the prototype Wellington, had an unorthodox fuselage in the form
a slender tail-boom and quickly acquired the nickname 'Flying Pan handle', but proved to have a remarkable speed
ge matching the Lysander though faster.

r some years the Blenheim, derived from *Britain First*, remained the RAF's fastest bomber and was a great
chnical advance on the Hind biplanes it supplanted, yet pilots quickly adapted to its complexities and fast landing
eed.

With the arrival of the Short Empire four-engined monoplane flying-boats, all others became outdated. *Canopus* was the prototype launched on 2nd July 1936. Above it is a Short Scilla landplane of only two years' earlier vintage.

End of the General Aircraft Company's hopes of a small airliner, for this is *Croydon* abandoned on a coral reef in the Indian Ocean.

The Armstrong Whitworth AW 29 single-engined all-metal day bomber first flew on 6th December 1936, but was overshadowed by the concurrent twin-engined Whitley which had secured a big production order requiring all available factory space.

)esigned as an eight-gun fighter to specification F 5/34, the Gloster fighter was radial-engined for competition against he Hurricane, but appeared too late and had slightly inferior performance.

As a safeguard before building a prototype a small balsa dynamic model was made to examine its free spinning characteristics in a vertical blast of air. A sequence of adjustments to the tail surfaces would then be made to ensure satisfactory recovery.

The two-seat Skua ighter-dive bomber was he first British deck-anding machine with ll-metal cantilever nonoplane wing, flaps, etractable indercarriage, and VP propeller.

Based on the Hurricane design and using the same outer wings, Hawker produced the Henley as a two-seat, high-speed monoplane light bomber, but instead 200 were built by Glosters as target-towing tugs.

Designed to replace the trusty Southamptons, the aerodynamically cleaner Saunders-Roe London was in first-line service from 1936 to 1939. Five with long-range fuel tanks, enabling laps of 2,600 miles, made a splendid 30,000-mile cruise from England to Australia and back.

The Blackburn HST 10 had a metal skinned fuselage and fabric-covered wing with single tubular metal spar. Intended as a fast twelve-seat feeder transport it was shelved after a single flight because of priority production of RAF aircraft.

he little single-seat Hafner AR 3 rotorplane differed basically from the Autogiro in its control system, and had the bility to take off with a dramatic upward leap.

he easy-to-fly Fairey Battle went into extensive production which continued until September 1940 by which time ,184 had been built, though they were already obsolescent in 1939.

arly production of Hawker Hurricanes filled the cramped assembly shops at Brooklands, but the main factory at ingston soon took over, and by September 1944, 12,780 had been built in Britain and 1,451 in Canada. Many referred the Hurricane to the Spitfire although it was slower.

Though it failed to secure production, the Hawker Hotspur two-seat fighter was based on Henley design but built to the same specification as the Boulton & Paul Defiant, both having a multi-gun turret as sole armament.

The DH 93 Don was a three-seat general purpose trainer powered by a 525 hp Gipsy King and originally mounted a gun turret but was converted to a normal cabin as a communications aircraft.

First flight of the aerodynamically refined Albatross mailplane at Hatfield on 20th May 1937 was watched by the entire Works employees. It was expected to carry 1,000 lb of payload for 2,500 miles flying at 200 mph against a 40 mph head wind.

and Sir John Salmond.' His knowledge of Service and civil flying was encyclopaedic. But he had enemies, for he was no respecter of persons unless they deserved respect, and it was even darkly hinted that Lord Swinton did not like Bullock and 'had been pushed into pushing Sir Christopher Bullock out.'

A White Paper made it clear that Sir Christopher's activities had not influenced negotiations for a projected contract between Departments of State and Imperial Airways, and that no attempt had beeen made to barter honours in return for personal advancement. Instead he had endeavoured to secure the best terms for the Government, resulting in a very satisfactory agreement with the company. The crux was: 'Sir Christopher Bullock is a fluent talker and not always a very patient listener, and, at all material times, he had constantly in mind the question of leaving the Civil Service and entering the world of commerce and business. He was impressed by consideration that he had already reached as high a position in the Service as ever likely to be open to him and that if he were to seek fresh fields he ought to do so while still young enough to have the prospect of a long business career.' Openly he had lunched with Sir Eric Geddes for that purpose, seeking appointment as a government director of Imperial Airways. 'This topic was distasteful to Sir Eric because he considered Sir Christopher's aspirations to be hopeless and he did not intend to support them. That could have been because in the spring of 1934, with negotiations about to begin on carriage of first class Empire mail, Sir Christopher Bullock had thought it an opportune moment for recognition of Sir Eric's services as chairman of the company by conferring a high honour, and he had put the idea to Lord Londonderry who asked Sir Eric whether an honour would be acceptable. Sir Christopher had interpreted this as authority to seek the answer, and Sir Eric indicated that one would be welcome. In the Birthday Honours List that year there was no mention of Sir Eric who angrily told his sponsor that if the reward was for negotiating the air mail agreement the Government could forget it. Matters went from bad to worse, each misunderstanding the other, and became impossible when rumours began to circulate that Sir Christopher was to succeed Sir Eric as chairman.' George Woods Humphery, as managing director, became involved in this at a lunch on 11th June when Sir Christopher referred to Sir Eric's recent indisposition, and said he thought that if he became chairman, with Woods Humphery as deputy chairman and managing director, that would be a very good working combination.

The report concluded: 'In 1928 the principles were formulated regulating conduct of Civil Servants in relations with the public, and these were commended by the Service generally as a true presentation of its traditions. They were enjoined by the Government and ordered by a Treasury circular dated 13th March 1928, to be incorporated with the rules of every department. We cannot escape the conclusion that Sir Christopher Bullock's conduct was completely at variance with the spirit of this code, which in our view clearly precludes a Civil Servant from interlacing public negotiations with the advancement of his personal private interests.' But as Sir Christopher afterwards said: 'I do not seek to burke responsibility for consequences which have flowed from my own actions. But it is easy to be wise after the event; and fortunate is he – be he politician, sailor, soldier, airman, journalist or civil servant – who can honestly say that if every private and informal conversation he had held was sifted and resifted months, even years afterwards in the rarified atmosphere of a solemn and formal inquisition, no passing phrase uttered in an unguarded moment could be held injudicious, no word or deed be called in question in some degree by absolute standards of tastes or propriety.' But that was the end of Sir Christopher Bullock as a public figure.

Colonel Sir Donald Banks, KCB, DSO, MC, currently Director-General of the Post Office, was appointed in his place. Previously he had been Secretary to a succession of Postmaster-Generals, but during the War commanded the 10th Battalion of the Essex Regiment and the 8th Battalion of the Royal Berkshires. He had gained experience of aviation as a member of the Council of Action correlating development of civil aviation and air mail requirements, and in November 1935 was one of the deputation which went to Canada and the USA to confer with their civil aviation authorities and Post Offices on transatlantic air services.

A great and historic pioneer of aviation died on 1st August in Paris. He was Louis Blériot, who for some years had suffered heart trouble. For all time he would be known as the first in the world to fly the Channel and thus breach England's age-long insularity. Manufacture of motor lamps brought earlier prosperity, and he became one of a clique of enthusiastic and far-seeing Frenchmen interested in achieving flight, and who, even had the Wrights never appeared on the scene, would have evolved the basic type of practical aeroplane. Behind most pioneers were clever but unrecorded craftsmen responsible for the construction of early machines, and with initial success came design assistants. Thus the earliest Blériots were built at the *Atelier* of Gabriel Voisin, but after disagreement Blériot opened his own business where Louis Peyret helped design and construct the Blériot No. VII tandem monoplane of 1907. Within a year Robert Saulnier became his designer of the successful Type XI with which Blériot flew the Channel on 25th July, 1909, winning the *Daily Mail* £1,000 prize. During the 1914–18 War Blériot's name disappeared as a type, but he bought the Deperdussin business, and its famous Spad fighter designed by M. Bechereau became world famous. After the War, Blériot employed such ingenious designers as Herbemont and Zappata, but was relatively unsuccessful in securing orders from the French Air Force, nor were his larger airliners and flying-boats much favoured. Financial vicissitudes were a consequence. As C. G. Grey commented: 'People said he was reckless of the lives of his test pilots, and quoted him as saying that "We must pay for progress in aviation" when several had been killed. That is the kind of hard thing a naturally kind-hearted man who is not afraid to face facts would say, and precisely the kind of thing people would ascribe to cynicism and carelessness for the lives of others. But France and England equally owe much to Louis Blériot.'

9

By Treaty of Alliance between Great Britain and Egypt, signed on 26th August, military Occupation of Egypt by the Forces of HM The King and Emperor was terminated. However, Great Britain and Egypt made a mutual pact of aid in the event of war, and it was conceded that the Suez Canal was an integral part of Egypt, but as it was an essential means of communication to the British Empire, it was agreed that until the Egyptian Army could ensure freedom of navigation, Britain would station forces there which would not constitute an Army of Occupation, though would extend to 10,000 troops and 400 RAF pilots together with ancillary personnel for administrative and technical duties. Land, barracks and technical accommodation, including water supply, would be provided by the Egyptian Government with a financial contribution from the British Government. Egypt also undertook the maintenance and constant availability of adequate landing grounds and seaplane anchorages to establish the passage of British Forces' personnel, aircraft and stores, and agreed to accord full facilities for the salvage of any aircraft in distress that might land in Egyptian territory. Regulations were defined controlling the passage of aircraft each side of the Canal, but because of the speed of modern aircraft the Egyptian Government accorded permission to the RAF to fly wherever necessary for training.

In England an unofficial body, the Aircraft Shop Stewards National Council, representing 40,000 aircraft factory workers, was endeavouring to negotiate recognition by aircraft firms, and claimed that the hourly rate for all workers should be increased by 2*d.* with a minimum rate of 1*s.* 4½*d.* for adults, and the 1931 cuts on piece work, overtime, and night shift to be restored. There was anger among the management, for in 1934, at the behest of the Amalgamated Engineering Union, wages were increased 2*s.* weekly, and as recently as June a further 3*s.* was conceded in stages of 1*s.* every two months until December. The Air Ministry took an equally serious view because costs must be rigorously controlled to keep within the annual budget.

The Soviet was showing how production could be tackled. Louis Breguet, *doyen* of the French aircraft industry, on return from a technical mission to Moscow said: 'With ten times as many workers as France, they are producing twenty times as many aeroplanes, motors and accessories. I estimate 200,000 are employed, and annual production is about 5,000. Three shifts work throughout the twenty-four hours. One factory produces Wright Cyclones; a second, Hispano-Suizas; a third, Gnome-Rhône; and a fourth, low power motors. At one factory 25,000 men produced four 250 mph light twin-motor bombers a day; another turned out five modern fighters a day. Two types of heavy four-engined machines are being built in large numbers. Some men are salaried, the rest poorly enough off. Work and leisure are organized. In fact this Soviet regime is far from communism as taught in Western Europe.'

However France was gratified that on 17th August, Georges Détré established a new world's record of 48,600 ft with a Potez 50 powered with the

Bristol-derived Gnome-Rhône – the same machine with which Mlle Maryse Hilsz a year earlier achieved a magnificent 47,000 ft. By now the Bristol high-altitude monoplane was almost ready for a British attempt. The special Pegasus with auxiliary blower and four-bladed fixed-pitch propeller was ready, and on 15th August the machine was returned to Filton for its installation, which was completed in a fortnight, and after a brief test by Uwins, was flown back to Farnborough on 5th September by Swain.

There seemed less serious intent with the combined manoeuvres of the three fighting Services in the last week of August, for it had the appearance of a great seaside party in Studland Bay where soldiers disembarked from troop-carriers guarded by the battleship *Iron Duke*, cruiser *Curacao*, four destroyers, and a monitor – the objective being an ammunition factory at Corfe Castle. The intention had been to land at Milford, east not west of Bournemouth, but fog rendered Studland easier until a stiff breeze forced essential heavy transport, guns and horses to be left at sea. Meanwhile Gauntlet fighters and Army Co-operation Audax disported above the troops wearily proceeding towards Corfe, and when the soldiers returned they found the sea too rough for re-embarkation. A hurried armistice followed while friend and foe shared what was left of the rations. As the doughty Mrs McAlery commented: 'If the combined Exercises of 1936 have shown the futility of this childish game they will not have been entirely a waste of time, money and energy.'

Much more successful was the fourth annual International Rally at Lympne organized by W. E. 'Bill' Davis and his dynamic little wife, managers of the Cinque Ports Flying Club. Harold Perrin was able to bellow in all his official glory, and there were races and displays, including Herr Paul Förster doing impossible things with his Focke-Wulf Stieglitz. There were dinners and cocktail parties, and a final evening at Port Lympne Mansion, home of Sir Philip Sassoon, set amid beautiful gardens bordered with high clipped hedges. Among the many visitors was tall Frank Courtney of the rimless glasses and wavy hair, revisiting the scene of former exploits after eight years in the USA where among other things he had developed an amphibian flying-boat for the Curtiss Aeroplane & Motor Corporation.

Concurrently the British Gliding Association held its annual soaring competitions on Bradwell Edge, Derbyshire; but conditions were indifferent, and nobody could beat Philip Wills and his *Hjordis*, holders of the Wakefield Cup for a flight of 105 miles earlier in the year. Unfortunately he badly damaged his sailplane during the meeting when he landed down-wind in Lincolnshire and turned over. The effervescent Dudley Hiscox with his efficient Kirby Kite had the highest scoring time, and the team prize was won by Dr Alan Slater and George O. Smith, who in turn flew the Golden Wren, the simple high-winger designed by Cpl Manuel. Despite several days of impossible weather while every one sat around in caravans and tents, the total flying of 151 hours was considerably greater than the previous year's meeting on the heather-clad top of Sutton Bank in Yorkshire.

A spate of transatlantic flights featured in the first week of September. On the

3rd two Americans, Harry Richman and Richard Merrill landed in Carmarthenshire after crossing from New York in their splendid Cyclone-powered Vultee low-winger in eighteen hours. Next day blonde Beryl Markham, who had replaced Amy Mollison as chief pilot of Air Cruisers Ltd, took off from the RAF aerodrome at Abingdon with her Gipsy VI Percival Vega Gull at 4.50 pm in an attempt to reach New York, but after twenty-one hours found Newfoundland swathed in fog, and fearing fuel shortage, landed on a swamp at Baleine, Cape Breton Island, and the machine tipped on its nose with too much damage to take off again, but she was duly rescued by an American aeroplane and taken to New York for the inevitable ballyhoo: nevertheless it took considerable grit to make that lonely flight over the desolate ocean.

By now interest centred on a race from England to Johannesburg with prizes of £4,000 for the fastest time, £1,500 for second, £1,000 for third, and a fourth of £500, offered by millionaire Isaac W. Schlesinger of that city in connection with the Empire Exhibition being held there. He had been a penniless young American insurance agent of nineteen when he arrived in South Africa by steerage in 1886, but now, at sixty-nine he controlled seventeen influential companies, including banking, newspapers, theatres, and had established the African Broadcasting Co.

Even before the race started there was tragedy. Tom Campbell Black, who had entered a Mew Gull, which at the best had limited visibility when tail-down, died from injuries at Speke, Liverpool, on 19th September after he taxied it into the fanning propeller of a slowly moving Hart and the blades scythed into the Mew Gull's left wing and across the cockpit. A foot either way would have saved him. Nobody could have been more experienced – but all who flew required a modicum of luck in the risky situations that inevitably occurred.

This thirty-seven-year-old ex-RNAS pilot had migrated to Central Africa after the War and presently was financially backed by Mrs Wilson of Nairobi in forming Wilson Airways there, and I first met him when he came to Westland to evaluate the Wessex which he selected for this service. Serious though he seemed, he had a great sense of humour and an unpretentious charm; but he could be quietly tough in business and had immense determination. Thirteen times he flew without publicity between Kenya and England, but fame came with his triumph as co-pilot with Charles Scott in the MacRobertson Race – an ideal partnership of mutual trust and outstanding ability.

Amy Mollison, with mistaken intent to honour Campbell Black's memory, offered to fly his Mew Gull in the race if repaired in time, but this was impossible. Many thought she was merely cashing in on tragedy – partly as a counter-blast to the news that Jim Mollison was in New York preparing for a transatlantic attempt with a Bellanca monoplane. Currently she was instituting divorce proceedings against Jim, prompted by her friend François Dupré with whom she was frequently seen hitting the highlights in Paris.

Of the fourteen aeroplanes entered for the Johannesburg Race only nine were pushed onto floodlit Portsmouth aerodrome for the start shortly after dawn on 29th September, watched by a considerable crowd in the lamp-lit public enclosure. As the machines taxied to the starting line, engines muttering and exhausts spitting red and orange and blue, the watchers surged onto the aerodrome, but were persuaded back – then in the cold grey light, with the West Sussex hills still shadowed, J. B. Reynolds, that veteran starter, raised his flag, and signalled Max Findlay away with the Airspeed Envoy. At minute intervals the others started, led by Capt Stanley Halse with his Mew Gull which, because of its variable pitch propeller, rocketed skyward. Tommy Rose with the twin-engined BA Double Eagle and Bagshaw his mechanic received a special cheer from the crowd. He was followed by the single-engined BA Eagle flown by George Allington accompanied by his brother and Lieut Booth. Then came Flg Off Clouston, with a learner's 'L' on the window of his Miles Hawk Speed Six, followed by David Llewellyn and Charles Hughesdon with their Vega Gull which also had a variable pitch propeller. Major Miller's white Mew Gull was signalled away and took off even quicker than Halse through using a small degree of flap. The blue Sparrowhawk piloted by Victor Smith should have been next, but his engine choked and it took time to restart, so Charles Scott and Giles Guthrie in a Vega Gull were next, and finally Victor Smith was off, and the crowd slowly dispersed in the sunrise.

The rest was disaster and fiasco. Max Findlay, a thirty-eight-year-old Scot highly regarded at Brooklands as Chief Flying Instructor, was killed with one of his crew when the Envoy failed to clear trees taking off from Abercorn. Max had served in the Black Watch and RNAS, and after hostilities remained with the RAF for three years in Afghanistan and Waziristan, but reverted to farming until formation of NFS in 1929, where he quickly made his mark as a sound, careful and sympathetic instructor before joining Brooklands Aviation Ltd. On 2nd October he was buried at Abercorn, and Llewellyn and Hughesdon, who were out of the race, went from Lake Tanganyika to join Findlay's remaining crew members, C. D. Peachey and Ken Waller, at the funeral. Well might Frank Bradbrooke 'regard the race as but a sorry procession; and that without the least disrespect to the sportsmen whose spirit endured through days and nights of weariness, difficulty and danger in no ordinary measure.' It became a chapter of misfortunes. Allington's Eagle damaged its undercarriage in a forced landing in Bavaria; Miller's Mew Gull retired at Belgrade; the undercarriage of Tommy Rose's Double Eagle collapsed at Cairo; an oil leak caused Victor Smith's Sparrowhawk to be withdrawn at Khartoum; at Abercorn, veldt fires caused such dense smoke that both Llewellyn with the Vega Gull and Halse with his Mew Gull failed to locate the aerodrome before running out of petrol and crashing in forced landings. Clouston with the Hawk Six had engine failure in total darkness and crashed, but luckily was unhurt. That left only Scott and Guthrie with the Vega Gull entered by Sir Conup Guthrie, and they completed the course in a little under fifty-three hours, greeted at Johannesburg by Mr Schlesinger accompanied by Oswald Pirow the Minister for Defence and Sir Pierre Van Ryneveld who in 1920 was the first to fly from England to South Africa.

On the day preceding the start of the race from Portsmouth, Britain at last re-achieved the altitude record. After taking off from Farnborough at 7.30 that morning, Sqn Ldr Swain climbed in wide circles, engaged the auxiliary blower at 35,000 ft, and at 40,000 ft flew south-west until 51,000 ft, at which point he was over the Bristol Channel, and later confessed: 'I felt small and lonely, and England seemed so minute a country that I could easily be blown beyond its borders.' By that time his pressure suit had ballooned out making it difficult to co-ordinate cramped arms and legs. There was fuel for a further forty-five minutes, but he seemed to have reached the ceiling, so began to glide down, but after losing 5,000 ft the cockpit windows and celastoid panel of his helmet hazed over, and he had to steer by the glare of the sun, unable to see his instruments.

Locked in his diving suit, to all intents blind, he continued to lose height in an erratic manner, and began to feel suffocated as though short of oxygen. He tried to press the release lever to open the cockpit cover but it would not move, nor had he the nervous strength to pull the emergency release. He tried to get at the ripping panel on his suit in order to get his head out, but cumbered with shoulder straps and parachute harness could not find it. In desperation he grabbed the knife hung within reach for just such an emergency and hacked out the window of the helmet. Immediately he felt better and found he was at 14,000 ft over Yeovil, so steered east, but after passing Salisbury there were only two gallons left, so he landed at Netheravon after two hours from take-off.

The sealed barograph when corrected for temperature and barometric pressure by the NPL confirmed 49,967 ft – but the designed ceiling was 54,000 ft, so it was decided to fit a finer pitch propeller, altered supercharger impeller, and small brakeless wheels to save weight – but this would take time.

The major factor in success was the rubberized fabric pressure suit and its helmet devised by Grp Capt G. Struan Marshal of the RAF Central Medical Establishment, and made by Siebe Gorman & Co Ltd, which overcame the physiological height limit of about 43,000 ft when breathing pure oxygen. Above that, means had to be found to increase the pressure in the lungs, and it was deemed necessary that the pilot should operate in a pressure equivalent of 32,000 ft. Since the lightest possible weight was vital in achieving maximum ceiling, a strong enough pressure cabin would have meant excessive weight, so the suit was the only practicable solution, but still required a closed oxygen circuit with a chemical cannister to absorb carbon dioxide and moisture. Protected by the suit, Swain had been experimented on in the low-pressure chamber at the Royal Aircraft Establishment from which air was exhausted to the equivalent of 80,000 ft at the – 60 deg C experienced on the Everest tests of the Westland machines.

On 5th October, a few days after the dismays of the Johannesburg Race had somewhat subsided and the height record still to be homologated, Jean Batten, latest recipient of the Johnston Memorial Trophy, left Lympne in the early morning darkness with her Percival Gull determined to fly home to her native New Zealand. It was the same long-range Gull which Lord Wakefield had helped to buy for her flight to Brazil and is preserved in the Shuttleworth Collection. When years later I asked about the Technical planning, she wryly remarked: 'I never really knew Edgar Percival very well, but found him most courteous and very businesslike, though he seldom answered a question without referring to the slide-rule which he always carried in his pocket! I have an idea that he, like the male members of my own family, considered that woman's place was the home, and not adventuring alone to distant lands in small aeroplanes – but with the Gull I was able to emerge from the "woman's category" and compete successfully with male contemporaries.'

Cruising at 150 mph she briefly stopped at Marseilles to top up, and reached Brindisi that afternoon having covered 1,205 miles. Then in a series of long stages starting at dawn and with mid-day stops for fuelling, sometimes flying blind in tropic downpours, she reached Darwin on the seventh day at 10.43 local time, having achieved the journey in a day less than Broadbent's 1935 record – but it was an emotioning arrival: the engine would not fully throttle, she overshot, had to make another circuit, and the ensuing landing required urgent braking, for apparently she did not think of switching off the engine. When she reached Sydney, the Australian Civil Aviation Board at first forbid her to cross the Tasman Sea to New Zealand, but after determined argument 'with what would be called obstinacy in less successful people, but distinguishes the great women pilots of all nationalities, she excused her rashness with the argument that the Tasman meant less to her Percival Gull than the Timor had done to the fifth-hand Moth with which she had first crossed it.' So she took off by the light of a flare path on 16th October, heading for New Plymouth, landmarked by the 8,000 ft peak of Mount Egmont visible from far at sea, thence she would turn northward along the coast to Auckland. The Tasman was at its worst with storms, but she reached her navigational target in nine-and-a-half hours, and an hour later received a tumultuous welcome when she landed at her home town.

Even *The Aeroplane* forgot its denigration of such flights, and was almost enthusiastic: 'Miss Batten's flying career has not blossomed into sudden perfection, but in five years she has built up, by study of navigation, blind flying, etc, the competence and airmanship with which her recent flights have been distinguished. Her performances have gained that elusive touch of reliability which makes them seem to follow an exact schedule even when she varies it, and is the mark of quality in long distance flying.' For her, as it had been earlier for Amy Mollison, the skies and beckoning distances had been a compulsive search, and her achievement was the endowment of courage, will-power, and technical skill.

10

Indicative of expanding techniques was the inauguration of the new premises at the village of Duxford for Aero Research Ltd, the brain-child of Dr N. A. de Bruyne, a Trinity don amateur flier who specialized in stress-bearing plastics and bonding solutions but was chiefly known as constructor of a 42-ft span low-winger of conventional appearance yet largely of plastic material. The opening ceremony was performed by David Pye, who quoted an injunction of the late Professor Hopkinson that 'the engineer in charge of research projects should carry out his work in daily contact with flying so that he may never get out of touch with the element with which he deals' – but as a note of warning added the comment of a RAF pilot: 'Poor old Hopkinson went up in bad weather and got into a cloud and into a spin from which he only emerged as an angel.' In more serious vein he said that what de Bruyne was now doing on synthetic resins might in a few years cause the sleek polished monoplanes of today look as old-fashioned as the biplanes of 1918 now seemed, for one of the objectives was to eliminate riveting by bonding metal parts together with the same adhesive strength as rivets.

Research and yet more research was now the accepted requirement everywhere. At the RAE the revolutionary change to metal-skinned construction had led to much mathematical and experimental investigation, and Dr D. Williams had contributed valuable R & Ms on such subjects as sheer lag, stress diffusion, box structures, and sandwich construction for which he devised a basic theory for designers' use. The new 24 ft atmospheric wind-tunnel was enabling full-scale investigation of air-cooled engines to provide adequate cooling with least drag, and similarly honeycomb radiators for liquid cooling which would blend aerodynamically without loss of thermal efficiency. With the increase in size of bomber and civil transport aircraft, development of servo controls was being studied as part of the general problem of stability and control. Metallurgists were devising improved protection for magnesium fuel tanks against corrosion from water and leaded fuel; the engine department was investigating vibration of propellers and aircraft structures; the wireless section was examining the German Lorenz and American Hegenbeurger blind-landing systems, the effect of noise and temperature on radio communication, and design of aerials and components; a new department under Dr Harold Roxbee Cox was involved with ground camouflage, balloon barrages, ground defence schemes involving cables, parachutes, rockets, and aerial mines released from aircraft; the armaments department, with R. Purves at its head, was devising fragmentation and incendiary bombs, release units and carriers, smoke-curtain generators, gun ancillary equipment, high-speed targets, and researching on bomb shapes. Under P. Salmon, the D.O. was designing a large hydro-pneumatic catapult to launch the projected heavy bomber aircraft of 65,000 lb all-up weight. The need for a high-speed wind-tunnel resulted in preliminary plans for one capable of 600 mph using a 4,000 hp electric motor, and was estimated to cost £200,000. Currently the RAF had requested authority to build a gas turbine, for Hayne Constant had reported: 'In the light of existing knowledge, an internal combustion turbine could be constructed using only types of components proved by past experience. Performance of such a turbine on the basis of specific weight and economy would be better than or substantially equal to the best modern water-cooled petrol engine except when cruising at comparatively low altitudes.' Meanwhile not only was there Whittle's project, but Professor Betz of Göttingen had been joined by Helmut Schelp on the application of jet propulsion to fighter aircraft, and Ernst Heinkel in conjunction with von Ohain was building a centrifugal type gas turbine.

More immediate day-to-day problems were involving designers of every aircraft company, and they attended in force at an RAeS lecture by Dr Gustav Lachmann on *Aerodynamic and structural features of tapered wings* on 8th October. Not surprisingly it stressed the value of slots-cum-flaps; but he also drew attention to the advantages of boundary layer control, a subject which the RAE was vigorously pursuing, and summarized the many aspects of aerodynamics and structural requirement which all firms were encountering. As he said: 'Most designers who applied the knowledge gained from biplanes to design of tapered monoplanes found difficulty in obtaining satisfactory longitudinal stability.' They also found many other snags, and were trying to unravel cause and effect. Thus behaviour at the stall varied between wings of the same taper and section depending on whether the mean chord line swept forward or back or was neutral. Wing fillets could have a destabilizing influence on overall pitching moment, and in some cases their removal might improve both stability and performance; in others they were essential. There were difficulties in trim when flaps were lowered, and variations of downwash sometimes had inexplicable results. Similarly with structures. As Lachmann said: 'Strictly speaking the term "stressed skin" could only be applied to structures where the skin takes practically the entire bending loads as well as the sheer due to torsion and anti-drag.' Such a wing could be lighter than one in which the spar took the bending loads and the skin only the sheer. But as he said: 'A dominating consideration is that structures must be simple to build in quantity. Riveting is the most expensive operation in building metal wings, particularly if done during assembly. Much saving is possible if the skin covering can be riveted separately, preferably by machine or spot welding, and the wing assembled with minimum riveting on the job.'

In the discussion, Teddy Petter advanced the case for the fabric-covered Lysander by skilfully arguing that use of a single spar wing with torsion nose cut the riveting cost in the ratio of 12:3 by reducing the distance between stiffeners compared with a fully stressed wing. There was even a hint of his next design, for he mentioned the destabilizing effect of engine nacelles, and added that full-span flaps seemed to be superseded by the wingtip slot and partial-span flap, for Westland found an increase of 20 per cent in total lift in that way. He did not disclose that the Lysander had been having problems. Though lateral control was now light, the automatic slots and flaps moving smoothly, and a larger tailplane had improved stability, the latter's big negative adjustment introduced a more serious snag, for now it was impossible to hold the machine level on opening the engine after an abortive landing, and even at half-power it required all the pilot's strength to hold the control column fully forward until he could re-trim. On the score that production was so urgent that no further time could be spent on development, a proposed automatic re-trimming device was turned down by Fearn who, like Petter, had no conception of piloting. Mistakenly the A & AEE agreed, and inevitably fatal accidents ensued.

Lachmann's twin-engined HP 54 to specification 29/35 was flown by Cordes on 10th October, shortly after Lachmann's lecture. Named the Harrow, a batch of 100 was being manufactured by the sub-assembly flow system recommended by Volkert on his return from the USA. Though Handley Page's gamble in switching from the C 26/31 competition had paid off, the new machine, with its fabric-covered fuselage, though a big advance in replacing the outmoded Virginia, was clearly only an interim bomber until the new metal-skinned high-performance equivalents were ready. Because of its derivation from the HP 51, only limited contractor's trials were necessary, and within six weeks the prototype K 6933 was flown to Martlesham for Service trials, equipped with prototype Fraser-Nash turrets.

The equivalent Bristol 130, similarly powered with two 900 hp Pegasus Xs, had been named Bombay, and though flown a year earlier, was suffering production delay becuase of the vital priority of Blenheims and the impossibility of quickly creating more floor space. However the Government-owned shadow factory on Queen's Island, Belfast, sponsored by Shorts and managed by the recently amalgamated Short & Harland Ltd was available, and a contract was placed for fifty Bombays to specification 47/36 – but it would take time to get so new a factory going despite the expertise of A. E. Bibby, a man who by his outstanding ability had risen from bench worker to works manager at Rochester where his latest production triumph was the line of Empire flying-boats.

Canopus had returned from the MAEE trials with its Certificate of Airworthiness on 1st October and on the 20th was handed over to Imperial Airways. Both Felixstowe and Martlesham seemed thronged with visiting pilots and trade representatives these days, so intense was the nursing of new types undergoing test. At nearby Ipswich, where RAF and civil pilots frequented the Pickwickian *White Horse*, there seemed to be unusual crowds of Pressmen on 29th October and we learned that at the local Assizes that day, in the case of Simpson v Simpson, that lady, already once divorced, had secured a *decree nisi* and therefore was free to marry again after six months. American newspapers headlined the news 'King will wed Wally'. British newspapers made only discreet reference to the divorce without mentioning the King – though a week earlier Baldwin had tendered his official advice to the monarch on how such a marriage would affect his constitutional position, the Commonwealth, and the Succession. The British people officially knew nothing.

Soon after Mrs Simpson's divorce, the King, pursuing his sympathetic course with the unemployed, made a long-promised tour of the depressed areas in South Wales where 60 per cent were unemployed. Next day the *News Chronicle* reported: 'The King's arrival at the great steel works which has been lying derelict for over six years was the most moving event seen by South Wales in living memory. When the people broke spontaneously into the National Anthem I thought the King was going to break down. Old men and young men, mostly ill-fed and thinly clad, cheered him again and again. One felt they were looking to him for salvation after years of political blundering. When the King looked at the tragic scene he suddenly stopped and said: "These Works have brought all these people here, and something ought to be done to find them employment." ' That comment was headlined in every newspaper. It triggered political controversy within the Government. What right had the King to implicate them in such matters? On 16th Nevember the position became worse, for the King informed Baldwin that he intended to marry Mrs Simpson and that though he wished to remain King he would abdicate if need be. Nothing would shake him. Baldwin, in his efforts to prevent both abdication and a marriage which would make Mrs Simpson Queen, even proposed a morganatic arrangement, but the Cabinet refused to agree to such legislation. The days mounted to an impasse.

Behind it all was the strategy of possible war. Singapore was recognized as a weak link, and on 14th October the Singapore III flying-boats of No. 230 Squadron skimmed thundering across the calm waters of Pembroke Dock to reinforce the Far East Command at the long-delayed new base of Seletar. No. 205 flying-boat Squadron was already there, and the 700-acre landing ground alongside accommodated two torpedo-bomber squadrons. An additional aerodrome was completed on 29th October at Tengah for further landplane squadrons which, with torpedo-bombers, were to be primarily deployed in the defence of Singapore.

Arrival of the Singapore reinforcements coincided with the issue of the *Report of the Royal Commission on the private manufacture of and trading in arms*. Bitterly C. G. Grey said: 'The Government appointed what amounted to almost a pacifist committee to investigate this important subject.' The recommendations amounted to a series of axioms, such as the best way of removing objections was the limitation of arms by international agreement; establishment of State monopoly would be impracticable and undesirable; civil servants and Service officers should not accept appointments with armament firms; profits should be restricted; conscription of industry in wartime must be without delay; private export of surplus and second-hand arms should cease and licences to export new arms be granted only to specially authorized firms; the Government should assume responsibility for organizing the arms industry and regulating the necessary collaboration.

The Government endeavoured to forestall criticism of the creation of their shadow industry by announcing: 'Arrangements have been completed for erection of six Government-owned factories for production of aero engines. The firms which have undertaken to manage them are: Austin Motor Co Ltd, Bristol Aeroplane Co Ltd, Daimler Co Ltd, Rootes Securities Ltd, Rover Co Ltd, Standard Motor Co Ltd.'

Securing the co-operation of these firms had been a gruelling task for Major Bulman. There had been many meetings before agreement was reached by all except the unpredictable, suspicious Lord Nuffield who considered he had been treated rudely by his fellow peer Lord Swinton. In an unpublished manuscript Bulman recalled: 'I found myself deeply moved by this strange, gnarled-handed man who had started as a mechanic and bicycle-maker and by sheer determination and shrewdness became throughout the world a synonym of success, and whose charities were in millions of pounds. Yet the battered old cigarette case he passed to me wasn't even silver, and contained a few cheap cigarettes and a half-smoked stub. To me he seemed a very lonely, very pathetic little man.' But there was nothing pathetic about the Press meeting he called to explain his exclusion, and he emphasized that the Wolseley Aero Motor Factory was initiated in 1929 because of his conviction that in time of emergency such firms would be needed, but there had been a complete lack of support from the Air Ministry, though he had spent £500,000 of his personal money on his aero factory at Birmingham which was separate from his other companies in which there were public shares. In July 1935 he had written to the Minister requesting an interview, but Sir Philip Cunliffe-Lister, before his elevation to the peerage as Lord Swinton, replied that it would be impossible to see him on the day mentioned and proposed no alternative. Nuffield as a practical engineer had been against the shadow scheme in which firms made components instead of complete engines, for he considered that in an undertaking of such technical precision there must be unity of control. He therefore offered to build 2,000 Bristol engines, but the offer was again turned down. To a man who had achieved world-wide success in mass production of cars this was gravely insulting. One fair-minded journalist commented: 'Disagreements in high places should not take on the atmosphere of Billingsgate, but the antipathetic temperaments of a highly successful and wealthy industrialist and a Minister of His Majesty's Cabinet seems to have generated a critical temperature between them. Lord Nuffield has been accustomed to dictate, and without that faculty probably would never have achieved his phenomenal success. His patriotism is not in doubt, but obviously he wishes to contribute to the national need either on his own terms or set limits on the degree of subordination to which he would submit. On the Government side he seems to have been treated with less than the tact due to a great industrial power. The Air Ministry were in a delicate position, but seem to have conveyed to Lord Nuffield with unflattering candour that they did not take him seriously as an aero motor manufacturer and wanted his co-operation only in his role of motor-car magnate. Whether sound or otherwise, the shadow scheme will hardly benefit from this kind of advertising.'

But Nuffield was not the only one aggrieved. T. G. John, chairman and managing director of the Alvis Car & Engineering Co Ltd which had established an aero engine department early in 1935 and held a manufacturing licence for the Gnome Rhône version of the Bristol radials, issued a statement: 'Our first knowledge of the shadow factory scheme was gleaned from newspapers. I immediately told the DAP that I could not understand why new factories were to be built at public expense while our own large and modern factory, built and equipped specially for making large-sized aero engines, was being officially ignored. At his suggestion I discussed this with Lord Weir and was amazed to discover that he did not know we had a factory ready for production, but he assured me that our resources were of considerable national value, and would be considered. I confirmed our conversation in a letter a few days later, but it was never acknowledged, and we have heard nothing since.' The doors undoubtedly were closed. Undeterred, John decided that his chief engineer, Smith-Clarke, should design an engine of their own. Meanwhile the Government issued a White Paper on its policy of aero engine production but seemed more concerned with excusing the exclusion of Lord Nuffield. It concluded with a letter from his arch-enemy, Lord Austin, chairman of the Aero Engine Committee, to Lord Swinton stating: 'We unanimously believe in the shadow scheme and have confidence it could be a success. Factories are rapidly taking shape, and I find no hold-up or stoppage in the work connected with the purchase of machinery, jigs, tools and fixtures. By the end of the year it should be possible to form a definite opinion as to when manufacture might commence.'

Parliamentary discussion on the White Paper followed after the Summer Recess and Viscount Dunedin emphasized that if raiding aircraft managed to cripple any one section of the aero motor scheme then every other section would be paralysed. Sweeping this aside, Lord Swinton said that the Air Ministry would have preferred each firm to make complete aero motors but the opinion of the firms was that the only practical operation was to sectionalize production: whatever the method, the Government deemed it essential that Britain should have air power second to none, and the new target for aircraft was now 8,000 in three years. Shrewdly an American visitor commented that mass production in England meant using more men to turn out more aeroplanes, but in the USA it meant turning out more aeroplanes with less men!

Lord Swinton did not mention the comprehensive engine works being erected at Crewe which Rolls-Royce would run on an agency basis. Arthur Wormald had recently resigned as general works manager, but remained a director, and dynamic, stubby Eric Hives was appointed in his place. Nobody was more fitted for the task, but for the moment he was in difficulty over shortcomings of the Merlin E which had failed its 100-hour type test due to distortion of exhaust valves, detonation producing erosion, and cracks in the asymmetrical cylinder head – so for the final design it was decided to use a scaled-up Kestrel type flat head; meanwhile a slightly improved version was put in production as the Merlin I, but though type tests were relaxed to permit replacement of the valves, not until November was it cleared for Fairey Battles. Tests with its successor, Merlin G, soon to be the Merlin II, using the integral Kestrel head, had succeeded a month earlier, and it was for this engine that Hurricane production was waiting while the new Crewe factory was being tooled up.

To help unravel complexities of supply, production, and industrial relationships Lord Weir, supreme organizer of production in the 1914–18 War, had accepted the Government's invitation to participate in an honorary advisory capacity. In the subsequent Debate on the Address from the Throne in relation to defence problems Winston Churchill, with rasping tongue, described the revelations of Lord Nuffield as a lightning flash revealing a dark and confused landscape of air and army production, and the position of Lord Weir as one in which he had neither formal authority not official responsibility. Criticizing progress, he reminded MPs that 124 Home Defence Squadrons had been promised by 31st March 1937, but at the end of the last financial year there were only fifty-three squadrons including Auxiliaries, and in the subsequent thirty-two weeks only twenty-seven squadrons had been added, so to maintain the progress we must have forty-three more squadrons within twenty weeks. If each of the current eighty squadrons averaged twelve aircraft we would have 960 first-line Home Defence machines, whereas Germany certainly had not less than 1,500 first-line aeroplanes and probably more. Stepping into the breech, the Prime Minister failed to point out that Britain had many squadrons abroad, but merely said that Mr Churchill's estimate of German metropolitan and first-line strength was definitely too high, though the Government could not give exact figures.

As a symbol of royal interest, and no less as the country's best PRO, the King visited the Home Fleet at Portsmouth on the 12th and 13th November, during which he inspected the FAA units on board *HMS Courageous*, comprising fighters of No. 800 Squadron, torpedo-bombers of 810 Squadron, and spotter-reconnaissance Squadrons 820 and 821. 'During the evening of 12th November, His Majesty attended a smoking concert in the lower hangar of the aircraft-carrier – and undoubtedly felt at home among other young men of his calibre.'

Two days earlier, while the House was earnestly debating Defence, forty-two-year-old Sqn Ldr de Haga Haig, that clever, tough lowland Scot who never quite grew up and had made the mistake in 1926 of abandoning his brilliant career in the RAF and joining William Beardmore & Co Ltd as test pilot, was drowned in heavy weather while taking his motor trawler from Plymouth to Southampton. He had made history on 4th December, 1925, as the first pilot to be launched in an aeroplane from an airship, the R 33, and hook on again, using a DH 53 light plane. He was a brilliant aerobatic pilot but also inventive, and with H. J. Stieger and Alan Chorlton, formerly Beardmore manager, had owned the Monospar patents but was persuaded to relinquish them to GAC when it took over; more recently he had been developing a system known as radiaura giving aircraft warning of others in the vicinity. To those who flew in the '20s this kind-hearted yet brusque man was remembered with affection.

A luckier man was James Mollison who without fuss, without newspaper publicity, arrived back at Croydon at 9 o'clock on the morning of 30th October, having flown from New York to Newfoundland and across the Atlantic with the Bellanca monoplane. 'Although we cannot award him a Blue Riband for doing so, he deserves credit for being one of the very few transatlantic fliers who has landed on an aerodrome,' sarcastically wrote C. G. Grey for whom Atlantic flying for publicity was anathema. A law suit followed disputing ownership of the machine on the part of the Irish Hospital Trust Ltd who originally had entered it for the England-Australia race piloted by Colonel Fitzmaurice; but it transpired that Jim bought it in America without knowledge of any defect in title, and he was declared innocent by Mr Justice Bennett. Amy's suite for divorce followed, and Mollison once again became a freelance, womanizing, but adventurous man.

11

Bombing was the order of the day in Spain both by Royalist and Communist forces using a heterogeneous collection of aircraft obtained wherever there was an aeroplane for sale. Mercenaries lured more by money than ideals were flocking in. The first British fatality was Sidney H. Holland, while flying a three-engined Russian bomber for the Communists. By now German pilots were operating as a Mission, commanded by Lieut-Gen Wilberg, and not only was the Luftwaffe participating in bombing but in November 4,500 German volunteers, known as the Condor Legion, arrived with twenty Ju 52s and twenty-two Heinkels of various versions which were steadily reinforced.

Jokingly the French explained that there were no German or Italian aircraft at the Paris Aeronautical Salon because all the new machines of those countries had been sent to Nationalists in Spain to try out against the latest types in the hands of the Reds. Designers and military men from all over the world were at Paris when the Grand Palais opened on 10th November, for this was an epochal year marked by universal change in design philosophy influenced by clean American designs. Among the thirty-two aeroplane stands the Bristol Blenheim was the sole British representative, for Hawkers at the last moment were debarred from showing the Hurricane. Most eye-catching of exhibitors was the twin-boom twin-engined Fokker G 1 bomber-attacker with two fixed cannon, two machine-guns in the nose, and a fifth rearward-pointing rotatable gun hydraulically controlled in the tail of the central nacelle – but it was not these details but its flaming orange and sulphur colour which made it so obvious despite the bright scarlet of Fritz Koolhoven's neat shoulder-wing fighter, reminiscent of the original Westland shaft-driven strut-braced monoplane submitted for the ill-fated F 7/30 specification – for the Koolhoven had the pilot forward above the engine, but was original in using shaft-drive for contra-rotating propellers. Biggest of all was the tandem-podded four-engined, strut-braced high-wing airliner which the pioneer Farman firm had built for Air France but earned the quick comment from a Frenchman: 'I perceive that M. Farman is still at war with aerodynamics.' Russia was almost as distinctive with the enormous span of the red-winged single-engined ANT-25 which held the world's closed-circuit distance record of 7,000 miles – but it had presented problems in trying to get into the building because the wing chord was greater than the huge entrance doors, so a deep trench had to be dug to move them in edgeways, and then it was found they had buried the wing-fixing bolts when the trench was filled in! Its famous designer A. N. Tupolev also convincingly proved that what seemed a cross between a Spitfire and Heinkel was contemporary with these designs and not merely a copy as many suggested. However his twin-engined ten-seater ANT-35 was certainly a replica of the DC-2. In fact, as C. G. Grey was quick to point out: 'The Grand Palais is full of children, legitimate and otherwise, of the Douglas airliner.' But Northrop was also an evident influence, for the Polish PZL 23A bomber-fighter with trousered and spatted undercarriage could have been from that stable; but the

PZL gull-winged monoplane alongside seemed dated, and France's Loire-Nieuport 46C 1 gull-winged fighter was of even more obsolete appearance compared with the sleek low-wingers everywhere in view.

In any case, France was immersed in political troubles, and though Germany was the recognized enemy there was an ominous feeling that Russia was the ultimate menace. 'France does not want Communism as commonly understood,' wrote a journalist. 'State Socialism of sorts they will have, just as we are letting it grow upon us in England, but the Frenchman is too much an individual to sink himself and his little properties in State ownership of everything. In France there is a swing against Communism, Bolshevism, Anarchism and everything for which the Union of Soviet Republics, under the dictatorship of Stalin, stands. And that will mean before long an end to the dangerous entanglement with Russia. So M. Pierre Cot, the Air Minister, whose political leaning seemed extreme Left of Socialism if not Communism, has not made himself any more popular by practically taking State control of the aircraft industry.'

But what did British designers learn at the Salon? That some of the French flush-riveted finish was superb and that of the Russians was awful? That there were differences of opinion as to whether a matt surface made resistance worse, for the shiny prototypes certainly would have to be camouflage-painted for war? There were ingenious details on many a machine, though nothing which seemed beyond the bounds of normal development. What seemed more evident was that the constellation of Bristol engines, admirably presented in starkly engineering manner, and the aristocratic Rolls-Royce stand with its glittering 1,065 hp Merlin II and 700 hp Kestrel XVI set in an alcove against a painting of the Hurricane, Battle, and Spitfire, were supreme.

So, on any basis, were the Short Empire flying-boats. While attention was still focused on the splendid French Salon, the prototype *Canopus* started its scheduled flights to Brindisi and Alexandria commanded by Capt F. J. Bailey after a month of demonstrations, practice, and proving flights by Imperial Airways pilots. *Caledonia* was being used by Brackley for instruction and clearance of further pilots prior to delivery of the machine to Hythe at the end of the month. *Cambria* would soon be ready, and the others were following in visible sequence.

Young Thurston James wrote at that time: 'No pictorial representation gives an idea of the size of the Empire boat. To say that the inside is 17 ft from bottom of the hull to top of the wing does not bring home its capacity. The vastness of the craft becomes more obvious when you wander around a smoking room as large as a modern flat and rather higher than some, and then hear a member of the crew walking above you on the upper deck. Then, consider that there are three more cabins, two of them even bigger and both longer, and they are being pushed through the air up to 200 mph – and you begin to see what Shorts have achieved. Yet this boat, and another twenty-seven, was built without there ever having been a prototype, though a great deal of experience was gained with the Short Scion Senior.'

On 17th November, Lord Strabolgi, formerly the Hon J. N. Kenworthy, demanded from the Lords that the Government furnish full information on the measure of rearmament achieved; but Viscount Swinton, with his customary manoeuvrability, said that the Government's foreign policy, whether agreed with or not, had been plainly stated, and he was surprised at Lord Strabolgi's indictment of dilatoriness in rearmament for there had not been a Service vote in the House against which the noble Lord's Socialist Party had not consistently and continually voted. Nevertheless discussion on defence continued for three days, in what became a comprehensive review of the whole question of air security.

'How little the present Government understands the people was demonstrated by Mr Baldwin's lamentable confession last week that he had not dared tell the people the truth or advocate rearmament because of the pacifist feelings which he and his colleagues imagined to dominate this country,' ran *The Aeroplane's* editorial. 'Popular opinion believes Mr Baldwin should have retired months ago but is determined to be Prime Minister at the King's Coronation. The Coronation will make practically no difference to the King, who has the devotion and admiration of his people to a remarkable degree; but so far as Mr Baldwin is concerned the sooner the Coronation is over the better, even if it means that Mr Neville Chamberlain is to be Prime Minister.'

But Baldwin was struggling with a situation that had got out of hand. On 16th November the King had firmly said he intended to marry Mrs Simpson. Matters came to a head on 2nd December when Dr Blunt, Bishop of Bradford, remarked at a diocesan conference that he hoped 'the King was aware of the need for God's grace. Some of us wish that His Majesty gave more positive signs of His awareness.' Newspapers seized on this to break their restraint and splash the news of the King's entanglement in great headlines. Controversy raged. The Bishops were bitterly criticized, and so was Baldwin; but many felt it impossible to crown a queen in a Church that disapproved of the marriage of divorced persons however innocent, for the King was head of the Church. In this crisis, Mrs Simpson diplomatically vanished to France. On 5th December the King told Baldwin that he would abdicate, but not until five days later did Baldwin inform the Commons in a skilful speech inferring that this was the result of his own careful advocacy. Next day *The Times* commented: 'Baldwin may have his defects as Party Leader or as an adminstrator, but in handling great national problems he has no comparable rival.'

On 11th December a Declaration of Abdication Act was hurriedly passed through Parliament. That evening HRH Prince Edward made his public farewell through the BBC radio network: 'It is impossible to discharge my duties as King without the help and support of the woman I love.' To everybody it was a tragedy, and the welcome to the slightly stuttering Duke of York as King George VI was subdued.

In his editorial greeting to the new King, who since 23rd June had been Marshal of the Royal Air Force C. G. Grey made valediction to the newly created Duke of Windsor 'whom as King endeared himself still more to aeronautical people by being the first English king to fly. Thereafter he treated air transport, in spite of its known dangers, as a normal means of locomotion. So on the meanest commercial grounds, as well as from the highest personal feelings, his loss is a blow to aviation. In this unhappy domestic affair Mr Baldwin's energy has belied the masterly inactivity which consistently marked his foreign policy and concealed from the people of the Empire our collective insecurity in the air, on the sea, and land. But to the Duke of Windsor we wish good fortune in whatever state to which he may be called. We have lost a king who would have done great work for the needy and oppressed. Our new King, George VI, has had more official connection with aviation than the Duke of Windsor, and those who served with him have the highest opinion of his understanding of air affairs. To him is due the same loyalty we gave to his brother – and in time, though in different ways, he will win the same affection from all ranks of his people.'

From the new King, the Secretary of State for Air received a message for all RAF Commands: 'I hasten to assure the Air Forces at Home and Overseas that I look forward to maintenance of my close connection with them, which has been such a happy feature of my life ever since I became a junior officer in the Service in 1918 and served with the Independent Air Force in France. I know full well that the Air Forces of the Empire will maintain to the full the great traditions that they have already established, combining with the highest efficiency and zeal a fine chivalry of service and a deep steadfast loyalty.'

12

On the eve of the abdication one of the world's greatest aeronautical inventors died. Señor Don Juan de la Cierva y Codorniu was killed flying as passenger in a KLM Douglas DC-2 which crashed into houses almost immediately after taking off from Croydon on a morning of fog. The usual technique had been used by Capt Hautzmeyer in following a white line laid from east to west on the more level portion of the sloping aerodrome. Having taken off, the pilot would continue by instruments, so when the airliner disappeared in the fog there was no hint of disaster until a phone call announced that it had crashed little more than 1,000 yards from the start almost at right angles to the take-off path. Fire broke out and the seventeen passengers, pilot and wireless operator, were unable to escape. Those familiar with fog take-offs at Croydon considered that Hautzmeyer lost sight of the white line early in the run, and instead of throttling for another attempt, continued, and the machine swung into the high ground to the left of the correct track.

'It is pure irony that Cierva should be killed in an accident of just the kind he spent his life trying to make unnecessary,' wrote the *Daily Express* air correspondent. As early as 1911, when a schoolboy of sixteen, Cierva and two friends built two gliders, the first to fly in Spain. Five years later he designed and built a three-engined aeroplane at Madrid Technical College where he graduated in Civil Engineering in 1918. While being tested, it stalled during a turn, spun, and crashed, so shaking his faith in fixed wings that he turned to alternatives such as the helicopter and ornithopter, having decided that he could only achieve safety if wings had a relative velocity to that of the framework independent of forward speed. As a mathematician he evolved a theory of mechanism resulting in his first co-axial rotor experiments, leading to understanding the dissymmetry of relative velocities on two sides of the rotor disc which caused lateral unbalance. By giving the blades a further degree of freedom, so that they could rise and fall about a hinge in response to variation of relative velocity, he found the essential solution. From that his series of Autogiros progressed to practicality.

Dr James Bennett, who worked with him on the direct take-off system of the W 3 and C 30, pointed out that: 'An immediate outcome of Cierva's work was the advent of the helicopter, although it was not Cierva's intention to use a mechanical power transmission to the rotor in steady flight. The possibility of attaining hoverability without the disadvantages associated with rotor torque was a problem that he thought capable of elegant solution.' Cierva in fact had only gone part way to the helicopter, and for the full solution it would have been necessary to follow either the Hafner or Kay systems where the blades altered pitch cyclically or asymmetrically and collective pitch could be superimposed to give increased incidence for take-off or set for auto-rotation.

In some newspapers announcing Cierva's death there was a small paragraph indicating that yet another great man had gone – Jean Mermoz, of whom Antoine de Saint-Exupéry wrote: 'Having taken off from Dakar bound for Natal he radioed briefly that he was cutting off his right-hand engine. Then silence ... We waited. We hoped. We were haunted for hours by this vision of a plane in distress ... Slowly the truth was borne upon us that our comrades would never return, that they were sleeping in the South Atlantic whose skies they had so often ploughed.'

These were still the formative years of air transport, still beset by many accidents, still expanding, and so in a flux of change with personnel and diversity of aircraft. Thus British Airways, who on 17th November had lost two pilots in a Fokker crash at night, had grown steadily from the simple establishment of Hillman's Airways to a business which was becoming runner-up to Imperial Airways, though still a long way behind. British Airways management recognized that metal-skinned American aircraft expressed the modern outlook, but the nearest British machine would have been the original Bristol Type 143 eight-seater, which was unobtainable because that company was far too involved with Blenheim production and experimental machines to tackle such a project in competition with the tooled-up production line of Lockheed. Negotiations were therefore in hand for several 225 mph ten-seat Lockheed Electras – which could be regarded as Douglas-influenced but of 49 ft 6 in span instead of the 95 ft of the latest twenty-one-seater DC-3. British Airways stated that they would 'acquire these machines with the approval of the Air Ministry and would continue to receive their air mail subsidy although previously subsidized airlines had been tied down to using British aeroplanes.'

That British Airways had great prospects was emphasized by the decision of Armstrong Whitworth's renowned test pilot, Alan Campbell-Orde, to resign from the Hawker-controlled Sir W. G. Armstrong Whitworth Aircraft Ltd to become operational manager of the airline. This tall, quiet, somewhat aloof, ex-Sherbornian had joined the RNAS in 1916, and his war-time service had been with Naval squadrons in France and Belgium, and in 1918 he was one of the original members of No. 1 Communication Squadron operating between London and Paris during the Peace Conference. On demobilization he joined Holt Thomas's Aircraft Transport & Travel Ltd, and when the company failed went to China as an instructor, helping to establish civil airlines in conjunction with Vickers; then in 1924 he returned and joined Armstrong Whitworth. Steadily Campbell-Orde had built a reputation as one of the most competent and trustworthy test pilots in the industry. As a colleague of his said: 'Anybody who knows Campbell-Orde knows that he is for efficiency above all things, in men and machines alike.'

He was succeeded by his skilful assistant, Charles 'Toc H' Turner-Hughes, who first came to notice as an aerobatic exponent of the Cobham displays and attracted Armstrong Whitworth's attention when he demonstrated the AW XVI fighter which Cobham took to South Africa. His latest task had been the initial flight on 6th September at Baginton of John Lloyd's bulky AW 29 rival of the Battle – but it was too late to consider seriously, for even the advantage of its rear gun turret was discounted because the Fairey was 25 mph faster.

Meanwhile there were appointments of significant interest in the Hawker-Siddeley Group. Thomas Sopwith was re-elected chairman, and Frank Spriggs became managing director of Hawker Siddeley Aircraft Co Ltd as well as the subsidiaries of Armstrong Siddeley Development Co Ltd, Armstrong Siddeley Motors Ltd, Sir W. G. Armstrong Whitworth Aircraft Ltd, A. V. Roe & Co Ltd, and Air Service Training Ltd. Their financial secretary, H. K. Jones became general manager of Hawker Siddeley and a director of all the subsidiaries. Few of the staff ever seemed to come in contact with Sopwith and Jones, who were primarily engrossed with business policy. In November the capital of the Group, then £2 million, was doubled, thus becoming the biggest aeronautical business in the world. As encouragement to worker participation £5,000 was allocated for key employees to acquire shares.

Similarly Oswald Short had firmly established Short Bros (Rochester and Bedford) Ltd, and at their first AGM a surprisingly high dividend of 30 per cent was declared, absorbing £45,000, but leaving a balance of £79,296. A. E. Bibby procured well-merited recognition by appointment to the Board, and it was agreed that Sqn Ldr H. W. McKenna, until recently chief engineer at Martlesham and the terror of all visiting industrial representatives and mechanics though off duty he was an amusing companion with a dry wit. Arthur Gouge was re-elected, and Oswald Short described this devoted and brilliant engineer as 'the man who has borne the greatest responsibility for the design of the

Empire flying-boat and to him goes the greatest credit.' But perhaps that was placatory, for often there was strife between them. On 10th December, Gouge read a paper to the Royal Society of Arts on *Recent Progress in the Design of Civil Flying-boats*, during which he confidently proved that there was nothing, except lack of an order, to prevent designing flying-boats double the size of the Empire boats and even bigger.

Optimism was now the keynote not only of the ring of SBAC major manufacturers but of non-members. Already a number of medium-sized and small civil prototypes had taken the air in expectation of a market next year. Nevertheless Fred Miles had a set-back with his twin Gipsy Six Peregrine low-winger which had a cabin big enough for two crew and six passengers. This was his first venture with bigger aircraft as well as a retractable undercarriage, variable pitch propeller, and a form of split flap extending across the fuselage from aileron to aileron – but the structure was above the estimated weight, so failed to go into production. However, Percival saw opportunity for a twin-engined machine of similar size and power, carrying only four passengers and two crew, and his small D.O. was designing such a machine, more elegantly.

Of newcomers, a single-seat, ultra-light, pusher high-winger on the lines of the 1933 Shackleton-Murray SM 1, was being offered as the Dart Pup by Zander and Weyl Ltd at Dunstable. More practically three young men, C. R. Chronander, J. I. Waddington and J. A. Heron made a breakaway from ply construction with a metal-skinned two-seat low-winger, the CWA Cygnet Minor of simple construction facilitated by the use of pop rivets which were inserted externally and expanded in position by a mandril drawn through the rivet until its head broke at a predetermined load, and was so simple that no particular skill was necessary compared with normal solid riveting requiring a 'dolly' held inside the structure while the head was being formed. As a result the three partners hoped to sell Cygnets at £800.

Even ultra-light, twin-engined two-seaters had been designed. The 29 ft 10 in span shoulder-wing Bee under construction by Carden-Baynes Aircraft Ltd at Heston and powered with two 40 hp pusher Carden-Fords was not yet complete, but an equivalent low-winger of 43 ft 4 in span and lighter loading, with two Continental opposed twins of similar power, had been built by Heston Aircraft as a ducal autoplane with folding wings to the joint specification of their directors, Edmund Hordern and the Duke of Richmond and Gordon, and was flying successfully though so over-heavy that with one engine stopped it could do little more than extend the glide.

In an attempt to set the price as low as the original Miles Hawk there was the strut-braced high-wing Wicko, with cantilever undercarriage, which Messrs Wickner and Foster had designed in wood for simplicity, powered with a 90 hp Ford V8 to which they had fitted Pobjoy reduction gear to reduce rpm from 3,300 to 1,500 – but although the bare engine cost only £60 from a Ford agent, it was heavy for its power at 450 lb, and with only single ignition would not be acceptable for a Certificate of Airworthiness until modified.

The last day of the year ended with historic changes at Vickers. Percy

Maxwell Muller, the general manager, who had joined Vickers at Erith in 1913 and thus was the link throughout the whole history of Vickers aviation interests, now at fifty-six was getting at loggerheads with the high-handed, forceful but brilliant Sir Robert McLean, and agreed to retire. His staunch companion, Archibald Knight, the works manager, originally a pre-war instructor at the Vickers Flying School at Brooklands, backed him to the hilt and also left. Thirty-six-year-old Trevor C. L. Westbrook, the brilliant organizing works manager at Supermarine, was brought back to Weybridge as works manager. Geodetic construction began to hum. By now the 1,200 workmen and staff of 1930 had increased to 2,690.

Westbrook had been with Vickers Ltd since 1916, and his dynamic energy was such that in 1919, aged nineteen, he was made production assistant and became assistant works manager in 1926. In 1929 he was put in charge of the Supermarine Aviation Works at Woolston and was responsible for the final Schneider racers, the big production of Supermarine flying-boats, and initiation of Spitfire construction. He believed that those engaged in aircraft design and construction should know how to fly, so he owned his own aeroplane. He keenly participated in local affairs, and was a member of the Management Committee of the Royal South Hants Hospital and Southampton Advisory Committee for Juvenile Employment. There was no doubt that Westbrook was making his mark and that Vickers appreciated his great organizing ability.

The end of the year also marked the death, at the early age of sixty-three, of Arthur Wormald, the recently retired general works manager of the Rolls-Royce plant at Derby where he was the longest serving member of the company, having joined Royce in making electrical equipment in Manchester at the close of the nineteenth century. He had quickly risen to management of the Works, helped in the early motor-car experiments, and in 1907 when Rolls-Royce Ltd was formed in Derby he became the general manager.

Though the passing of Wormald made no impact except for paragraphs in the aeronautical and automobile Press, the death of Lady Houston, DBE, two days earlier had been world-wide news, for the publicity accruing to her from the 1931 Schneider Trophy Contest had been enormous. To most people it was astonishing that this much photographed eccentric widow, first of Lord Byron and then of Sir Robert Houston, was eighty, for she dressed in the smartest fashion and beauty parlours had done their best. 'I was immensely impressed with the intensity of her patriotism and the fervour of her feeling for Great Britain and the Empire,' wrote Lord Clydesdale, who had succeeded to the Marquessate of Hamilton. 'I felt there was no sacrifice she would not be prepared to make in order to strengthen British prestige throughout the world and set the British Empire more firmly on its foundation.'

C. G. Grey added: 'She was not well educated or even very intelligent, but had the right ideas as represented by Victorian patriotism. People laughed at Lucy Houston as a crazy old woman, but we have not noticed that anybody except our highly conservative Chancellor of the Exchequer objected to taking her good money.'

CHAPTER III

1937: *The Veiled Threat*

'The English have never been good at preparing for a merely possible war; they are apt, indeed, to regard such preparation as ill-omened and impious.'

Sir Walter Raleigh (1924)

I

Among the New Year's Honours a knighthood was conferred on Henry Tizard for 'services as Rector of Imperial College', but basically because of his extensive voluntary work on aeronautical research committees. Acknowledgment was made with a CBE of Arthur Hall's quiet management work as chief superintendent of the RAE. Nominally he was responsible to the quietly authoritative, monocled Major Harry Wimperis, the DSR who currently retired. His able Deputy Director, fifty-year-old David Pye was promoted to Director, and bespectacled forty-five-year-old William S. Farren, of swift chatter, became Deputy. Both Pye and Farren had served as pilots on experimental work in the RFC and RAF, and were Fellows of Trinity, Cambridge, as well as members of the ARC.

January 1st also marked the promotion of Sir Edward Ellington to Marshal of the Royal Air Force; Air Marshal Sir Hugh Dowding to Air Chief Marshal; Air Vice-Marshal Wilfrid Freeman to Air Marshal – and in due course they were formally presented to HM the King by the Secretary of State for Air at the customary levée at St James's Palace. Even more significant for the aircraft industry was the appointment of L. V. Meadowcraft as Director of Contracts following the retirement of Basil Holloway who had so successfully and rigorously controlled the financial benefits of manufacturers.

The industry was facing up to a tremendous task. Recruitment was intense. In the past two years the number of employees had increased tenfold. More would be required as soon as the acute shortage of materials was relieved. Meanwhile factory space was a pressing problem, and there were not enough houses in the vicinity of any of the major aircraft manufacturers for the new input of employees.

There were rumours that General Aircraft Ltd, Aero Engines Ltd, and the recently formed British Marine Aircraft Ltd which held the Sikorsky flying-boat licence, would amalgamate with Westland Aircraft Ltd – for rather than build more new shops at Yeovil for Lysander production, Sir Ernest Petter hoped to use the big premises British Marine were constructing on the shore of Southampton Water near Hamble. Authorized capital of British Marine was £500,000, and the Board comprised Air Marshal Sir John Higgins, Lord Willoughby de Broke, William Craven-Ellis MP, George Handasyde, and two others. This suited neither Peter Acland nor the rising Teddy Petter, who coldly rebuked his father. Two months later at a shareholders' meeting convened to increase Westland capital to £750,000 by the creation of one million 5*s*. shares, Acland proposed and young Petter seconded a resolution to adjourn on the grounds that British Marine had an agreement with British Aircraft Manufacturing Ltd by which the latter received a management fee of £5,000 per annum for four years and was unacceptable to the majority of the Westland directors. Sir Ernest admitted that this unfortunate circumstance had arisen, causing a difference of opinion, but explained that British Aircraft Manufacturing had consented to waive the agreement for £5,000, so it was not in the interest of shareholders to hold matters up and he was therefore against the adjournment. On putting the amendment it was lost by nine votes to ten, so the original resolution to increase capital was carried *nem. con.* But Acland had won. Three days later a circular was sent to Westland shareholders stating that the Board had decided not to take over British Marine Aircraft Ltd.

The latter's directors considered their only course was to liquidate, but the shareholders' committee secured a stay in proceedings and succeeded in replacing the directors with swarthy thirty-five-year-old Alan P. Goode, an astute financier and chairman of Heenan and Froude Ltd, Lagonda Motors Ltd and other companies, together with Edgar Granville director of British Airways Ltd and other companies, Capt J. Dawson Paul a director of Boulton & Paul Ltd, Alfred Charles Kimpton delegate director of several ICI subsidiaries, and Henry Folland, formerly technical director of the Gloster Aircraft Co Ltd. Goode was elected chairman, and said that the advantages of an association with the Folland technical group had been recognized by the original Board and they now generously agreed to co-operate with the shareholders' committee.

Folland had found it impossible to accept subordination to the Hawker Siddeley masters of the Gloster Aircraft Company. As Hugh Burroughes, a Gloster director, explained: 'He naturally thought the new owners would give preference to Camm's designs. Harts and Furies had put Camm on the top of the world, and the Hurricane was at hand – yet Folland had come to Glosters with a well established reputation as a first-class designer. The SE 5s of his days at the RAE and the Nighthawks of his Nieuport period provided ample evidence of his ability and that of his chief lieutenant H. E. Preston. He gave GAC outstanding service and two long spells of success with the Grebe/Gamecock and the Gauntlet/Gladiator. He was naturally conservative, but all in all was one of the most successful designers of his period.' Whether he quite matched up to his contemporaries, Barnwell and Pierson, or the later Mitchell, is doubtful, but as Roy Fedden said: 'He had the great gift of leadership,' and it is significant that on resigning from Glosters he wasimmediately joined by little Henry Preston, his talented chief assistant who except for a brief sojourn with de Havillands in 1920–21 had been with him since SE 5 days. He was followed by Frank Radcliffe of the genial grin, who had been in charge of Gloster's airworthiness department after earlier experience at Blackburn and Armstrong Whitworth, but who had joined Airspeed as chief designer in 1935, and now brought with him his assistant F. A. Hamlin.

George Carter, who had joined Sopwiths in 1916 and for five years was chief designer of H. G. Hawker Engineering Co where Camm was one of his juniors, now became chief designer of Glosters – for he had been with that company since transferring from de Havillands in 1931 with the sub-contracted DH 72 bomber project. His latest effort, the Gloster F 5/34 Mercury-powered low-wing fighter, was under test by Gerry Sayer, but like its uncompleted Bristol 146 rival, was outdated by the more aerodynamically perfect Hurricane and Spitfire. Currently the shops were full of Gauntlets of which 204 had been ordered, but long before their completion a production line of the succeeding F 14/35 Gladiators was well underway. Meanwhile Carter was completing the design of a twin-engined, turret-armed, two-seat fighter to F 34/35 specification.

Neither he nor other designers were prominent at the seventy-first birthday celebrations of the RAeS at the Science Museum where the 400 elegantly attired members, wives, and guests were received by the Society's President, Major Wimperis, and his wife. Aeroplane designers were modest and somewhat socially shy except when it came to business, and they rarely attended the superb series of lectures organized by Laurence Pritchard, the beetling-browed secretary. But as one critic reported: 'The alleged *pièce de résistance* in the programme of the conversazione was a film on *Methods of Aeronautical Research at the RAE*. A lot of us had been anxious to know just how that department spent its time, so the title scored a record attendance of members and the vituperative Press. The film, recorded on inferior stock with ancient cameras and disregarding cinematographic lighting techniques, showed no item of technical research which could be dated later than 1926. Photos of airflow were the sort associated with school text books of the pre-Georgians. Semi-recognizable shots of the vertical spin tunnel brought suppressed laughter, for the model bore a resemblance to a pre-war Fokker, and the subject would have been interesting around the time of the Meopham accident circa 1928. As for the hydroplane tank, was it not one opened by Queen Victoria herself? We were shown that vicious complaint known as porpoising – an epidemic studied and cured around the time of the second Schneider Trophy. Though an impressive prologue introduced the big wind-tunnel, the shots might have been scale models rather than full-sized aircraft, and we were struck by the atmosphere of leisurely peace and grandeur which permeates the building, for the only human action was a sort of verger in overalls picking up such parts of the building as had fallen during the night. Or was that a blind for foreign spies? We suspect his real duty was to rouse any scientist who had fallen asleep while contemplating their problems. Despite all this, the party went with a swing. So many eminent persons had come to see each other, and so many others had come to see them, that a social success was assured from the start. The drinks were soft, as became the setting of the Science Museum, and the dresses displayed a similar veneration for the great past. Certainly the committee scored a magnificent joke in choosing such a setting for a meeting of modern aeronauts, forcing to-day's big industrialists face to face with the prior art of nature and of the scientists of the nineteenth century, thus meeting the original sources of their present royalty incomes!'

While the RAeS was modestly feasting, the Berliners were finding that shops were short of butter. There was a suspicion that the German Government was stock-piling against emergency, and this was underlined when Herr Rudolph Hess, a great Hitler supporter, defensively said: 'Germany prefers guns to butter.' Anthony Eden made the *riposte*: 'We are prepared to do our best by economic co-operation, and by working for European appeasement, to ensure that others have butter in a world which has no need of guns.' Few followed the example of the oil mogul Sir Henry Deterding, who sent several hundred thousand pounds to relieve the German poor.

Said C. G. Grey: 'To-day we are doing precisely the opposite to Germany. In our passion for guns, which Mr Eden denies, we are refusing to supply aeroplanes to countries that want them, and we are losing our markets to Germany and America. In the meantime we are not encouraging our farmers, but are limiting the supply of milk instead of subsidizing them to make butter and cheese which instead we buy from Denmark, Holland and Soviet Russia.'

That the implicit warnings of Hess and others were not empty threats was emphasized by Germany's willingness to show British officials a great deal of her aeronautical, naval, and military hardware. At the invitation of the German Air Ministry, Air Vice-Marshal Courtney, the Deputy CAS who was also Director of Operations and Intelligence, accompanied by a senior Staff Officer of Bomber Command, Wg Cdr R. V. Goddard, and bomb expert Wg Cdr M. Moore, flew to Berlin on 17th January for discussions with General Milch and General Kitzinger. Although shown a number of new military aircraft, including the Me 109 equivalent of the Spitfire, the twin-engined Junkers Ju 88 was not revealed for it had been designed largely by an American, Al Gassner, as the German answer to the Bristol *Britain First*, and only eleven months had elasped between initial drawings and the first flight late in December.

2

When in the previous year, the Air Navigation Bill became law it provided for the formation of an Air Registration Board (ARB), based on the 1934 Gorell Report proposal that the Certificate of Airworthiness of civil aircraft should devolve upon 'a statutory autonomous board' formed from the British

Corporation Register and Lloyds Register. Though Air Ministry officials said this was acceptable in principle they insisted that official control should still apply to design and construction of commercial aircraft seating more than ten passengers. The next move was for the Air Council to invite the SBAC, the two insurance Registers, commercial operators and the RAeC, to form a draughting committee in conjunction with an Air Ministry committee to devise a scheme within the provisions of the Act. Their proposals were accepted and the Board was now registered as a company on 26th February with membership drawn from the constituent groups.

From the construction and maintenance interests, the SBAC appointed Frederick Handley Page (Handley Page Ltd), Charles Walker (de Havilland) and G. H. Handasyde (British Aircraft Manufacturing), together with Henry N. St Valery Norman (Airwork) for maintenance and repair firms. The underwriters' group delegated Capt Alfred G. Lamplugh (British Aviation Insurance) and L. Murray Stewart (Aviation and General) together with a deputy chairman of Lloyds, Arthur J. Whitall, and Ernest R. H. Hill representing Lloyds underwriters. The commercial operators selected George Woods Humphery (Imperial Airways), L. T. H. Greig (Jersey Airways), and Eric L. Gandar Dower (Aberdeen Airways), and Harold Perrin (RAeC) was appointed to represent private operators, owner-pilots, and clubs. The convened Board then added Major R. H. Thornton, a marine transport operator and experienced private pilot, together with G. F. Johnson, a barrister-at-law of the Accident Offices Association. In due course the Secretary of State for Air appointed two 'independents', respectively Flt Lieut Rex Stocken as a 'person with not less than five years professional experience as pilot of civil aircraft' and Sir Laurence Philipps with the somewhat anomalous qualifications of chairman of the United British Steamship Co Ltd, member of the Executive Committee of the Wye Fishing Board, Lords, Turf Club, and Jockey Club. Meanwhile the Board was empowered to appoint an outside chairman, and chose Sir Maurice E. Denny chairman of William Denny and Brothers, shipbuilders of Dumbarton, who 'should before long be building on the Clyde flying-boats developed from the Blackburn flying-boats designed by Major J. D. Rennie.' But Sir Maurice was not without aircraft knowledge, for his firm had pioneered two huge pre-war helicopters, and during the Great War, as sub-contractors, had built numbers of aircraft as well as ships, and he had been chairman of the British Corporation Aviation Committee of 1927.

For revenue, the Board would have a government subsidy equivalent to the costs hitherto incurred by the Air Ministry for the same functions, together with subscriptions from the SBAC, Imperial Airways, British Airways and unsubsidized operators, the aviation insurance companies, and revenue from the issue of C of As, licences, and surveys.

Concurrently the Report was issued of the Maybury Committee formed in June 1935 'to consider and report upon measures which might be adopted by HM Government or local authorities for assisting the promotion of civil aviation in the United Kingdom.' Brig-Gen Sir Henry P. Maybury had been chairman, and the members were Colonel Sir Donald Banks, J. A. N. Barlow, Sir Cyril Hurcomb, Sir Arthur Robinson, and Lieut-Col Sir Francis Shelmerdine. In the course of twenty-three meetings they had heard from municipalities, air transport companies and allied interests. The three-part Report summarized the history of civil aviation in the UK, examined conditions in the light of the terms of reference, and offered recommendations – all at considerable length. As C. G. Grey, said: 'At first sight it seemed a blinding flash of the obvious. One might describe it as a collection of historic platitudes. Its mass of sententious remarks seems based on the analogy that if you don't stand up you'll fall down.' Nevertheless it revealed pertinent lessons of ill-fated and badly financed enterprises. Aerodrome location was discussed at length, and proposals made for a central junction airport between Liverpool and Manchester with radial services to Belfast, Glasgow, Newcastle, London, Portsmouth, and Bristol to serve a third of Britain's forty-two million population. There were recommendations that a single company should operate all internal services, and local authorities provide aerodromes for private flying and clubs, but aerial displays, of which there had been a plethora, should be replaced by demonstrations showing the practical application of flying. Rigorous licensing of operators and airline pilots was recommended, improvement in ground organization considered essential, and limited relaxation of aircraft maintenance proposed so that reconditioning could be at more economical periods and possibly not all at once, though in fact Imperial Airways to some extent were already operating such a system.

Coinciding with this wave of civil aircraft interest came the annual conference of the Aerodrome Owners Association (AOA) opened by Lord Swinton at Central Hall, Westminster, where an exhibition of appropriate equipment was staged. Prime mover was its past chairman, Councillor Dick Ashley Hall, mainstay of the Bristol and Wessex Aeroplane Club at Whitchurch. Prospects for aerodrome operators and the establishment of regular air services were less rosy than the Maybury Report indicated, even though the Government might take over all radio services at a cost of £300,000 and an annual upkeep of £240,000. Thus Hull's Municipal Airport no longer had potential traffic, for KLM had replaced it by Doncaster as their intermediate stop on the Liverpool-Amsterdam route in order to serve a greater industrial and agricultural area. At a gloomy meeting of Hull's Aerodrome Committee, Alderman F. Hill moved that the Airport manager be given warning of notice, for the previous year's income of £148 from landing fees had become negligible, and it was felt that the land should be sold as building sites – so it seemed that Councillor J. W. Bateman of Hull, newly elected chairman of the AOA in succession to Dick Hall, would find himself without the background authority of an aerodrome. A few municipalities were more optimistic. Cardiff City Council considered their airport might prove a paying proposition, and were encouraged by an Air Ministry offer to take over seventy-six acres of surrounding allotments to increase the area so that an RAF squadron could be established there – but Socialist members of the Council were siding with disgruntled allotment holders, and declared they would rather face machine-gun bullets than establish the aerodrome for military purposes.

Contention of a different kind was echoing through Whitehall – for instead of supporting the British aircraft industry, the Australian Government, to the tune of £250,000, had ordered forty North American NA 16 low-wing general purpose two-seaters powered with the 550 hp P & W Wasp. The bulk would be constructed under licence by the Commonwealth Aircraft Corporation Proprietary Ltd at Fisherman's Bend, Port Melbourne – a Government-backed organization run by Wg Cdr Wackett, an enterprising man who had built various experimental aircraft of modest success, and on behalf of the Australian Government had recently visited manufacturers in Britain and the USA to study manufacturing techniques and the capabilities of their aircraft. His selection of the compact North American design was undoubtedly justifiable, for the engine was remarkably reliable, the fabric-covered welded-steel fuselage easy to construct, and the metal-skinned wing a good introduction to completely metal-skinned machines. North American Aviation Inc. of Los Angeles was a new and go-ahead concern whose president and general manager was J. H. Kindelberger, formerly vice-president in charge of engineering at the Douglas Aircraft Co Inc, and he had been joined by ex-employees of the former Fokker Aircraft Corp in which General Motors held 40 per cent shares. To British critics it seemed a breach of faith that a big order should go to an almost unknown firm, and American at that, for it had been agreed that RAAF and RAF equipment should be identical to make war-time co-operation easier. However, Sir Archdale Parkhill, the Australian Minister of Defence, had clearly stated that the deal was made 'in view of certain special considerations which had been the subject of consultations with the British Government.' That was the genesis of the Harvard.

The nearest British equivalents were more powerful and costly and either were still experimental or had not yet flown – though it had been hoped that the Australian Government might consider the Hawker and Fairey monoplanes to specification P 4/34 for a Merlin-powered light bomber capable of tactical support. The Fairey was an elegant re-design of the Battle with 7 ft 6 in less span and 12 ft shorter, weighing 2,000 lb less; but though 43 mph faster it failed to achieve the specified 300 mph. Chris Staniland gave the prototype K 5099 its first flight on 13th January, and as expected found it handled as easily as the Battle except for a heavier rudder and more emphatic stall. Because Battle production was considerably advanced, there was little hope for the P 4/34 despite improved performance.

The equivalent Hawker Henley had a few inches greater span but looked much smaller because its fuselage was shorter and bulkier. Construction had started in mid 1935, but in company with the Hotspur had been subordinated to the Hurricane, and basically was a slightly scaled-up version of it, with standard flitch-plate jointed fabric-covered fuselage structure to which a deep ventral coaming was added, and wing span extended to 47 ft 10½ in by a longer root section incorporating central internal bomb stowage. George Bulman made the first flight on 10th March, and found handling was viceless. Top speed, using a Hamilton hydromatic three-blade propeller because the equivalent de Havilland version was not yet available, proved almost as fast as the prototype Hurricane. Orders seemed certain in view of the considerable interchangeability with Hurricane components.

Blackburns also had been flying a two-seat fighter/dive-bomber, the B-24, later known as the Skua. Built to FAA specification O 27/34, it was a flush-riveted, Alclad-skinned low-winger which had a buoyancy compartment under the pilot's cockpit, another behind the gunner's cockpit, and a watertight box between the main spars forward of the ailerons. George Petty, the chief designer, neatly solved the folding problem by adapting the inclined hinge used on the Blackburn Airedale monoplane of 1925 which enabled the wing to be turned leading edge upward while moving back until parallel with the fuselage and was then latched in position, giving a folded width of just under 14 ft. Unlike the Fairey and Hawker machines it was powered with a radial – the 840 hp Bristol Mercury IX driving a Hamilton three-bladed propeller. Choice of the circular-section fuselage was aerodynamically questionable because to avoid wing-root flow breakdown it required a massive fillet extending far along the rear fuselage, and the narrow enclosure for pilot and gunner had to be made high because the top curvature of the fuselage had minimal cut-away. The maiden flight by 'Dasher' Blake proved straightforward, but the centre of gravity was near the aft position, so longitudinal stability was marginal, leading to insertion of a bay in the engine mounting to bring the engine 2 ft 5 in further forward. Subsequent tests showed the wingtip slots were unnecessary so they were eliminated from production machines, of which 190 to specification 25/26 had been ordered 'off the board' in July 1936, so urgent was the need for a naval dive-bomber.

Brough since February 1934 had been the home of Cirrus-Hermes Engineering Co Ltd, successors of the original Cirrus engine built by the Aircraft Disposal Co for the first DH Moth, and Robert Blackburn had become chairman when the business transferred from Croydon. Blackburn Aircraft now took over the entire business as a branch of the Brough factory, and Carol S. Napier, formerly technical director of Cirrus-Hermes, was appointed chief engine engineer, thus continuing the tradition of his father's involvement with aero engines.

Production concentrated on his 90 hp Cirrus Minor which had a considerable export market, and recently Minors had been installed in a fleet of fifteen BA Swallow low-wingers for ab initio trainees at Blackburn's steadily expanding RAF Reserve FTS, now registered as Flying Training Ltd, managed by Flt Lieut N. H. Woodhead as CFI. Also a new 150 hp Cirrus Major had been bench tested and was now installed in a Percival Gull bought as a flying test bed from Blackburn's North Sea Aerial & General Transport Co. Development was in charge of Sqn Ldr J. L. N. Bennett-Baggs, the new director, who did some of the flying, though this was chiefly undertaken by Flt Lieut H. Bailey, a civil engineer who came from the Reserve School as assistant to Blake to deal with Shark production test flying. The Shark already was obsolescent, but as an interim measure until later aircraft were available the biggest order to date had been given in January for ninety-five, designated Shark III, and in fact was the last, bringing the total to 237 in addition to PV prototype, six for Portugal, nine for the RCAF, and seventeen built by Boeing Aircraft of Canada Ltd.

The old controversy between the Admiralty and Air Ministry had not only revived but strengthened, for the Royal Navy required complete control of the FAA, its manning, and ordering of aircraft, as in the first years of the Great War. Admiral of the Fleet Sir Roger Keyes declared: 'The Navy cannot afford to be dependent on the Air Ministry any longer, for they have utterly failed to keep pace with development of aircraft in other navies and have never been able to fulfil the Admiralty's requirements.' In a letter to *The Times* Admiral Sir H. W. Richmond wrote: 'A navy of to-day consists of all those ships, vessels, and craft whose duty it is to preserve the country from isolation. Whether they move on or under or above the surface of the water is immaterial. They are engaged upon one task, and they must be part of one single whole. By no possibility can they be subdivided under different Ministries in separate categories of waterborne and airborne instruments.' It seemed surprising that the Senior Service could not enforce its will, but Lord Swinton was against it. The *Morning Post* joined the fray with the argument: 'During the past two years much has been heard of the marvels of new machines for the RAF. Not once has mention been made of new types for the FAA or a new type of flying-boat. This is because no new type has been produced except one which is a combination of some old types. Thus the FAA has scant possibility of getting up-to-date types so long as the present system of supply maintains. The Navy is also anxious about flying-boats as these are controlled and manned by RAF personnel without naval experience, and all our Service flying-boats are obsolete. Every other important nation, even British private enterprise, is producing flying-boats capable of a performance 60 to 100 per cent better than those in service to-day.'

There was some justification, for although the Saro London and Supermarine Stranraer to specification R 24/31 had entered service respectively in October and December their biplane structures seemed antique compared with the latest American Consolidated PBY twin-engined high-winger – later known as the Catalina – but it was not generally known that Shorts were busy with a military version of the splendid Empire flying-boats with bow and stern turrets. The first of a batch of eleven of these S 25 Sunderlands was already far advanced and revealed an ingenious hull stowage of bombs which were run out on rails through hinged side hatches to their operational location under the wings.

The Empire flying-boat fleet was steadily increasing. On 4th February two return services a week from Hythe to Alexandria had started, with an overnight stop at Brindisi, and from the beginning of March was increased to four. An immediate difficulty was embarking and disembarking in rough weather, for what was needed was a wet dock into which machines could taxi after alighting on Southampton Water. Even Woods Humphery got into difficulty when he came to Hythe to inspect the Empire boats moored off the pier, for it coincided with an exceptionally low tide and the launch was unable to return to the quay near the Supermarine sheds used by Imperial Airways, and the last seen of him for some hours was while he was drifting downstream in a dinghy.

The Navy with ships, the Army with guns, and the RAF with flying-boats were on combined manoeuvres at Singapore early in February testing its defence. The attacking force comprised two cruisers, five destroyers, several submarines with their depôt ship, the aircraft-carrier *Hermes* and well over a hundred aeroplanes. Defenders were the regular garrison, eight RAF squadrons of bomber-transports, a flight of the Singapore AAF torpedo-bombers, four destroyers, four submarines and a submarine base ship. The Army's impressive installation of fixed guns, designed to fire across the sea on a prescribed arc of fire, pointed over the Johore Straits so that attacking ships could be hit with armour-piercing shells. Nobody foresaw the contigency of attack from the landward rear, for the defence of which these guns were impossible, so the attack by Blueland against Malayan Redland was unimaginatively conventional, nor had the secret experiments with radar yet been applied to operations by the Navy and Army or the RAF Overseas.

On the first day various ships were located far at sea by the RAF, one of whose flying-boats crashed next morning while taking off from the Straits, and although five crew were saved, the second pilot was fatally injured. Rain closed in that afternoon, and enemy destroyers and submarines were later reported heading for the Straits and other submarines were east of Singapore. During the night Redland patrols and coast defences drove off invading destroyers. On the third midday the Blueland carrier was attacked before she could launch her aircraft, and a second attack followed an hour later when nine fighters took off and chased Redland bombers and torpedo machines to Singapore and shot them up while they were landing at Seleter. At nightfall the Blueland fleet approached Singapore from three directions, preceded by aircraft from the carrier, but were stopped in Singapore Strait by continuous attacks from air and submarine, and presently withdrew using smoke screens. Presumably Singapore was safe!

3

The pioneer aircraft industry was suffering the pangs of forced growth against competition from a nationalized aircraft industry in the form of shadow factories. Sir Thomas Inskip explained in the Commons on 17th February: 'Of these firms Messrs Austin Motor Co and Rootes Securities will undertake the manufacture of airframes as well as aero engines. During the first year each airframe firm will be paid management fees of £50,000 for erection, equipment, and production planning of their factories, and similarly £24,000 for each aero engine factory, and £20,000 in the case of the six responsible only for assembly. When production is started the firms will receive management fees of £225 or £200 per airframe according to type and £75 for engines. After the first year, advances on an agreed scale will be set off against management fees. Provision has been made whereby a percentage of savings in manufacturing costs below a figure to be determined will accrue to the firms concerned.'

Already Rootes Securities Ltd had been in difficulty, for their preferred site was a factory at White Waltham near Maidenhead which, as Baldwin explained, was chosen because production could begin there more quickly than other areas as local labour could be found without building new houses. There was the outcry that rural amenities would be ruined. *The Times* had columns of letters. In deference, Rootes Securities turned to Lancashire with proposals for a factory at Speke.

The rest of the industry seemed disturbingly embroiled in disputes with employees. There were strikes at Fairey, Vickers, Boulton Paul, A. V. Roe, and Rolls-Royce – all described as 'unofficial' because they were not supported by the unions concerned, though backed by the aircraft shop stewards' National Council which was in arms against the Amalgamated Engineering Union (AEU). The trouble stemmed from the 1931 economic depression cutback of rates for overtime, piecework, and night shift. Employees had been quick to note that eleven companies in the motor and aircraft industry almost quadrupled their profits in 1935, and judging by 1936 returns might now be four times as much again. Therefore why not pay more to the workers who had remained loyal to the aircraft industry through the lean years? Yet the principal pressure was coming from the massive new entry of semi-skilled workers paid 1*s.* an hour and unskilled at 9*d.*, all of whom registered with the National Union of General and Municipal Workers, which led to friction with the skilled workers.

When the AEU held its annual conference at Scarborough the president, Jim Little, said that the Employers Federation had refused an advance of 2*d.* per hour on the standard basic rate of 1*s.* 4½*d.* per hour for skilled adult aircraft workers, amounting to £3. 6*s.* for a 48-hour week or £4. 1*s.* 6*d.* with overtime. Employers explained that the demand raised vital issues, and though the industry had progressed, this was offset by the long years of depression, and they did not see why the aircraft section should be treated differently from the engineering industry as a whole, but though national negotiation was the only solution they nevertheless had increased the wage of labourers by 3*s.* 10¾*d.* per week and up to 8*s.* 6*d.* for tool fitters. The conference therefore voted on a resolution to take a ballot of all skilled aircraft workers for or against strike action, but surprisingly this was defeated by a large majority – though, as a supporter said: 'Heavy taxation for rearmament means higher cost of living. Coal is up 4*s.* a ton. The workers rightly expect a higher standard of living, not a lowered standard, in prosperous times. Speculators in shares get big profits. Well-managed businesses get increased dividends. Surely the workmen, whose work produces all these rises, are entitled to a rise themselves?'

At Vickers Ltd there were signs of different unrest. General the Hon Sir Herbert A. Lawrence said he wished to be relieved of his duties as chairman after the forthcoming AGM, and it was agreed that Mr A. A. Jamieson would succeed him. Within a few weeks this was followed by an announcement that Vickers Ltd had decided to transfer to Vickers-Armstrongs Ltd the operation of Vickers (Aviation) Ltd and Supermarine Aviation Works (Vickers) Ltd, and that Commander Sir Charles Craven, chairman of Vickers-Armstrongs, would therefore become chairman of the two aviation companies, and Sir Robert McLean, as their managing director, would be appointed to the Board of Vickers-Armstrongs Ltd. The course was set for a struggle between two strong personalities.

More astonishing was the news that Arthur Hagg, chief designer of de Havillands, had resigned and given up the directorship held since May the previous year. This recent relationship had been an uneasy one. More and more had Geoffrey de Havilland relinquished participation in design because of his involvement in policy and business affairs, and consequently he could do no other than give unqualified support to his chief designer 'backing him to the hilt against all comers,' as his aerodynamicist Richard Clarkson recorded. At forty-nine, Hagg was both a creative artist and a practical engineer of long experience, for he had joined Capt de Havilland as assistant designer in 1915 at the Aircraft Manufacturing Co, and held the same position when de Havilland Aircraft was formed, yet even as chief designer he had never been fully accepted as a member of the inner ring comprising de Havilland, Walker, Hearle, and Nixon, with St Barbe as an honorary member. Since achieving his directorship Hagg had naturally used his authoritative position, and his complete concentration on the unique and fantastically advanced design of the all-wood Albatross airliner had increasingly overridden the technical advice of Walker and Hearle, and even de Havilland himself. Tension increased. Ultimately came the breaking point. The senior directors decided that he had become too difficult to deal with, and he was sacked.

In defiance that his world had collapsed around him, Hagg went to his office next day and carried on with his usual tasks – but it was soon made clear that he must go. As he told me many years later: 'I had not been able to get any co-operation from them for a long time, and they expected me to undertake other designs while I was fully engaged on the intricate Albatross, so I told them they must find someone else to handle the extra work. Nevertheless I already had made plans for the future, and had bought a small boatbuilding company at Walton-on-Thames where I proceeded to develop a new line of motor-boats.' In fact, these craft were the most distinctive of their time, and particularly reminiscent of earlier DH cabin aircraft, such as the DH 61 and its cabin appointments. Within a few months he was also appointed consultant to Napiers 'to vet their proposals from an aircraft designer's point of view.'

His departure left an awkward gap in the de Havilland set-up. De Havilland himself had been too long out of touch to take control. To promote from within his technical department seemed preferable to engaging an outside designer unfamiliar with the long-established de Havilland techniques. De Havilland and Walker considered the merits of their three senior section leaders: steady Percy Bryan; the inventive ex-Vickers draughtsman Charles T. Wilkins; Ronald E. Bishop who had beaten me to the post in securing the only available DH apprenticeship in 1921, and it was he who now was made chief designer. His immediate task was finalizing the almost ready Albatross on which Imperial

Airways were pinning high hopes, for despite wooden construction it would be their earliest answer to fast American airliners because the all-metal AW Ensigns were still further from completion, and even the impressive Empire flying-boats now entering service were a year later in operational use than the equivalent big flying-boats of the USA. Not only Imperial Airways in the larger class of airliners, but smaller companies were limited by lack of British equivalents to the new generation of fast American low-wingers, for the nearest approach had been the twin-engined, twelve-seat Blackburn HST 10, which after its initial flight the previous year had been standing abandoned in the Brough flight shed because the FAA re-equipment production programme was too urgent and profitable to risk a private-venture series construction of an expensive civil aircraft. British Airways had therefore bought the well proved Lockheed Electra, powered with the almost unfailingly reliable Wasp. At a Press demonstration on 24th March not only reporters but several pilots who briefly handled the little airliner were greatly impressed. 'Those few minutes at the wheel aroused an enthusiasm which is very difficult to keep within patriotic limits,' said one. Another more critically reported: 'By comparison with the Douglas DC-2 the Electra is much lighter to handle in proportion to size, and considerably more stable directionally. The ailerons are light but seem unnecessarily geared down. Relative to response, a little more resistance in the wheel would do no harm. The fore-and-aft control is considerably stiffer than the others, which is as it should be in a really stable machine, and the crank on the roof which works the trimming tab is all that is needed to set it in a steady climb, glide, or level course. The electric flaps produce a satisfactorily steep glide and an easy landing at about 60 mph.'

When Major Greaves, the airline's commercial manager, explained the circumstances forcing the company to import foreign aeroplanes a reporter noted: 'The visible conflict between regret at stern necessity and beaming pride in this new possession would have been amusing but for the light it sheds on our foreign market problems.' How many other airlines would follow suit? There were seventeen serving Great Britain and Ireland; sixteen in Australia; four in New Zealand; Europe exluding Britain had twenty-five; Asia thirteen; Africa, including Egypt, eight; South America eleven; Central America seven; the USA topped all with thirty-five.

Equally impressed with American aircraft was Sir Hugo Cunliffe-Owen, chairman of the Scottish Aircraft & Engineering Co Ltd, who shrewdly secured the British licensee of the 37-mm Hispano cannon, and less wisely had constructional rights for the long obsolete American twin-engined Burnelli 'flying wing' with deep centre-section which seated the passengers in remarkably cramped conditions. The British-built prototype, named *Clyde Clipper*, projected with two Rolls-Royce Kestrels, was about to be put in hand, and with the remarkable optimism was stated to be entered for a proposed air race from New York to Paris in four months time. Amy Johnson had also entered a foreign-built machine, believed to be a twin-engined Caudron.

4

This year the Air Estimates of £56,500,000 net was not only £17,500,000 greater than in 1936 but for the first time exceeded those of the Army, though the Navy was still well ahead. An Admiralty grant-in-aid of £4,200,000 for the FAA, various appropriations and a projected armament loan, brought the total to £88,588,600 gross. 'Why they bother about that odd £600 I cannot imagine,' said one commentator. 'We cannot even visualize the total – but during 1918 we were spending roughly £6 million a day on the war, so this very Gross Estimate means roughly a fortnight's war at full throttle.'

The House debated the Estimates with all due solemnity and even acrimony after a typically brilliant presentation by Sir Philip Sassoon who had memorized all the figures and juggled with them without using notes before turning to a wide variety of subjects involving men, material, and the war potential in which he emphasized that the expansion programme was without parallel in times of peace in the history of Great Britain and was being accomplished by voluntary effort, not emergency powers, based on good will and co-operation and not dictatorial decrees. 'To expedite production,' he said, 'the Air Ministry aim at reducing the number of types, and making as few modifications as possible. Both Air Ministry and main contractors are doing their best to widen production by sub-contracts. Accelerated production can only be secured by more skilled personnel, but to draft picked men from other engineering industries would dislocate those businesses at a time of great activity, and many unskilled workers whose occupations are dependent on skilled workers would be thrown out of employment.

'The shadow factory scheme is a practical policy which gives us a war potential and reinforces production without interfering with normal industry. Manufacturers have been persuaded to make large extensions of factories and plant, and to safeguard them against being left with redundant buildings they can claim compensation if on the completion of expansion they find themselves burdened with these buildings and plant, but any excess profits will be taken into account in assessment.

'Where new types are concerned and the contractor does not know their production cost, the Air Ministry will give an Instruction to Proceed and wait until enough information is available for a fixed price to be agreed for the contract, and this is called a basic price. If by increased efficiency the contractor delivers the bulk order at less than the basic price the saving is shared between him and the Air Ministry – but if the price is exceeded, the contractor receives what he has actually spent and his profits are then based on the basic price, though he has the right to go to arbitration.'

Churchill growled his say, batting for Admiralty control of the FAA: 'Because of this unhappy controversy the Admiralty and the Air Force are fighting each other so keenly that the Air Ministry, in the natural desire to placate the Admiralty, is taking an undue proportion of senior officers and sending them to the Fleet Air Arm. The RAF would benefit enormously if the

FAA were developed by the Navy itself. The Navy should be given all it wants, but only for its own specific and highly specialized function.'

Turning to expansion progress of the RAF, he said it was not true that we had a hundred squadrons, for twenty two were skeletons of only a single flight and equivalent merely to seven squadrons. 'Adding them to the established seventy-eight, we have eighty-five, but deducting fifty-four which existed two years ago we have an increase of thirty one squadrons of the seventy one promised. If the full expansion of the RAF projected in 1935 had been executed by 31st March we would have had 1,500 machines and 123 squadrons, yet it still would not give parity with the leading air power within striking air distance of these shores. The effective fighting force of Germany must be at least 1,700 or 1,800 machines capable of going into action and being continuously maintained in action during the course of a war.'

Undoubtedly Churchill was accurately advised by his German contact, for information available a quarter of a century later indicated that early in 1937 the Luftwaffe had at least 200 operational Staffeln, each of which would have nine aircraft, so there were 1,800 not counting reserves, trainers, or aircraft in Spain. Those squadrons were being rapidly re-equipped with more modern aircraft, for the philosophy had been that 'obtainable equipment of limited usefulness is better than no equipment.' Thus the Heinkel He 51 was outdated as a fighter and being superseded by the Messerschmitt Bf 109; the bomber JU 52s by Heinkel He 111s and Dornier Do 17s, and the Ju 87 dive-bomber, fully proved in the Spanish War, was already in service with the Immelmann Gruppe 162, replacing the Henschel Hs 123 biplane. Type by type, German re-equipment was proceeding substantially faster than equivalent British.

The Debate re-exhumed the old arguments against rearmament. Inevitably Labour members moved a motion for abolition of aerial warfare and the need for international control and ownership of civil aviation, but were defeated with 175 votes for the Government and 119 against, and the Debate drably continued into the small hours to little purpose other than airing ill-found views. The great British public remained serenely indifferent; there were jobs in plenty, money was circulating, and anyone could see the future was brighter than for years.

The day after the Debate a Select Committee of the Commons considered the Southern Railway's Bill for a proposed airport at Lullingstone, between Sevenoaks and Dartford, to which Woods Humphery said his company hoped to move in due course. Meanwhile the provision of new headquarters for Imperial Airways on the Southern Railway's property in Buckingham Palace Road was planned but Parliamentary authority was needed to remove a wall screening the railway from adjacent houses. After giving evidence, Woods Humphery hurried to an inquiry on the crash of one of his DH 86s, *Jupiter*, during an experimental night flight from Croydon to Cologne, where the burned out wreckage was found twenty miles away, together with the bodies of the three occupants – his close friend Capt C. F. Wolley Dod, OBE, European manager of Imperial Airways; Capt George B. Holmes the pilot; and Charles Langman the radio operator. They had been flying in sleet and snow in a high wind, so it seemed that ice formation must have forced a landing or caused a stall, for both Wolley Dod and Holmes were pilots of long experience. Worse followed eight days later. The Empire flying-boat *Capricornus*, piloted by Capt A. Patterson, left Southampton at 11.30 on 24th March for its first commercial flight, and shortly after radioing Lyons Airport that it was lost in snow-storms, flew into a mountain twelve miles south-west of its destination Mâcon, and was totally wrecked, killing four of the crew and the only passenger, but the radio operator, despite a broken arm and other injuries, managed to make his way through the snow to a farmhouse two miles away where he summoned aid.

The very remoteness of crashes such as these meant little to a public long-conditioned to the weekly succession of RAF fatalities, but one and all were dismayed by news on the eve of the *Capricornus* disaster that the sporting seventy-two-year-old Duchess of Bedford was missing with her Moth after setting out from Woburn alone with the intention of fifty-five minutes flying to complete her total of 200 hours solo. Her private pilot, Flt Lieut 'Chev' Preston, had set a triangular course across country she knew intimately to Huntingford, Cambridge, and back. To many she was an even greater heroine than Amy Mollison or Jean Batten, for she had first come to popular acclaim with her record flight to India and back in 1929. The first identifiable clue that she had died was ten days later when four interplane struts were washed up on the East Coast, and a day or so later a broken propeller drifted ashore, indicating that she had hit the sea hard. Some believed it intentional, for her marriage to the 11th Duke was not over happy, and she was deaf and hated the thought of growing old. Even those who knew her combination of forcefulness and gentleness, and her devotion to helping those in sickness or trouble, thought she probably preferred this sudden end; but it could not be assumed that the Duke was unmoved, for in token of the services by all ranks of the RAF in the search for her aeroplane he sent £1,000 to the Air Council for the RAF Benevolent Fund.

That springtime saw also the death of forty-nine-year-old 'Bill' Kenelm Lee Guinness, ex-RNVR sporting grandson of the founder of the brewery, and known throughout the motor-racing world as largely instrumental in the success of Bentley racing cars in the famous team which included Malcolm Campbell and Henry Segrave, though few realized that his initials KLG indicated he was the originator of the famous sparking plugs of that name.

His friend the Rt Hon Frederick Edward Guest, CBE, DSO, MP, died a few days later on 28th April at the age of sixty-one from cancer of the liver. Universally this brother of Viscount Wimborne was known as 'Freddie'. He was a cousin of Winston Churchill, and started his political career in 1905 when both switched from Liberal to Conservative. After the 1910 election he became Churchill's assistant Private Secretary successively at the Colonial Office, Board of Trade, and Home Office. In 1921 he became Secretary of State for Air, appointing Sir Sefton Brancker as the first Director of Civil Aviation, but he never again held high office. Earnest though he was, but no real politician, he often got facts and figures wrong. As one of his supporters said: 'The consequence was that nobody took him very seriously in the Commons – but his friends did not mind that, for they regarded him as a great sportsman, an entertaining companion and a thorough good fellow all round.' Certainly No. 600 (City of London) Squadron of the AAF, of which he was Hon Air Commodore, regarded this valiant soldier of the South African and Great Wars as a commanding and inspiring figure, and particularly appreciated that he had learned to fly when over fifty. His marriage to the daughter of an American iron-master millionaire led to secondary fame, for she was a great traveller, and in particular initiated elaborate hunting camps in Kenya which became known as **'going on safari'**.

5

Britain was agog over the forthcoming Coronation. Even the Japanese were involved. On the score of bringing an illuminated congratulatory Address to the British Nation, the president of Japan's most prominent newspaper dispatched the latest Mitsubishi monoplane from Tokyo to Croydon, where the *élite* of London's Japanese awaited it with John Galpin representing the DGCA and Harold Perrin of the RAeC, as well as a huge crowd who almost converted the welcome to Masaki Iinuma, the pilot, into a riot. *The Aeroplane* reported: 'The pilot brought his monoplane to Croydon at 15.30 hours on 9th April to a brick-like landing with the bounceless plop of a mashed potato.' Nevertheless he had achieved 9,900 miles in just over ninety four hours, which was outstanding, for he had no experience outside his local territory except a single flight between Korea and Formosa.

None could have imagined that his aeroplane's name *Kamikaze*, the Heavenly Breeze, was significant of the future, but the machine showed that the Japanese could now match construction skills with any country though it was of American inspiration, with fixed spatted undercarriage and a Japanese-built 550 hp Wright Cyclone fitted with a Japanese-built version of the Hamilton. On 16th April Masaki and his observer flew to Brussels escorted by a squadron of the Belgian Air Force and were received by King Leopold. They then headed for Germany where they were received by General Göring and the Lord Mayor of Berlin, thence in stages to Moscow, across Siberia to Vladivostok, and so to Tokyo where they were honoured with the insignia of the Rising Sun.

Three days before the *Kamikaze* left England Lady Maud Hoare launched the latest aircraft-carrier *HMS Ark Royal* of 22,000 tons, 685 ft long with 94 ft beam, and able to accommodate seventy-two aeroplanes. Two more aircraft-carriers of a thousand tons greater displacement were being built by Vickers-Armstrongs, and two more had been authorized under the 1937 Naval Estimates. They were as fast as the scarlet-funnelled 81,000-ton *Queen Mary* which on its maiden transatlantic crossing in May gained the Blue Riband by averaging 30.63 knots, burning 6,000 tons of oil in the process, but she only precariously held the record against France's *Normandie* which had one-fifth less shaft horsepower because of her superior hull form creating less wave resistance.

Germany, holder of the Blue Riband from 1929 to 1933, was out of the race but remained faithful to the airship for transatlantic flying. On the evening of 6th May came terrible disaster to the hydrogen-filled *Hindenburg* at Lakehurst, USA. Some twelve hours late because of head winds, she had just completed the first crossing of the year, and on approaching the mast and dropping the handling lines there was an explosion at the stern. Within a minute the flames were roaring. The ground team rushed to right and left as the burning hulk dropped down but several failed to escape; yet before the flames consumed the new outside cabins built onto the ship during the winter, a number of passengers and crew managed to jump and run clear – but sixteen of the thirty-three passengers and seventeen of the sixty-one crew were killed, and Capt Ernst Lehmann, famous for many airship flights, subsequently died, and Capt Max Preuss, the commander, was gravely burned. One newsman observed: 'The number killed was much the same as in accidents to American airliners a few weeks earlier – but the world's newspapers were more impressed by the wreck of the *Hindenburg* than any aeroplane crash. Perhaps a sense of its size had something to do with it, yet so many millions travel every day by motor-car that the killing of a few thousand passes almost unnoticed.'

Static discharge was Dr Hugo Eckener's view of the cause of the accident, but a contributory factor was a hydrogen leak in an aft gas cell probably punctured by one of the ballonets' stabilizing wires. It meant the end of airships as passenger-carriers, unless non-inflammable helium was used, and then payload would be only 50 per cent of what was possible with hydrogen. Even that former great advocate of airships, Sir Dennistoun Burney, wrote in the *Daily Telegraph*: 'One is forced to the conclusion that the airship has been developed as an ocean transport serving only for a short era. That era seems to be the period required for flying-boats to develop sufficiently to become capable of fast, regular and reliable oceanic service, after which the airship's chance of survival appears precarious, for top speed is unlikely ever to be more than 90 mph.' But the USA had a monopoly of helium supplies, and despite the Zeppelin disaster they would not abandon their own naval airships. In a burst of generosity Hitler telegraphed the Zeppelin Airship Works that he had set aside £2,500 for the families of the crew. Göring told them: 'The destruction of this proud ship should be only a spur to the will of German workers to build another and better one to replace it in the shortest possible time,' and he ordered work on the airship LZ 130 to be pushed on with all speed.

Though the papers were full of the *Hindenburg* disaster they continued building up excitement over the Coronation. The streets of London were gay with bunting and flags. As a contribution to the jubilations the RAeS held its annual Garden Party on 9th May. *Flight* reported: 'That term suggests sunshades, silk hats, and strawberries. Last Sunday, at Fairey's Heathrow Aerodrome, it implied gumboots, galoshes, and gamps. Yet, to the credit of all those modestly anonymous people who stage-managed this notable fixture, it went off more or less to programme. Over 3,000 guests, including many distinguished overseas visitors here for the Coronation, sheltered in hangar doorways or under accommodating wings while various stout-hearted pilots, penned beneath a 500-ft ceiling, peered through opaque windscreens and drizzle while they made their aeroplanes do all those things that should be done beneath a blue sky. In spite of those weeping heavens, some thirty widely assorted aeroplanes were present on parade.' But of modern military aircraft the new Fairey P 4/34 was already housed there, and only the Swordfish and Wellesley managed to get through the murk – though not the Blenheim, Battle, Lysander, and Bristol Bombay; and both Spitfire and Hurricane were regarded as too secret for foreign visitors. Many of the latest light aircraft were missing, such as the twin 40 hp Carden-engined Baynes Bee which had its maiden flight by Hubert Broad a month earlier, but among the many dripping Hawks and Gulls and Dragons could be seen several single-seat ultra-lights such as Kronfeld's Drone de Luxe, little Tipsy and Chilton low-wingers, a big-span Dart Kitten, the Luton Minor high-winger designed by Latimer-Needham of earlier Halton fame, the somewhat similar Carden-powered Broughton-Blayney Brawny optimistically offered at £195, and the funny little Scheldemusch tail-boom pusher biplane with which its designer demonstrated the effectiveness of slots and tricycle undercarriage by letting it sink to the ground with almost no forward run.

Raoul Hafner's AR III giroplane drew more popular interest, for its leap into the air seemed fantastic after the more laboured take-off of conventional aircraft. His earlier partner had returned to Vienna where he was experimenting with a helicopter powered by a Pobjoy which drove both the two-bladed rotor and a pusher propeller. Hafner had the more practical solution, and the display of his machine by Flg Off Clouston even dimmed the appeal of Reggie Brie's skilled efforts with a standard C 30 Autogiro.

Thirty firms exhibited in the big Fairey shed, and the rain certainly ensured that people milled around them, but essentially the occasion was one of socialities rather than technics, though hats and mackintoshes enabled many a well known personality to mingle unrecognized. Thus the Australian, Harry Broadbent, disguised beneath a trilby hat, was there, having flown his Leopard Moth from Australia the previous week in the fastest solo time of six days eight and a half hours. A rivalling record-breaker was H. L. Brook who had landed at Heston on 5th May having flown his Percival Gull from the Cape in four days eighteen minutes, brilliantly beating Amy Mollison's time by sixteen hours; but he was somewhat disgruntled, saying: 'When a man risks life and money on these sporting efforts one would think he was entitled to some practical recognition, but unless one is willing to hang over the bar in the RAeC and buy people drinks one gets no bouquets.'

No flight could match a Coronation. The sun shone brilliantly on the morning of 12th May, when the King and Queen in the great gilded state coach drawn by four pairs of caparisoned white horses drove down the crowded and decorated Mall to the Abbey and were crowned amid the aura of pageantry and pomp typifying the great ceremonies of England. The entire nation rejoiced, for this seemed a happy conclusion to a difficult period which had seen the end of George V's reign and the abdication of his son King Edward VIII. All the world heard the radio broadcast of the ceremony and the triumphant fanfares, and, for the first time on such occasions, television sent pictures of the procession and the traditional balcony appearance of the new monarch and his consort with their heirs, thirteen-year-old Princess Elizabeth and little Princess Margaret – and below them the huge crowds cheered and cheered. There might be difficult problems ahead, but there would be no thoughts of dictatorships and revolutions. Three weeks later, the Duke of Windsor quietly married Mrs Wallis Simpson at Monts, France.

The Coronation Honours List did not entirely ignore the aircraft industry, for Sir John Davenport Siddeley was elevated to the Baronage as Lord Kenilworth – the first to gain such an honour as a member of the aircraft industry. Since retirement in July 1935, Siddeley had dropped every commercial interest, busying himself with local government matters, and as a near-millionaire had recently purchased Kenilworth Castle and presented it to the nation, together with money for upkeep.

As part of the many Coronation events a Royal Review of the Fleet at Spithead on 20th May took precedence. Prominent were the aircraft-carriers *Courageous*, *Glorious*, *Furious*, *Hermes* and the previous *Ark Royal* renamed *Pegasus*, but – after the King aboard the *Victoria and Albert* sailed down the serried ranks of battleships, cruisers, and destroyers – hazy weather limited the massed squadrons of Nimrods, Ospreys, Swordfish and Sharks to a single fly-past led by Rear-Admiral N. F. Lawrence, AOC Aircraft-Carriers, aboard a Shark piloted by Lieut B. W. de Courcy Ireland of the Royal Marines. Among visiting foreign warships was the German *Admiral Graf Spee* with a seaplane on its catapult, two small aircraft on the Japanese *Asigara*, and a floatplane on the Portuguese *Bartolomeu Dias*.

Next came the Royal Tournament at Olympia on 27th May, where the King in uniform of Marshal of the RAF, accompanied by the Queen and the two Princesses, inspected a Guard of Honour drawn from the three Fighting Services. As one newspaper reported: 'The Royal Tournament is a display of skill at arms for the benefit of Service charities and has been going on for fifty-four years but never better than this Coronation year. All branches of the Services excelled themselves to provide a pageant of colour, movement, precision and beauty which is unrivalled.'

Empire Air Day, organized by the Air League, was on the following Saturday. Fifty-three RAF Stations, and many civil aerodromes were open to the public with the main purpose of showing RAF squadrons operating, but also to raise money for the RAF Benevolent Fund. The day was sunny; the crowds turned up in their hundred thousands – but the most modern aeroplane among the host of obsolescent RAF biplanes was a solitary Handley Page Harrow. The only other big monoplane in service, the Fairey Hendon, totalled only fourteen, for a subsequent contract for sixty had been cancelled. Pilots preferred by far the more nimble and resilient Heyford.

Civil aviation scored with new aircraft compared with the RAF, for the almost completed 123-ft span Armstrong Whitworth Ensign prototype was visible in the Hawker Siddeley-owned AST hangar at Hamble to which it had been transferred because the Armstrong factory was full to capacity with Whitley bomber construction. The beautiful but smaller and lighter rival, the 22-seat DH Albatross, was similarly visible at Hatfield, where young Bob Waight had piloted its first flight on 20th May after days of system testing, propeller and engine adjustments. The novel reverse-flow air cooling devised by Richard Clarkson with ducted intakes in the wing leading edge had proved successful, but there were problems with the characteristically shaped twin fins and horn-balanced rudders inboard above the tailplane, for there was directional wandering, and with an outboard engine cut the rudder was too heavy to hold the machine straight, nor was the elevator generally satisfactory. In the next few months many flights were required and both Geoffrey de Havilland and his son tried the controls. None could foresee that it would be autumn before the decision to make a new dihedral tailplane with inset balanced elevators having servo and trimming tabs, and fit new fins with similarly balanced rudders at the tailplane tips.

6

Already there were indications that the King's reign would not be easy. On 27th May Baldwin resigned in favour of Neville Chamberlain, who took office the following day. The new Prime Minister had shown no interest in aviation, although he had joined the move to install Lord Swinton, now reported ill, as Secretary of State for Air.

Said one of Baldwin's critics: 'He was the most astute and circuitous politican of the century.' Others bitterly said that no prime minister ever gained so much support with so little accomplishment, and that he was a master of dialectic enabling him to confuse any issue while giving the impression of clarification. In his own words, his fundamental outlook was: 'Give peace in our time, O Lord!', and when he scrapped the Geneva protocol many agreed with his comment: 'I do not know what the word "internationalism" means. All I know is that when I hear it employed, it is a bad thing for this country.' To most people his air of a stolid, pipe-smoking country gentleman seeking the moderate course carried conviction, and at least women could be grateful that in 1928 he extended the vote to those under twenty-five, thus rounding off the battle for which their mothers and grandmothers struggled in pre-war days.

Fifty-eight-year-old Neville Chamberlain, until now Chancellor of the Exchequer, was very different. Brusque in manner, harsh of voice, but with great will power, he was a masterful administrator – though his foreign policy was doubtfully regarded, for it was both too innocent and too commercial because of unfamiliarity with foreign peoples.

Changes followed affecting aeronautical affairs, and though no colleague liked him, Swinton remained Secretary of State for Air. However, Sir Philip Sassoon, who for thirteen years had been the most effective Under-Secretary this country had known, was translated to First Commissioner of Works. As an admirer said: 'For sheer intelligence he is unsurpassed in contemporary politics, and if he devoted himself wholly to such affairs he might rise to great eminence in history.' Succeeding him was Lieut-Col A. J. Muirhead, an Etonian who graduated at Magdalen, and as Conservative Member for the Wells division of Somerset had been Parliamentary Secretary to the Ministry of Labour since 1935. Gloomily C. G. Grey remarked: 'He has no aeronautical record that we can discover' – but that at least meant he might approach aeronautical problems with an open mind.

Sir John Simon succeeded Chamberlain as Chancellor of the Exchequer. Anthony Eden remained at the Foreign Office though it was unlikely his views would agree with Chamberlain's. Sir Thomas Inskip continued as Minister for Co-ordination of Defence, in which post he had earned great respect, but Sir Samuel Hoare changed from the Admiralty to the Home Office and was succeeded by Alfred Duff Cooper, formerly of the War Office to which position Leslie Hore-Belisha was transferred, leading a cynic to comment: 'He cannot put the Army in a greater mess than it is already, nor do anything more foolish than he has already done as Minister of Transport.'

As part of the process of up-grading 'consonant with the promotion of Sir Francis Shelmerdine to be Director-General of Civil Aviation instead of the less glorious title of Director', the Department of Civil Aviation was physically divorced from the Air Ministry and removed from Gwydyr House, the Air Ministry's headquarters in the lower corner of Kingsway, to the nearby historic Ariel House, previously Marconi House, in the Strand. 'Which,' as Lord Swinton said at its opening, 'will no doubt inspire air navigators sitting for examinations in the rooms where their grandfathers frolicked in the days of the Gaiety Restaurant.'

Newspapers were urging attention to the big increase in air accidents allegedly due to the increased number of pilots rushed through the flying schools in the last twelve months – though all pilots, military or civil, whatever their experience, had to face this occupational hazard in which there were at least ten major accidents that were non-fatal for each one where one or more crew were killed.

The DGCA was concerned specifically with civil aircraft crashes, whether those of airline or private fliers, and was considering establishment of an Accident Investigation Board. The Manx Air Races at the end of May had added another victim, for Sydney Sparkes, despite long experience dating to pre-war days at Hendon, stalled in a steep down-wind turn when taking off from Hanworth at the start of the race, hit the roof of a house, and the machine dropped into Hounslow Road, burst into flames and set fire to the building, so badly burning the houseowner's wife that she as well as Sparkes died in hospital next day. There was concern that the RAeC had permitted the race in such poor weather, for despite sunshine at Liverpool, fog eliminated all visibility across the sea, and there was every risk of collision or flying into the invisible water because most competitors were skimming the waves. A visitor, Herr Seidemann, flying a Messerschmitt Taifun, was adjudged winner as the only machine officially observed to round all the turning points.

To help British air crews 'overtaken by misfortune, or the dependants of those who may have lost their lives', the GAPAN had instituted a charitable Trust by Deed on 14th January which already was aiding dependants of ten Guild members lost in seven fatal accidents the previous year. The honorary treasurer was that faithful friend of pilots 'Lamps' Lamplugh of insurance fame, and Farey W. Jones provided office and staff to achieve a fund of £50,000 to which it was hoped the industry would contribute despite its parsimony in effecting adequate insurances.

British aviation received a great blow at the announcement on 11th June that Reginald Joseph Mitchell, the brilliant Supermarine designer, had died. He was only forty-two, and had joined his company in 1916 when it was directed and largely owned by Hubert Scott-Paine, becoming its very youthful chief engineer in 1920. Tutored on Admiralty designs of early flying-boats, he first came to public notice in the summer of 1919 with his modified version of the N 1b Baby flying-boat, renamed Sea Lion I, but was unlucky in the Schneider Contest that year, though redesigned as the Sea Lion II (Sea King) it won the Trophy in 1922. From that time a string of successes led to the conception of the Spitfire with John Faddy as section leader, and initiation of the four-engined bomber B 12/36 on which his unassuming assistant, Joe Smith, was currently working. Even between July 1934 and September 1936 Mitchell had been responsible for twenty eight designs and variants. As with every other manufacturer, the number accepted by the Air Ministry for construction was minimal, but twenty three were agreed, of which fifteen were successful, eight went into production, six were failures, and two not completed.

Joe Smith wrote: 'The facts are tragically simple. In 1933 he was taken ill and an operation was only partially successful. Although apparently completely recovered, he was increasingly subject to pain during the next three years. In March 1937 there was a further operation, and the disease was found incurable. He put his affairs in order, made provision for his wife and son, and in April was flown to Vienna to the Clinic of Professor Freund, but the diease was too far gone and he came home at the end of May. Three weeks later he died, having been unconscious for several days.' But at least he had seen his Spitfire fly, and knew that it would prove triumphant – a belief fully shared by Joe Smith whose unswerving policy was to develop the Spitfire, confident that 'it would see us through any war.'

Mitchell, rather shy with strangers but always preserving an easy manner, had both outstanding capacity for leadership and unique creative genius – but the popular conception that chief designers physically design their machines is incorrect. They are chairmen of discussions on solutions to specifications, for which a section leader, later more fittingly described as project engineer, draws a series of general arrangements meeting parameters defined by stressmen and aerodynamicists, and after broad agreement of the appropriate envelope other sections leaders ingeniously contrive the structure and fitments of which details are drawn by their draughtsmen. But all chief designers intensely continue to consider and alter such work. As Joe Smith said of Mitchell: 'He was an inveterate drawer on drawings, particularly general arrangements. He would modify the lines with the softest pencil he could find, and then re-modify with progressively thicker lines until one would be faced with a new outline of lines about 3/16 in thick. The results were always worth while, and the centre of the line was usually acceptable when the thing was re-drawn. When construction began he would assess the result and if not satisfied, discussed it with the people concerned, always bearing in mind the practical aspect of any alteration in relation to the state of the aircraft and the ability of the Works to make the change.'

Alan Clifton, Supermarine chief technical assistant, told me: 'After R. J.'s death we stopped sending in designs for a bit. Our small team of about fifteen technicians and sixty or seventy draughtsmen was fully occupied productionizing the Spitfire and designing the Sea Otter amphibian and the B 12/36 bomber. Joe, who came from the Midlands like Mitchell and was an Austin Motor Co apprentice, had the same common sense, and was by nature a conservative and sound engineer, but experience with Mitchell led him to support innovations as a matter of course.'

Already the four-engined Supermarine strategic bomber design in company with the equivalent Short, had been awarded a prototype contract, and A. V. Roe and Handley Page had contracts for the corresponding medium range twin-engined P 13/36 requirement, but the Vickers tender was declined together with Pierson's proposals for a six-engined 235 ft span geodetic bomber which he argued would have the offensive power of a squadron of Wellingtons and cruise 55 mph faster.

The crucial question was whether to continue the big Supermarine bomber now that the guiding hand of Mitchell had gone. In later years, Air Commodore Verney, still DTD at that crisis, told me that the Supermarine revealed all the technical insight acquired by Mitchell in his Schneider designs, and was estimated to be 50 mph faster than the rival tenders, so he was strongly in favour – but his powerful superior, Air Marshal Freeman, after discussion with Sir Robert McLean, decided that neither Vickers nor Supermarine factories had sufficient design and production capacity because of involvement with the tremendous requirements for Wellingtons and Spitfires. This infuriated Sir Charles Craven, for he hated McLean's assumption of plenipotentiary powers and was trying to knuckle him down – a state of affairs by no means peculiar to the aircraft industry. Inevitably this meant that big Rex Pierson had to satisfy two masters, but he had a calm persistence. As Roy Fedden recorded: 'He was a grand person to work with, and without doubt his handling of such situations and of technical discussions was masterly. Basically he was always constructive, and we invariably arrived at a satisfactory conclusion, but he was at times a hard taskmaster for an engine-maker, though a most loyal and valuable friend if one had the sound sense to listen to him. I have heard criticism that he kept too much to the centre of the road, and that some of his designs were not as clean aerodynamically as they should have been, but he was in command of Vickers technical side at a very difficult formative period and had great responsibility to keep the production side going while designing what were in those days large machines. He always favoured light wing loading and high aspect ratio, but was not a subscriber to the adage "that which looks right, is right" and was interested mainly in producing a practical aeroplane to do the required job, backed by his thorough engineering knowledge and innate sense of what would work. Nevertheless, when he produced the Wellington it was a world beater, and the fact remains that R. K. P.'s long family of designs were for the most part successful and built up an excellent and lucrative business for Vickers.'

At this juncture Volkert was having second thought about his P 13/36 big twin-engined bomber, the specification of which had been frequently discussed with Verney, for Volkert considered the four-gun tail turret was too great a drag handicap, and by replacing the R-R Vultures with the 2,000 hp Sabre engines which Halford was developing, he could offer an unarmed bomber capable of 400 mph – but this was rejected. Volkert remained dissatisfied. He knew the Vulture was suffering delay, and therefore re-schemed the HP 56 as a four-engined machine, powered with Fedden's new Taurus engines, with wings of 20 per cent increased area to compensate for the greater weight, extending the span from the original 80 ft, which Chadwick also had used for his Manchester, initially to 90 ft, whereas the Short four-engined bomber was 99 ft and only just met the requirement of hangarage through the standard 100-ft doorways. A fortnight later a four-engined Merlin-powered version was also submitted, and within a few days the contract was amended to accept this version as the HP 57 in lieu of the twin-engined bomber, although slower.

Concurrently the US Army Air Corps was testing the new and imposing 150-ft span Boeing XB-15 bomber with 4,400 hp for take-off from its four P&W two-row Wasps, and an all-up weight of 35 tons compared with $26^1/_2$ tons for the four-engined Short bomber to specification B 12/36.

The complexity of the four British bomber designs was far greater than anything previously attempted, for the Air Ministry tried to include too many alternative purposes. Thus the B 12/36 four-engined machines not only were required to accommodate a reserve crew but also twenty-four full-equipped troops, which fundamentally affected structural design; but the P 13/36 required only seating for twelve troops and eliminated the reserve crew. 'A significant difference,' recorded Chris Barnes, 'was that the largest "stores" specified for the B 12/36 were 2,000-lb armour-piercing bombs, whereas the smaller P 13/36 had to take two long torpedoes, with the unforeseen result that the B 12/36 had divided bomb compartments suitable for nothing larger than those bombs, while the P 13/36 had uninterrupted bomb cells of maximum length and width.' However, Volkert succeeded in having the torpedo and dive-bombing requirements deleted – but in addition to the involved equipment design, in turn depending on development of such items as gun turrets, the Air Ministry required the airframe to be readily disassembled into components fitting RAF packing-case sizes devised to suit standard railway waggons, and all such components had to be interchangeable. 'Finally, the bombers must have short take-off and landing runs matched to existing grass airfields, and also good ditiching characteristics with ability to float level for several hours.' Despite the complication, 'design, construction and satisfactory operation of all services must proceed with the greatest possible speed'. For Gouge's bomber, Oswald Short agreeed that a half-scale model would be built to gain early experience of any aerodynamic problems following the precedent of using the Scion Senior as guide to the Empire flying-boat design. Designated S 31, the fuselage and wings would be conventional spruce structures with plywood skin, and the four engines were 90 hp Pobjoy Niagara IIIs in long chord cowls.

7

The work of the ARC in England and NACA in the USA was substantially contributing to the aerodynamic and constructional advances of all new designs. Though much was too secret to mention in the annual report of the ARC, published modestly by HMSO at 4*s*., many a visit was paid by the industry's technicians to the RAE and NPL for discussions with specialist scientists on particular problems. Facts and figures were beginning to emerge on the effect of air compressibility. Totally confused, the *Manchester Guardian* in reviewing the ARC Report, commented that at over 400 mph the compressibility of the wing covering must be taken into account!

Even the ARC had discovered that longitudinal stability of low- and mid-wing monoplanes was still a difficult problem, and required either the centre of gravity some 10 per cent further forward than with conventional biplanes or the tail volume must be considerably greater. Directional control behaviour of the latest aircraft was optimistically assessed as good. 'Previously when one engine of a twin-engined machine was cut out, nearly all the available rudder was required to keep a straight course; with many modern low-wing twin-engined monoplanes practically no rudder movement was required' – but that was generalizing too far and failed to take into account the instinctive side-slipping which pilots employed. There was even a light knock at Handley Page in the statement 'there should be little difficulty in designing a wing with flaps which will give satisfactory lateral control without wingtip slots', but despite three decades of experience, ailerons still presented problems.

Much full-scale experiment had been conducted with flaps of various types, and the deterioration of lateral control at low speeds was less unfavourable than expected. Continuous flaps across the bottom of the fuselage caused tail-heaviness on deflection, and it was again recommended that the tailplane should be located as high as possible to avoid the wake. The Report stated there was little to choose between plain and split flaps for lift and profile drag, but as flap size increased the Zap gave the highest lift coefficient but, like the Schrenk, had larger profile drag than plain or slotted flaps. The latter could offer high maximum lift with relatively low drag for take-off, and then with increased angle gave as high a drag as split flaps for landing and seemed the best solution.

The long, largely mistaken, concentration on spinning at last had ceased. The panel devoted to its study had been dissolved, for the Air Ministry had seen the light and decided that prolonged spinning should cease as a regular Service manoeuvre. There was also better understanding that concentrated masses in the fuselage had an unfavourable effect on spinning recovery, but when along the wings had a favourable effect. Slots, while greatly decreasing liability to accidental spinning, might increase the severity once the spin was established. Experiments showed that multi-engined monoplanes were unlikely to have a dangerous spin, but multi-engined biplanes were liable however good the design of fuselage and tail. Autogiros were also investigated, and the report considered two accidents, in one of which longitudinal control was lost in a dive due to rotor blade twisting, resulting in a dive into the sea with the pilot unable to escape by parachute because the hanging control column pinned him in his seat; and in the other a seaplane Autogiro suddenly executed a half loop, righted, but remained very tail-heavy, and dropped into the sea in a steep spiral – the wreckage disclosing that during flight some of the blade covering had parted from the ribs because of inadequate gluing.

Marine aircraft were still believed to be a valuable line of development, and the ARC endorsed that no new problem would arise in designing and constructing flying-boats of 100 tons. Considerable research comparing stub wings and wingtip floats surprisingly showed that the former did not adversely affect take-off, but their weight was 3 per cent greater and lateral stability on the water was less.

Aero engine improvement had received considerable attention, particularly fuels, sparking plugs, superchargers, and cooling systems. The ARC opined there were advantages in the DC-3 bi-fuel system by which 100-octane was used for take-off with reversion to 87-octane for cruising. There was little to choose between carburettors and petrol injectors except icing prevention, though this could be achieved by heating. Inter-coolers for air under compression were being studied, and the introduction of ethylene glycol cooling liquids instead of water had reached the production stage, but the new regenerative cooling system reduced the need for liquids with still higher boiling points. Work on single-stage superchargers indicated that if developed to higher pressure ratios, then two-stroke compression-ignition engines, for which Ricardo & Co had developed a sleeve valve requiring no sealing rings, might give as much power as petrol engines of equal size and weight. In connection with material research, the ARC drew attention to the possibilities of X-rays for the detection of flaws in metals.

Variable-pitch propellers were also extensively investigated. The two-pitch propellers currently in production for the new line of fighters and bombers were already outdated though would remain in use for some years, but de Havillands had secured a Hamilton licence for the later oil-operated variable-speed design which had a hydraulic piston within the hub controlled by valves governed by 'constant-speed' units mounted on and driven by the engine. Roy Fedden had long urged Bristols to manufacture just such a propeller, and proposed acquisition of the Hele-Shaw which Glosters had dropped through lack of an official contract. Fedden found backing from Sidgreaves and Hives at Rolls-Royce, and this led the Bristol directors to form a joint Bristol-Rolls enterprise, for which Bill Stammers, general manager of the Bristol aero division, proposed the name 'Rotol'. With the two major aero engine firms thus involved, the Air Ministry almost eagerly promised support, and the company was duly registered on 13th May with nominal capital of £250,000 in £1 shares 'to deal in aircraft, parachutes, gliding machines, and all parts thereof, in particular airscrews and propellers.' Temporary premises were obtained in Gloucester while a special factory was being built opposite Staverton aerodrome where experimental flying would be conducted. 'Although the main production of Rotol Airscrews Ltd will be of the hydraulically operated type,' ran the prospectus, 'the electrically operated airscrew will also be made and is a development of the well known Curtiss-Wright airscrew of the USA particularly suitable for heavy multi-engined machines as it can be "feathered". Initially all propellers will have magnesium blades, but the company will conduct developments with blades of wood and other materials.' Bill Stammers was appointed general manager, and in charge of workshops was Bob Coverly, a stolid smooth Yorkshireman of devious machinations and great engineering skill; tall thirty-three-year-old Leonard Fairhurst, latterly of Rolls-Royce and formerly of Bristols, was appointed chief designer.

Aircraft and component factories seemed to be springing up everywhere. In a burst of flattery C. G. Grey wrote: 'One of the most remarkable achievements in the history of British aviation has been equipping the new Fairey Works at Stockport and getting it to a high state of production in little more than twelve months.' Major Tom Barlow had so extensively multi-jigged the big floor area that Battles were already streaming from the production line and went by road to Ringway where they were assembled in the company's erection shop and test-flown by Flt Lieut Duncan Menzies, the amiable chief production pilot. Although Battles were to be built by the Austin shadow factory – where the motor-car experts considered that they alone knew how mass production should be organized – complications were so involved that while there might be Battles in name, no aircraft would emerge for some time.

Rootes Securities of Speke were more successful, for they used Bristol-built production jigs and there was effective liaison between the two companies, just as with Faireys between their Stockport and Hayes factories. A. V. Roe & Co Ltd at Chadderton were building 250 Blenheims, but under Roy Dobson's fiery and experienced guidance had no problems. Even more important was Spitfire production at Supermarine in charge of H. B. Pratt, under whom

Barnes Wallis had gained early airship experience, but the real fire-cracker was Trevor Westbrook who, as Victor Gaunt recollected 'was a lean, dark, wiry type, and the real driving force of Vickers-Supermarine. He rushed around to keep foremen, inspectors, AID, and everbody in sight hard at their jobs to get each machine finished for tests by his personal forecast date and thus inspired everyone to tremendous effort.' Westbrook immediately set up a production office to design and construct jigs and tools in great number at the Woolston Works, expand the labour force as quickly as possible, and organize an extensive system of sub-contracting whereby Supermarine initially built only fuselages but took responsibility for eventual assembly and testing at nearby Eastleigh where Jeffrey Quill was chief production pilot. Fuselage frames were built by J. Samuel White of Cowes; Folland Aircraft Ltd manufactured the tail end; Westland Aircraft built wing ribs; the Pressed Steel Co manufactured wing leading-edge sections; General Aircraft and Pobjoy Air Motors made the wings which were fitted with ailerons and elevators from Aero Engines Ltd; and Singer Motors Ltd constructed the engine mountings. By March, production had begun to flow, and as soon as it became clear that Rolls-Royce would meet demands a further 200 machines were added to the initial 310.

Morris Motors also had been designated Spitfire constructors, and their huge shadow factory at Castle Bromwich, near Birmingham, was being prepared. Alan Clifton told me: 'To our surprise they drew every thing ten times full size, and forecast that some parts would not fit together. It took a long time to work out the unfamiliar aircraft problems, and eventually the authorities got tired of waiting and an offer by Vickers to run the factory was accepted.'

This was an incongruous period between the old and the new. Germany was already forming Jagdgeschwader 132 to which the first production Messerschmitt Bf109B fighters were being delivered with a machine-gun firing through the propeller hub, and two on top of the engine synchronized to fire through the disc. But in Britain the fighter just going into service was the Gladiator biplane, of which the first three production machines emerged in January and were delivered in the last half of February to No. 72 Squadron at Tangmere as an improvement on their Hawker Furies. Delivery of Hurricanes was still some months distant owing to delayed engines. Nine squadrons were therefore to be equipped with Gladiators as an interim measure. In April, No. 3 Squadron at Kenley next began to replace their Bulldogs with them, and a Mk II version with three-bladed metal propeller was already under test and speed had increased to 253 mph. Glosters were working on a further 180 to be delivered by the end of the year.

The Fleet Air Arm seemed even further behind the times. While Battles were emerging from Stockport, the main factory at Hayes was filled with Swordfish production. So suited was this machine to the Naval outlook that the original contract for 201 was increased to 692, and Blackburns were being considered for further orders eventually amounting to 1,700. That spoke volumes for operational requirements conceived in 1930 and handling characteristics which were far from perfect yet which made pilots believe the Swordfish was easy and safe because of its general steadiness, ease of recovery from a dive, and sense of solidity given by its heavy ailerons and rudder.

Faireys were also busy with a batch of Sea Fox biplanes constructed at Hamble, and on 23rd April the first production machine flew, but it was not so pleasant as the Swordfish, and there was considerable criticism at the MAEE that it touched down at 50 knots instead of the 40 knots specified. Nevertheless water behaviour was remarkably good, even in a choppy sea and 20 mph wind, and alighting with 35 degrees flap was particularly easy; but it took sixty-four seconds for take-off with a wooden propeller, and though it was hoped a two-speed metal propeller would give satisfactory results, it still took thirty-six seconds, and that was considered too long. Meanwhile there were problems with excessive cylinder and oil temperatures of its Rapier engine.

The Navy seemed solidly wedded to biplanes, for Lobelle currently was working on specification S41/36 for a replacement TSR three-seater, and although he offered both monoplane and biplane, the latter was chosen in the form of a 50-ft span single-bay machine with cabin extended to fill the entire centre-section gap. Nevertheless, the P4/34 was now interesting the Admiralty. The second prototype was flown on 19th April, and was now being modified by removing eight inches from each wingtip and converting into a two-seat eight-gun fighter, thus leading to the eventual Fulmar specification O8/38. 'It was outstanding,' Marcel Lobelle told me, slyly adding, 'but of course all my machines were.'

The Hawker P4/34 Henley was less fortunate despite earlier optimism. The second prototype had the new Merlin II, and was tested by Lucas on 26th May – but unlike the Fairey, this easily handled compact semi-mid-winger was relegated to target-towing trials, for which purpose it was delivered to Glosters for suitable modification. Meanwhile the first prototype was fitted with metal-skinned wings similar to those prepared for the Hurricane as a replacement for the original fabric-covered versions.

It was at least encouraging that early representatives of the clean aerodynamic formula were beginning to infiltrate initial RAF squadrons. First had been Blenheims which went to No. 114 Squadron at Wyton in January, replacing its Hind day bombers; then in March the first Battles went to Nos. 52 and 63 Squadrons at Upwood in Huntingdon, again replacing Hinds, and the first Whitleys, which had no dihedral, went to No. 10 Squadron at Dishforth, Yorkshire, where they began to supersede the much loved Heyford biplanes. RAF pilots took conversions in their stride.

As a sign of the times, Hawkers and Glosters, like de Havillands earlier, announced a contributory pension scheme of 1*s.* weekly administered by the Eagle Star Insurance Co, offering a death or total disability benefit of £100 and minimum pension of £26 a year; and there was a rising scale of contributions to 15*s.* per week for those with over £1,000 per year, affording a death or disability pension of £1,000 or an annual pension of £15 for each year of future service – and at twenty years accumulation that was enough to live on in those days. At the AGM of Short Bros later in the year, Oswald Short announced that to augment the £10,00 staff pension fund, he would personally give £30,000 to found a Trust which increased pensions in cases of hardship.

Imperial Airways were more concerned with increasing the company's capital to £5 million rather than pension schemes, as shareholders found at an Extraordinary General Meeting on 18th June with Sir George Beharrel presiding in the absence of Sir Eric Geddes who had been unwell for several months. The big increase, the chairman explained, was necessary because all first-class mail would in future be carried by air to and from Empire territories, requiring nine return services a week to Egypt, five to India, three to East Africa, three to Singapore, and two to South Africa and Australia under an agreement which would run for fifteen years. All would be operated by *Canopus* Class flying-boats except two per week between London and India for which AW Atalantas would be used. The Singapore-Sydney section would be operated by the Australian associate Qantas Empire Airways by agreement with the Commonwealth Government. Those to India would be jointly run with an Indian subsidiary, and auxiliary services in Africa would be by two associated companies, Rhodesia & Nyassaland Airways, and Wilson Airways. Subsidies and Post Office payments would total £1,650,000 per year for the first three years, progressively reduced to £1,350,000 for the ultimate three years, adjusted for fluctuation in cost of fuel and oil or changes in the basis of the scheme.

'This,' said the chairman, 'brings a new era in world transport. The carriage of all letters from the United Kingdom to Empire territories for a 1½d stamp for a ½oz letter will make an enormous difference to commerce and general intercourse,' and he mentioned that eight years ago the total route mileage had been 6,400 miles, whereas currently it was 28,000. To meet the capital requirement four million shares would be created, for which the Board would invite immediate subscriptions for one million ordinary shares and issue the balance as preference or ordinary shares according to market conditions.

Four days later Sir Eric Campbell Geddes died, aged sixty-two. 'In his own way, he was a great man,' said C. G. Grey, 'and we have so many little men these days that the Empire is poorer by his death.' Geddes had been a man of iron will and strong prejudice, but was no scholar and at seventeen had induced his father to send him 'steerage' to the USA, where he worked as a labourer, logger, and on railroads. Four years later his father paid for his return, and he worked on forestry in India, then railways, rapidly made good and became traffic superintendent. On leave in London, he met Sir George Gibb of the North Eastern Railway, joined the company, and by 1914 was deputy general manager. War brought his great chance, and he was made Deputy Director-General of Munitions from 1915 to 1916, knighted for his services, made Brigadier-General on the staff of the C-in-C British Army in France, and was soon Director-General of Military Railways and Inspector-General of Transportation in all war theatres with rank of Major-General. His progress was fantastic. In 1917 he became Comptroller of the Navy, temporary Vice-Admiral, and First Lord of the Admiralty. Next year as Unionist MP for

Cambridge he joined the War Cabinet. In 1919 he was Minister without Portfolio; then Minister of Transport until 1921, and in the recession of 1922 became chairman of the Economy Committee, resulting in the notorious Geddes Axe, under which expenditure of the Fighting Services and Civil Service was cut to the bone. Thereafter the shadowy Mr Szarvasy, a Hungarian financier with a mandate to put the Dunlop Rubber Co on its feet, brought in Sir Eric as chairman to reorganize the business – and it was also Szarvasy whom Holt Thomas after the war had interested in air transport, leading to the amalgamation of the struggling pioneer airlines into Imperial Airways Ltd and appointment of Geddes as Chairman. As one obituary mentioned: 'He could not be called a leader of men – he was a driver of men. Those who came in close contact developed an intense admiration for his business ability, his capacity to make others work, and his foresight, but they feared rather than loved him – and in the last resort people will do more for a leader than they will under orders.'

Meanwhile the Imperial Airways issue had become a matter of contention. Certain shareholders sought to have the capital authorization rescinded because the new shares were quoted at 30*s.* whereas the old ones were almost double. Sir John Mellor asked the Chancellor whether the Government, as a substantial shareholder, had consented to an offer so much below the market price, but Sir John Simon replied that the Government agreed the company must get its money on the most advantageous terms practicable. In fact the issue was below market price because the latter was unduly inflated, so the hardest hit were those who had recently been speculating by buying at a high price in the expectation of a rise, for the expanding aircraft market now always seemed worth a gamble.

There was therefore a ready response when de Havillands sought to increase their capital from £600,000 to £1,200,000 by creating 600,000 ordinary £1 shares – a steep climb from their initial little capital of £50,000 with which the business was founded in 1920. At an Extraordinary General Meeting to approve the issue it was stated that work in hand was nearly £4 million – enough to keep the aircraft, propeller, and engine divisions fully occupied for a considerable time.

8

Earlier in June there was an international gathering at York Municipal Airport of private pilots from Germany, France, Belgium, and Holland, and there followed two days of gaiety and flying despite dull weather. Among those from Germany was Kurt Tank; but none knew he was technical director of Focke-Wulf Flugzeugbau who were designing a radial-engined single-seat fighter to supplement the Messerschmitt Bf 109. He arrived in a Stieglitz piloted by Friedrich von Braun, who won the York Handicap race next day, watched by a vast crowd. A high spot was an aerobatic display by Henrich Wendel, CFI of a flying school at Jena in Thuringia, the German equivalent of RAF civil operated schools in England.

Among later aerial junketings that summer was a reciprocal meeting at Frankfurt to which British, French, Belgian, Swiss, Austrian, and Polish private pilots were invited to meet their German equivalents, and the ninety-four visiting aircraft of all makes and colours were parked for the night in the vast *Graf Zeppelin* shed alongside the airship emptied of gas and arranged for free inspection by the public who crept through it in thousands daily at a charge of DM 1 per head. 'Thanks largely to the British *Week-end Aériennes* and other international rallies, there is a growing body of aviators who know each other well, and at Frankfurt there was a general atmosphere of friendly reunion rather than the half-formal meeting of strangers determined to be genial,' wrote a visitor.

Some 275 guests attended the official reception and ball at the Frankfurter Hof, welcomed by Herr von Gronau, famous polar air navigator and president of the German Aero Club. For the next two days it was sightseeing, including the Wasserkuppe to watch the summer's gliding competition where they found Philip Wills with his *Hjordis* and a number of other prominent British pilots including Mrs Joan Price whose gliding at Cobham's displays had brought her renown. There was also diminutive Hanna Reitsch, Germany's leading airwoman, now licensed to fly the Focke-Wulf experimental helicopter. They also met 'Papa' Ursinus, whose part in creating the Gotha bombers of the Great War was submerged in his fame as instigator of the German air renascence with gliders. There was no doubt of the cordiality of German friendship. The farewell party at an old hostelry with tables on vine-canopied balconies overlooking the Rhine 'was unforgettable as the four-day-old moon sank behind the mountainsides and the beacon fire blazed in our honour on the island in midstream. The French spokesman in graceful sentiments acknowledged that we had been shown several different Germanies: the factories; the scientific in a marvellous exhibition of technical craftsmanship; the youthful on the crest of the Wasserkuppe; the ancient in the buildings and treasures of Frankfurt; and the hospitable in the inns and guest houses where we had been entertained. The beauties of the country and the friendliness of its people left no other course than declare choice of Germany as a second country of adoption after one's own.'

There were Germans too at the annual RAF Display on 26th June, intent on assessing British air strength. In some ways they were misled, for most of the participating aircraft were biplanes, such as the Gloster Gauntlets and the Hawker Hectors built by Westland making their first and last Display appearance – but there were also five squadrons of Ansons and several squadrons of monoplane bombers. 'The great event of the day was the mass formation. The 260 aeroplanes in five columns gave everybody a greater thrill than anything ever seen at earlier Displays. The throb of all those motors overhead seemed to beat into one's brain until one felt oppressed by the mass of metal in the sky,' said one of the visitors. Others said: 'That shows what to expect in the next war! Fancy raids of that size coming over all day long.'

It was a Royal occasion. The King and Queen were received by Viscount Swinton, Sir Edward Ellington as CAS, and Sir Hugh Dowding, AOC-in-C Fighter Command. With them in the Royal Box were both Royal Dukes and their Duchesses with Prince Arthur of Connaught and Prince Chicibu of Japan. Members of the Air Council, officers of the RAF, and representatives of the air industry were presented to His Majesty, and among those so honoured were Frederick Handley Page, Arthur Sidgreaves, Sydney Camm, Arthur Gouge, and Frank Barnwell.

Unquestionably this was the most successful of the eighteen Displays since the War. Some 200,000 paid for admission. Across blue skies high clouds drifted, making it easy to watch the antics of aeroplanes performing in formation or with individual demonstrations. Certainly it was a great day for the Hawker-Gloster-Armstrong Whitworth-Avro armada forming the bulk of the RAF.

Several hitherto unrevealed types were among the ten in the New Type Park. First to taxi out was the DH 93 Don GP trainer to specification T 6/36 – a wood-structured low-winger with 450 hp Gipsy Twelve, side-by-side dual control, radio trainee accommodation, rotatable dorsal turret, retractable undercarriage of considerable resistance, and chines preceding the tailplane similar to those being fitted to Tiger Moths to improve spinning recovery. Its rival at Hendon was the Miles Master, developed from the earlier private venture Kestrel trainer which had failed to interest the Air Ministry despite a top speed of 295 mph. The new version was a much more practicable tandem two-seater than the Don, and though of wood, was smaller, neater, and more conventional, and its 226 mph was 35 mph faster. After it came the Airspeed Oxford trainer, with two 350 hp Cheetahs – a straightforward development of the Envoy to meet specification T 23/36, but it nearly missed the occasion for it was only ready the previous Saturday for its maiden flight in the hands of Flt Lieut C. H. A. 'Percy' Coleman, the firm's chief pilot.

By virtue of its Swan Hunter backing, Airspeed had become respectable in the eyes of the Air Ministry, and accordingly Tiltman had been given the opportunity of designing a radio-controlled, pilotless I-strutted taperwing biplane of wooden construction. George Errington, who had joined the company as an inspector in 1934 and turned to test flying a year later, made the initial flight on 11th June, and it appeared at the Display as the Queen Wasp powered with a 350 hp Cheetah. The five other machines in the New Type Park were the Fairey P 4/34, Hawker Henley, Gloster F 5/34, and Blackburn Skua, all of which flew, though the latter could not retract its undercarriage, and last was the splendid great DH Albatross, but it merely ignominiously taxied because only three engines could be started.

For old and young alike, it was the display of Great War aeroplanes that enthralled more than these later manifestations of aerial art. The old two-seat pusher Horace Farman, flown by Wg Cdr D. V. Carnegie, attempted to shoot down grotesque balloon monsters; then came a Bristol Fighter, Sopwith Triplane, and SE 5 trinity attacking a kite balloon, with a lone German LVG defending – but it was with the aid of electrically-fired patches that the balloon at last began to blaze. There was also an evocative scarlet-painted Vickers Virginia of 1924 vintage which was attacked by Gladiators and fled while discarding its crew by parachute.

The great battle of the splendidly fabricated seaport of Hendon concluded the Display with the usual impressive set-piece, tremendous bangs and a general flare-up following an attack by two torpedo-bombers, and dramatically a large ship in dock heeled over. Squadrons of Whitleys, Wellesleys, Blenheims, and even Hinds and Vildebeests came streaming over in successive formations and were attacked by Gladiators, Demons and anti-aircraft guns – and with a final last explosion the great day ended. None except RAF Command realized that it signalled the last of the RAF Displays ever held.

The customary SBAC Show followed on 27th and 28th June at Hatfield. A banquet on the first evening to celebrate the Society's coming of age was attended by 450 guests from forty-seven countries, making more than 800 at the tables. Frederick Handley Page, as chairman of the SBAC, welcomed the new Under-Secretary of State for Air, Lieut-Col Muirhead, and also Sir Henry White-Smith the Father of the Society. Hiding his gratification at enormous contracts, Handley Page mournfully said: 'We aircraft constructors are more concerned with the aeroplane as a means of bringing peace to the whole world than as a weapon of destruction. The aeroplane is the best peace-bringing machine that has ever been invented. By it we can wipe down barriers between races and so bring widespread happiness.'

Muirhead received tremendous acclaim when in toasting 'The Society' he said· 'Although we now have machines which can fly by themselves, and gramophone records which will play themselves, we can not escape the need for human initiative, and can not dispense with after dinner speeches. Except for a scrappy breakfast I have fed exclusively this day at the expense of the SBAC. I was met by Mr Handley Page, who seemed to me to be an independent gentleman interested in aircraft as a hobby. As we went round the exhibition I learned that Mr Handley Page held strong views about aircraft construction. I noticed that he lingered long by two big aeroplanes which, curiously enough, bore the same name as his own!'

At the static exhibition that first day, many guests closely inspected the impressively large covered exhibition of components on uniform stands, and examined the forty aeroplanes displayed on the Hatfield turf with civil aircraft intermingling with those of Service type. Next day most of these machines were evoluted by their test pilots while Major Oliver Stewart provided admirable commentary. Even the Empire flying-boat *Calpurnia*, piloted by Lankester Parker, came roaring in to skim low across the aerodrome several times. The one machine visitors wished to see above all others was not there – the Spitfire; but its stable-mate the Venom was present, as well as the Hurricane, and their aerobatics were of high order. Even the Avro Anson in the hands of Geoffrey Tyson was nonchantly looping. As Frank Bradbrooke commented: 'One lesson recurs at these displays – yesterday's "hot stuff" is treated with casual disrespect as the pilots become accustomed to it.'

Seal was triumphantly set on the achievements of the British aircraft industry when Flt Lieut Maurice J. Adam of the RAE established a new world's height record at the end of the month by climbing the slightly modified Bristol 138A monoplane to 53,937 ft, thus defeating the Italians whose Lieut-Col Mario Pezzi six weeks earlier had achieved 51,348 ft with a large span Caproni 161 biplane powered by a Piaggio fourteen-cylinder radial. Adam, a pleasant, tall, curly-haired young man, wore the same helmeted pressure suit as Swain, and in the preceding few weeks had made six flights to 50,000 ft to try several propellers and changes in supercharger.

The first transatlantic flight of an Empire flying-boat followed on 5th July when, in the presence of President De Valera, the unfurnished *Caledonia*, piloted by veteran Capt A. S. Wilcockson with a crew of three, left Foynes at 8 pm, and flying at only 1,000 ft because of a 25 mph westerly, headed for Botwood, Newfoundland, arriving next day at 11 am after a flight of fifteen hours nine minutes having averaged 133 mph. At 3.30 on 8th July *Caledonia* left for the St Lawrence River, alighting off Boucherville, Montreal, five and a half hours later, and next day left at 4 pm for Port Washington, New York, arriving two hours twenty five minutes later.

'Early in the morning of 6th July two aerial merchantmen spoke to each other in mid Atlantic, thereby making a significant piece of history, the more so as one was British, the other American, and both running to a mutually agreed schedule,' reported the *Daily Mail*. The meeting was not unexpected, for the Pan American *Clipper III* Sikorsky S-42B had taken off from Botwood on the same day that *Caledonia* left Ireland, piloted by Capt Gray with crew of four, and flew at 10,000 ft on course for Foynes to take advantage of the tail wind, and alighted on the Shannon at 10.44 on 6th July after flying twelve hours thirty four minutes and an average of 157 mph, though her true air speed was 19 mph less than the cleaner Short. De Valera was there to welcome them, and the ubiquitous William Courtenay was also there, having stayed up all night in the control tower with Major Brackley and Capt Entwistle the meteorologist to receive direct reports of progress. Said Courtenay: 'Brackles, whose hard work for his pilots, and whose inspiring leadership is not so well known as it ought to be, went out in the rain on the water at night to lay a flare path in case Wilky had to come back. After a four-hour break for breakfast we gathered at Foynes at 10.00 hours to await the *Clipper*. In she came at 10.50 hours, alighting beautifully near the jetty, though alarming some of us in case she hit it. De Valera presided at the official lunch for one hundred in an old barn decorated with bunting. Sumptious courses were served with Champagne and cigars which would have done credit to the Ritz, but there was not a single Union Jack about though the Stars and Stripes waved gaily overhead and many Free State Tricolours were flying. Capt Gray proved the very opposite of American movie pilots. He was modest and reticent and his well chosen words ending with a thousand thanks in the Irish language was a nice bit of diplomacy that brought down the house.'

Next day the *Clipper* left at 9 am, and reached the Imperial Airways base at

Hythe shortly after midday, greeted by the US Air Attaché, Colonel Scanlon with the Mayor of Southampton; the Admiral of the Port; Sir Francis Shelmerdine; and nautically clad Hubert Scott-Paine. However Sir George Beharrell, a sixty-three-old Yorkshireman of Hebraic appearance who was now Imperial's chairman, was not there nor Woods Humphery, for they were reserving their publicity guns for the return of *Caledonia*.

But world interest centred on another American trans-oceanic attempt which ended in tragedy. With a fanfare, Mrs Amelia Putnam, better known as Amelia Earhart, set off in her twin-engined Lockheed Electra on a much publicized round-the-world flight with Harry Manning of the US Navy as navigator, but due to a delay in repairing the machine after an accident at Honolulu, his leave expired and Fred Noonan took his place when she set off for Howland Island in mid Pacific on 2nd July. They failed to arrive. Ships were searching. Amateur radio operators reported they heard her call sign. Nothing was found. Though she was world news, a caustic reporter could say: 'After the first aeroplane has flown an ocean just to show it could be done, further flights are mere suicidal gambles. All who have drowned have only themselves to blame. Mrs Putnam's much advertised flight proved nothing until she went into the Pacific – but the spirits of Mrs Putnam and Capt Noonan may be consoled that at least they gave the US Navy's Air Service a search exercise which may prove of high value when the great war in the Pacific starts.'

That one could encircle the world by air with safety, was shown by Mrs Walker-Sinclair in a series of flights by regular airlines in an extensive and expensive air tour starting from Croydon with a flight to Brussels and Cologne in a Sabena machine, then from Frankfurt to Lakehurst, USA, in the ill-fated *Hindenburg*; thence to South America with Pan American; across the Pacific in a Martin Clipper, to China where she spent a month; finally Australia, returning by Imperial Airways across Arabia to Croydon, having covered 50,000 miles in five months. 'For traveller like myself, whose time is of no value, and who is interested in the constantly changing sky and the countries and seas flown over, may I plead with designers of aircraft to let us see out and watch the view. I think we should marvel, while it is still marvellous, that it is possible to fly round the world as ordinary fare paying passengers not only in safety but mostly in considerable comfort.'

9

Germany had just put in service a fourth capital ship, the *Friesenland*, capable of launching a bomber of 37,500 lb gross. For that purpose two Bloehm und Voss four-engined Ha 139 gull-winged monoplane seaplanes had been built. which, because of their great span, were able to take off in nil wind at an all-up weight of 12½ tons, soon extended to 15½ tons. For long-range reconnaissance the Royal Navy had nothing to compare with this, for the

Admiralty was more interested in the Short-Mayo composite and its top four-engined seaplane component *Mercury* which might prove suitable for long-range reconnaissance as its still air range was 3,800 miles.

Germany was also pulling ahead with helicopters, though France had hoped to obtain a lead with the Breguet-Dorand which had oppositely revolving blades on a central spindle to balance out torque. Germany's new Focke-Wulf was very different, and carried two contra-rotating rotors each side on an outrigger structure. Already it held the height record of 8,200 ft, a speed of 76 mph, and duration of one hour twenty minutes fifty seconds. This at least impressed Britain's popular Press and also Jimmy Weir and his team in Scotland. There was a feeling that such a machine might be useful for air-sea rescue.

The use of conventional aircraft was being examined in a three-day Coast Defence Exercise in mid July in which all three Services participated. The last had been a flop due to separate control of the units. But this time there was a combined Staff working from a single operation room. The attacking Blue Force, consisting of four battleships, two aircraft-carriers, three cruisers, twelve destroyers, two escort vessels, and five submarines, supposedly based 600 miles west of England, was to attack the Red Defence which comprised twelve destroyers, five anti-submarine trawlers, and six submarines. The attackers had seven squadrons of FAA on the carriers and one amphibian on each 8-in gun cruiser except *Rodney* and three on each Southampton Class cruiser. The defenders had a flying-boat squadron at St Mary's Scilly, Mountbatten, Portland, and Pembroke Dock; two torpedo squadrons at Exeter; two reconnaissance squadrons at Woodsford, Dorset; one general reconnaissance and one heavy bomber squadron at Boscombe Down, and one day and night fighter squadron at Tangmere.

The object was to test co-ordination of the defences of the fortresses of Portsmouth, Portland, and Plymouth. The Army co-operated with the Navy in defence and the RAF aided with anti-aircraft guns and searchlights. Heavy rain and fog made matters difficult on the opening day when four Saro Londons left Mountbatten and four from the Scillies for a preliminary search to the south-west, but within forty-two minutes three hostile cruisers and five destroyers were located sixty miles west of the Scillies and, soon after, an aircraft-carrier, five cruisers and four destroyers thirty miles south-west of the Bishop's Rock. Out came the light bombers, and the war was on. Most of the ships seemed, in the Navy's opinion, to have got through as they pursued their zig-zag course. Eventually six attacked Portland, and the carrier aircraft attacked Plymouth and Portsmouth, others later adding their bombardment. What the final results were remained a mystery to the public, though a great deal appeared in the Press written by naval officers who were anti-Air Force judging by homilies on when and when not to attack, and that 'it would be advisable if the RAF participants had some knowledge of the more elementary principles of naval warfare.'

On 30th July Neville Chamberlain announced that all shipborne aircraft of the Royal Navy would come under the Admiralty for operations and administration, but shore based aircraft, including flying-boats, employed in naval co-operation would remain under Air Ministry control. That left a number of things uncertain, such as responsibility for training naval pilots, supply, and control of aerodromes used as FAA shore bases. When Attlee asked the Prime Minister to prevent matters continuing as an internecine warfare between Air Ministry and Admiralty he was told that though individuals could not be prevented from firing shots, he hoped they would refrain because it was in the public interest that this controversy should be settled. The *Daily Telegraph* did its best to renew it by interviewing four Admirals. Sir Roger Keyes fulminated: 'The dual control under which the Navy has suffered has hampered development of Naval aviation and had been extremely unfair to officers and men. As to the Prime Minister's statement about shore based aircraft, any arrangement which does not give the Admiralty control of flying-boats and aircraft which the Navy needs for security of our sea communications – which is entirely a Naval responsibility – can only perpetuate the system of dual control. This will be unsatisfactory in peace and is bound to break down in war.'

Ten days later the RAF had it on their own when the annual Air Defence Exercises commenced on 9th August at 18.00 hours with thirty raids on targets near London, during which half the raiding formations were attacked by defending fighters before reaching their objectives. Heyfords, Battles, Blenheims, Demons, Gauntlets, were the mainstay. 'By day and night, almost without exception, the bombers had to confess they were intercepted by fighters. If these Exercises had been real war, some of the targets might have been hit and some of the fighters would have been shot down, but there would not have been one bomber left.' That was the pattern of all three days, whether daylight of night attacks. Since the new location systems were secret the Press alleged that: 'Fighters at night were practically helpless without searchlights to show them the enemy. Nor could they do much at any time without information supplied by the Observer Corps.' Nobody except those intimately concerned knew that three radar stations (Bawdsey, Cahewdon, and Dover) had in fact proved invaluable and led the Air Ministry to put in hand seventeen more from Dry Tree at Land's End to Ottercops Moss north of Newcastle. Location ranges of 100 miles had been proved practicable.

Though the Hawker Demon two-seat fighters had only mild success against the bombers, there was still the traditional belief that the more advanced two-seat fighters to specification F 9/35 had a useful purpose. Concentration on the Hurricane was delaying the Hotspur contender, but on 11th August the Boulton Paul Defiant was given its maiden flight at Wolverhampton by Flt Lieut Cecil Feather, formerly a specialist armament pilot at the A & AEE and now chief pilot of the company. Merlin-powered, with pointed nose, it was smaller and neater than Camm's massive-looking Hotspur and entirely of stressed-skin construction, whereas the Hawker had a fabric-covered rear fuselage and basically was a variant of, but not identical with, the Henley light bomber for which an initial order of 350 was now reduced to 200 as a target-tower and was sub-contracted to Glosters at Hucclecote.

Already the Air Ministry had promised an extensive order for the Defiant, and production was officially sanctioned a few months later. In Parliament it sounded well that another fighter was being manufactured – yet the Defiant was the result of misjudgment based on earlier methods of converging and beam attacks on bombers so that a 4-gun power-operated turret could be brought to bear as in the days of the Bristol Fighter with its Scarff-ring, but unlike that machine there was no fixed forward gun because the turret afforded a forward arc of fire above the propeller. Turrets had become the mainstay of the Boulton Paul business, for only twenty-four Overstrands had been ordered and the last was delivered at the end of 1936. With the advent of Defiant production the company's prospects seemed financially encouraging, and led John North to concentrate increasingly on power controls for aircraft, marine, and automobile purposes. 'It was the development of North's ideas on control which led to his deep interest in control mathematics and cybernetics,' wrote Lord Kings Norton (Harold Roxbee Cox) many years later. 'I remember an article he wrote discussing the operation of controls as extensions of the human anatomy to perform the elaborate functions required in controlling an aeroplane. From this simple start developed his wide ranging knowledge of operational research, ergonomics and statistics.'

Every aircraft firm was breathing a sigh of relief that their path at last seemed profitable. Blackburns announced that profits for the past year were £120,480, with adequate general reserves, enabling a 12 per cent dividend on ordinary shares. Bristols were even better with £295,088, enabling over £100,000 to be added to the general reserve, which coupled with the premium on £1,200,000 shares issued in July brought the reserve to £1,100,000. De Havillands with a profit of £71,709 were £6,000 down on the previous year but maintained their 10 per cent dividend as the company now had doubled captial due to the recent creation of 600,000 £1 shares. Even General Aircraft, chaired by Sir Maurice Bonham-Carter, was able to say that their miniscule profit of £3,239. 6*s.* 5*d.* indicated an improvement in affairs, though it had been a bad set-back when their ST-18 Monospar was lost in the Timor Sea; however, contracts were such that the staff was being doubled and the Works would be fully employed all next year even if no further orders were accepted. They were also inaugurating a new RAF Volunteer Reserve School at Fairoaks, Woking.

At the Handley Page AGM, Worley explained that accounts could not be presented because the Inland Revenue might appeal against a recent court decision that money received from patents was capital and not revenue; but if no liability was established the general reserve would be £140,000, and the directors might consider capitalizing a portion by another bonus issue to ordinary stockholders. Gleefully Handley Page added that agreement had been reached with the Air Ministry on excellent compensation if the Cricklewood and Radlett extensions were redundant at the end of the expansion scheme 'though my policy is to increase output by increased efficiency rather than size, and I must remind shareholders that the technical characteristics of the company's products are not exceeded anywhere in the world!'

Yet it was the threat of war that was building up a prosperity greater than anything since the previous war – with cheap goods, cheap transport, massed produced entertainment, inexpensive houses and easy mortgages. Most men still walked or cycled to their factories, but draughtsmen and clerks, let alone the well-to-do, were beginning to have cars and discover the pleasures of the countryside and shores. Though people ignored the breath of war, the revolution in Spain had expanded to a struggle on an international scale with Italy and Germany as accomplices on one side and Russia on the other.

By this time the *Legion Condor* had extensively organized airfields, anti-aircraft units, supplies and medical units, and was operating some 200 aircraft comprising fighters, dive-bombers, reconnaissance aircraft, and bomber-transports, and now began sending some of the latest fighters such as the Messerschmitt, and also Dornier bombers, to study their effectiveness. By comparison the gentlemanly British air defence exercises were merely scholarly; but here the Germans in detachments of a six-month posting were coldy assessing in a miniature theatre of manoeuvres with death and destruction as the score. The little town of Guernica had been obliterated one spring afternoon by waves of their dive-bombers, and 1,654 Spanish civilians were killed and hundreds gravely injured – a symbol of Nazi frightfulness. A month later 4,000 Basque children evacuées were landed at Southampton. The military situation was near stalemate, but Franco as Generalissimo was fighting on. The *Sunday Chronicle* published an article purportedly by him stating that the majority of Red volunteers enrolled by the USSR in the name of Popular Front were French, Belgians, Czechs, Germans, and anti-Fascist Italians, but though most of the war material was Russian the majority of the aeroplanes were French, and the machine-guns were either French or Czechoslovak. He alleged that the Red Government at Valencia was existing on gold reserves stolen from the Bank of Spain. Such was the attitude of the British Press that recently Franco had broadcast a protest from Seville against its Bolshevik bias. Already he had secured three-fifths of Spain, including most of the mines and the best agricultural land. 'What impresses most as one enters Nationalist Spain today,' wrote Richard Findlay, 'is the sudden translation into harsh and terrible reality of the whole grim story, entirely credible and yet infinitely remote when read at a distance. The Nationalist-Bolshevik war is being fought to-day on the Spanish front. When General Franco has won on that soil, all the signs point to shifting the battle ground to France. Must we in England, blind to the alien forces which sway our destiny, pass through suffering to salvation, through darkness to the light?'

From the Far East came news of intensified fighting by Japan in the undeclared war on China, resuming on still larger scale the battles of 1931. Peking was evacuated. Swiftly the Chinese were driven back to the line of the Yellow River. There was no hope, particularly from the League of Nations who merely formally condemned the attack. Nor did the USA take action, for the Americans had long declared they would not resist by force any action of Japan short of an attack on Hawaii or Honolulu.

In England a few Parliamentary pacifists might mutter that their own country had been militarily engaged in the Near East ever since the Great War. Nevertheless the Foreign Office was still hankering for armament limitation, and on 17th July had signed a Naval Convention between England and Germany and England an Russia limiting aircraft-carriers to 23,000 tons, and their guns to not more than 6-inch calibre. By agreement in 1935, Germany had permission to build up to 35 per cent of the British tonnage of aircraft-carriers – but there was no limit to the size or number of her mercantile fleet of aircraft catapult launchers.

10

By one of Parliament's ingenious quirks of procedure a Debate on National Defence resulted from the consideration of the Civil Estimates with special reference to the salary of Sir Thomas Inskip as Minister for Co-ordination of Defence, cleverly initiated by the Liberal Sir Archibald Sinclair moving a reduction of the vote by £100. C. G. Grey shot his usual barbed shaft: 'This insistence on National Defence may be only another manifestation of the English non-conformist conscience, or the perfidy which our Continental friends regard as our natural characteristic. When we invented the science of boxing, the nobility and gentry who supported it called it "the noble art of self-defence." So if our grandfathers regarded pounding one another to jelly as self-defence, we may come to regard bombing other nations as a form of National Defence considering that the Air Ministry programme seems to consist of colossal orders for bombers and those for single-seat fighters are much smaller. The little affair in Spain is apt to give wrong ideas on air tactics, for distances are short and targets strictly limited to outskirts of towns along main roads which the Communists are defending, and the old part must be missed at all costs because the bulk of the population is Nationalist and not Communist. In a full-out European war most of the bombing will be at night from a great height, and the targets will be the whole of what Mr Hore-Belisha chose to call "built-up areas" in the neighbourhood of munition factories or troop centres.'

It was an entangled Debate raising many pertinent points to which only disguised reply could be given lest Defence techniques be revealed, but Sir Thomas Inskip was able to give authoritative if generalized assurances and explain that the Committee of Imperial Defence comprised only the Prime Minister and himself but called to the deliberations a wealth of experience from over 500 soldiers, sailors, airmen, industrialists, and politicians. He said that nothing, even feeding the children, had been omitted, but no minister could undertake responsibility for answering all the many topics that might be regarded as Defence. A great deal had been done with air raid precautions and the whole country surveyed and divided into zones and when financial responsibility was agreed these plans could take effect. In due course the vote earned a Government majority of 165.

With Parliament's summer recess a more peaceful Britain seemed to reign. There was Cowes, with its armada of glittering sails, including the great J Class yachts owned by Sigrist and Fairey, but Sopwith's beautiful *Endeavour* was still in the USA after inglorious defeat in the America Cup races when it lost all four, and Sopwith as helmsman was unkindly blamed by pundits who did not have to bear the cost of this brave expedition. But for the masses it was County cricket which held their gambling interests. For some there were the club air days, flying meetings, and races, as well as considerable touring to the Continent, including the Netherlands Aero Show which had opened at The Hague on 31st July. Britain was well represented by Rolls-Royce, Bristol, Armstrong Siddeley and de Havilland aero engines, but lacked actual aeroplanes except for a Leopard Moth. Germany with a beautifully arranged mass exhibit occupied half one of the two Halls. Biggest was that of the Netherlands Defence Department, backed with impressive stands by Fokker and Koolhoven, those one-time opposing designers of the Great War but now Holland's leading constructors. Rivalling the Netherlands show internationally was the great air race from Paris to Damascus and back organized by the French Air Ministry; but Amy Mollison was a non-starter, and the solitary British entrant, Flg Off Arthur Clouston of the RAE and his friend Flt Lieut George Nelson flying the Comet which had won the England-Australia race, came fourth, while the well-organized Italian team, undoubtedly helped by the British-built Smith automatic pilots fitted to their powerful machines, won tremendous prizes totalling £23,000.

The Auxiliary Squadrons were having their Summer camp, and so were the public schools Aviation Group with members from Uppingham, Eton, Malvern, Millhill, and Imperial Science College, most of whom intended to make the RAF their career – and the King travelled by air for the first time since his accession, using the Royal Airspeed Envoy piloted by Wg Cdr 'Mouse' Fielden, from Smith's Lawn in Windsor Park to Martlesham Heath, thence by road to the camp he had sponsored for Eastenders at Southwold. Concurrently Lieut-Col A. J. Muirhead, Under-Secretary of State for Air, and Sir John Maffey of Imperial Airways flew by the regular flying-boat service to Mombasa to meet Oswald Pirow, Minister for Defence for the Union of South Africa who arrived the same day from Nairobi in his Airspeed Envoy. An Imperial Conference, attended by Air Marshal Sir H. Brooke-Popham, Governor of Kenya, began next day to review co-ordination of the defence of British Africa. *The Aeroplane* speculated: 'Attack by whom? Not Germany, for Mr Pirow is a German. Hardly Portugal, our oldest ally; not Italy with its new African Empire in Abyssinia; not Russia which can never be a sea-going power; not Japan unless we lose a big war in the Far East – most likely the ever increasing black population which foolish people are educating to ideas beyond their mental capacity and proper station in life.'

At the end of August, Lympne's fifth International Air Rally was held and a great number of Continental private owners came flying in, including the two German aerobatic stars Oberleutnant Fischer and Herr Förster, the latter startling everyone with a dive across the aerodrome ending in an unexpected flick roll into a smooth climb. There were two days of races and demonstrations, and the RAF with three Gauntlets introduced looping and rolling in line-abreast as well as the usual 'vic' and line-astern displays. Some 500 guests attended the great barbecue supper and dance organized by Ann and Bill Davis. 'Festoons of lights led to their mansion known as Berwick Cottage, and the scene was illuminated by glowing pits of logs surmounted by sheep carcases on great iron grills with flames leaping a yard or two in the air. A bar in the orchard and tables and chairs under the lighted trees, with those fires in the background, provided a unique scene for the feasting and dancing. Viscount Willingdon, Viceroy of India, was there as President of the Cinque Ports Club, and Noel Coward, a Vice-President, presented the prizes.'

The quiet infiltration of Germans to England's air meetings expressed only the close bond between pilots whatever their nationality – but in the Fatherland, Göring's intention could have a deeper meaning than implied in a speech to overseas Germans attending the Congress of the Foreign Organization of the National Socialist Party at Stuttgart: 'I take a special pleasure in the pride you all must feel when the new German warships sail the seas and visit your countries and harbours bringing greetings of the Homeland and demonstrating the unity between Germans at home and abroad. In similar manner the time is not far distant when I may send our air squadrons to friendly countries and nations as tokens of greeting from the Homeland to Germans abroad.' Was that a threat, or his usual *bonhomie* in trying to unify his people with the vigorous spirit of the new Germany? 'Your great Führer,' he said 'never despaired in the darkest days of the Nazi Movement and has succeeded in rescuing his nation of 60 million people so that they once more can boldly sing Deutschland über Alles.'

Strictly British was the theme of the King's Cup Race for a circuit of Britain starting from Hatfield on 10th September, since pilots and aircraft alike had to be of British design and construction. Of the thirty-one entrants the new TK 4 from the DH Technical School and the Miles M 13 Hobby had been specially built to beat the Mew Gulls, the sixth and latest of which, flown by Edgar Percival, was even more refined and had slightly smaller wings, while the other, flown by Charles Gardner, was a re-build of the one in which Campbell Black was killed the previous year. Despite frantic all night work on the Hobby it was not ready until the night before the race and difficulty with the retractable undercarriage and lack of a Certificate of Airworthiness prevented it competing. However the miniature TK 4 with only 19 ft 8 in span was even smaller than the slender DH 71 Tiger Moth racing monoplane of ten years earlier and had made its first flight as E-4 in July, but with a wing loading of 23 lb/sq ft was regarded as somewhat dangerous, though as a design exercise its cantilever wing planked like the Comet and Albatross, semi-monocoque fuselage, and balsa wood stabilized flying surfaces, were regarded as outstanding. Except for the Foster-Wikner high-winger with inverted Cirrus Minor replacing its unapproved Ford, in company with the fully metal-skinned CWA Cygnet seen statically at the RAeS Garden Party some weeks before it flew, and the neat 95 hp Pobjoy-powered enclosed Moss low-winger, the rest were well known and overwhelmingly Percival, with Miles productions in second place and a significant absence of de Havilland machines which in earlier years had been the mainstay of Britain's pilot-sportsmen. 'The dullest King's Cup Race,' lamented C. G. Grey. 'Some 20,000 went to Lympne on Saturday and Sunday and saw jolly good air racing and fancy flying. But who on earth is going to be interested in a race around all these towns and seaside resorts? There is no interest in seeing one small aeroplane flash past without anything to explain why and what it is. For 95 per cent of the population, aeroplanes flying over their heads does not amuse them. The only good the aircraft industry can get out of this race is to advertise the performance of any machines which fly particularly fast.'

Nevertheless he found a great number of people 'trekked to Hatfield on Saturday and stood in an icy blast with the sun in their eyes, when there was any, to watch aerobatic displays by the RAF's Gauntlets and Herr Fischer and Herr Förster in their Bücker biplanes, followed by cavorting models which strike me as dangerous uncontrolled projectiles; and lastly the finish of the King's Cup Race which was the cause of all the pother and fuss. Charles Gardner thoroughly deserved to win his second King's Cup; he is the star pupil of AST and is not merely a good pilot but a navigator and engineer and sportsman. General Lewin's win of second prize with the Miles Whitney Straight was most popular; for that hardened sixty-one-year-old campaigner the bumpiest flight on record round Britain was merely a joy ride. Edgar Percival's third place was a curious combination of good and bad luck; he is so valuable to British aviation that I doubt whether his co-directors ought to allow him to fly, for what would happen to Percival Aircraft if he broke his neck?'

That is what happened to one of the most popular men in the RAF, thirty-seven-year-old Wg Cdr Ted Hilton from Martlesham, and his friend and co-pilot Wg Cdr Percy Sherren, and both died. The weather was appalling, with low cloud, drizzle, and tremendous turbulence. The first day's course was along the east side of England to Aberdeen thence diagonally across Scotland and the Irish Sea to Baldonnel near Dublin. At Scarborough it was roughest of all. As a competitor explained: 'Above Scarborough Castle Rock we hit a bump that was like flying into Niagara Falls. I was heaved a foot off the seat, and my pilot's belt creaked under the strain and my co-pilot thudded back to his seat with his forehead laid open. Ballast weights broke through the floor of the locker and fuselage, though we did not know until later.' Hilton, piloting a Falcon loaned by Miles, similarly met the bump at speed, and despite his harness hit the cabin top, fracturing his neck, and Sherren, a heavy man, was catapulted through the roof and out, and the machine crashed into the sea. Others were luckier, or more judicious in flying higher, but even the experienced Wally Hope arrived home bandaged and bloody having hit the roof of his BA Eagle near Montrose hard enough to stun him for a few seconds, and both business man Sqn Ldr A. V. Harvey of the AAF in his Whitney Straight and Wg Cdr Stent of Phillips & Powis in his Sparrowhawk had been near to catastrophe.

Return next day was initially northward via Belfast to Portpatrick on Scotland's Rhum of Galloway, thence Carlisle and southward via Leicester to Cardiff, and east again to Hatfield. Several competitors had dropped out, and by the time the field reached Cardiff only thirteen were left. So to Hatfield and Gardner winning by two minutes, and the next three within a second of each other but Lewin just in the lead, and Percival and Harvey so close that many thought the latter had third place, but the judges on the winning line ruled he was fourth. Last was Charles Hughesdon described as 'something in the City and honorary instructor to the Insurance Flying Club at Hanworth.'

When visiting Martlesham a few weeks later, the tragedy of smiling Hilton, renowned for his tatty cap and careless uniform, was still evident. 'Because it was private flying there is no pension for his dependant,' explained his oldest friend, 'nor was he insured. His beautiful widow is at her wits' end for money.'

While I was there, Harold Piper, the New Zealander assistant to Lankester Parker, passed overhead with a seaplane version of the Short Scion Senior mini-airliner on its delivery flight to West of Scotland Air Services at Greenock. Oswald Short's gamble in building six had not proved very profitable, though three were bought by the Irrawaddy Flotilla Co at Rangoon, and the Air Ministry was interested in purchasing the company's blue-painted commuter with which Piper nearly experienced disaster in the King's Cup Race. Parker and he had been busy with the production run of Empire flying-boats, as well as *Maia* which made its first flight on 27th July, and since then had been fitted with the central pylon structure to which *Mercury*, the pick-a-back machine, could be attached. Parker had flown *Mercury* on 5th September, and considered it sufficiently satisfactory for a demonstration to the Press three days later. Since then Piper had concentrated on the *Mercury* trials, and Parker with the *Maia* to determine their respective stabilities for Major Mayo on whose calculations the success of the composite aircraft depended.

Both Mayo as technical adviser to Imperial Airways and Major Brackley as air superintendent, had been co-operating closely with Arthur Gouge, who recently had been made a Freeman of the City of Rochester. There was a great banquet given in his honour by Oswald Short, and among the civic dignitaries was George Gouge JP, father of the recipient and pillar of the local Labour Party. Toasting the health of the new Freeman, the Mayor paid tribute to the energy and enterprise shown in rising from bench worker to general manager of so famous a firm, but seemed more intent on flattering Oswald Short to whom he proposed a second toast. Replying, that tough man said: 'I felt I could not let this important occasion pass without celebrating it with a dinner which would give an opportunity not only of congratulating Mr Gouge publicly, but to thank the Mayor and Councillors for having conferred that honour upon a member of my firm, though it could not have been on a more deserving man. He had been with my firm a good many years before I came in close contact with him. I quickly realized his abilities at a time that he, in his modesty, did not realize himself. He needed opportunity, encouragement, and backing for his talents to be exercised to the full, and it was in my power, as head of the firm, to see that he got it because he was the man I was looking for. But there are also other members of my firm here to-night, all very talented in their respective spheres in the art of designing and constructing aircraft. Unfortunately Mr Lankester Parker, skilled in the courageous art of taking new monsters into the air to see if they are what the designers profess them to be, is on holiday. But we have Mr Kemp and Mr Bibby, my co-directors, whose names I link as men of great achievement as production managers. We have Mr Jackson and Mr Lipscomb, wizards of technical knowledge and mathematics as applied to aircraft design. When I question how I have shaped this world of ours I say with humility that I have at least gathered around me many brilliant men. That lifts me from the depression into which I occasionally descend. Thirty years have passed since my two late brothers and I founded this business which is now a major industry of this city. Being young we plunged into it through scientific interest and hoped it would pay its way and even bring fame and fortune. It had done both – but it was a hard struggle. That we survived is due to those who stood by the firm and worked for salaries much below that which their merits entitled them. I hope I inspired these brilliant men, and particularly in all-metal construction of the form now rapidly superseding all others throughout the world. I claim that revolution in design began at Rochester nearly twenty years ago and so is entirely British in origin.'

Encouragement of young designers was the tenet of most aircraft firms. De Havillands had their outstandingly successful Technical School. The SBAC was offering annual Scholarships. Currently Arthur Sidegreaves of Rolls-Royce at the opening of their Technical College extension at Derby announced that an annual Travelling Research Fellowship of £450 would be established in memory of Sir Henry Royce, open to University graduates or equivalent qualification, so that a course of advanced study could be pursued at home or abroad, but recipients on completion must spend at least two years with Rolls-Royce at a salary not less than £300 per year.

II

Military interchanges between Germany and Britain continued. A party of British Staff officers, including General Sir Cyril Deverell as Chief of the Imperial General Staff, and Sir Arthur Longmore as Commandant of the Imperial Defence College, flew in a Bristol Bombay to see the German autumn manoeuvres accompanied by Grp Capt F. P. Don, the British Air Attaché. Within a few weeks the British Government reciprocally invited General Milch, Maj-Gen Udet and Major Stumpf, and they arrived at Croydon on 17th October for visits to several key aircraft and engine factories and RAF establishments at Odiham, Mildenhall, Cranwell, Halton, and Hornchurch where they watched displays by Nos. 54 and 65 Gladiator Squadrons as the most modern operational British fighter. But the Gladiator was obviously obsolete, for the first production Merlin II had been delivered to Hawkers at

Brooklands in August, and on 13th October the first production Hurricane was pushed out for Philip Lucas to test. A re-designed hood, full equipment and engine-driven hydraulic pump for flap and undercarriage systems had been installed but no instrument vacuum pump, nor were the two-speed metal propellers available so the initial batch had wooden airscrews. Between April and October there had been so many snags with the original Merlin of the prototype that only twenty-three hours testing had been accomplished in twenty-nine flights, but this included spinning, which became stable and flat after three turns. Though the elevators remained effective the rudder was not, and 2,000 ft was lost in recovery, but model tests conducted in the RAE vertical wind-tunnel led to an improvement by adding a small ventral fin and leaving the tailwheel extended, and it was advised that instead of the usual stick forward and opposite rudder for recovery it was better to keep the elevator up before applying rudder.

In November, Lucas had a narrow escape when his production Hurricane emerged through cloud almost into a wood although the altimeter indicated safe height. Only much later was it discovered that these Kolsman instruments read incorrectly due to diminished cockpit pressure, and were then cured by connecting to the pitôt static. Later still the wooden propeller blades failed when pulling out from a TV dive. As he laconically told me: 'All we did was land in a field with undercarriage down, fit a new propeller and fly home!'

Though production of the very much more elaborate Spitfire was in full swing there were problems of construction and co-ordination and still no sign of the first machine. In fact Supermarines were to find that it took over 300,000 man hours, or three men for a working lifetime, to produce each airframe, and 100,000 man hours for the engine. Already it was clear that if war came the Hurricanes must hold the fort, and No. 111 Squadron at Northolt was awaiting them.

Just as Supermarine had lost their great innovator by the death of Mitchell, so did Ernst Heikel Flugzeugweke GmbH through the death in a motor smash on 1st October of Walter Günter, their chief engineer, who had been responsible for the He 70, He 111 and He 112 – three of the most efficient aeroplanes so far built.

That same day the latest Imperial Airways Empire flying-boat *Courtier* was wrecked when attempting to alight on the shimmering glassy surface of mist-obscured Phaleron Bay, Athens. Four passengers lost their lives, four were slightly injured, but four others and the crew of four were unhurt. The crash coincided with a new regulation making lap straps for passenger seats compulsory in British aeroplanes, and at least two fire extinguishers must be carried in every passenger aeroplane with more than ten seats. But *Courtier* had no straps.

There was yet another fatality that day, for keen Bob Waight, the de Havilland chief test pilot who in five and a half years had accumulated 2,150 hours, crashed while steeply turning during a demonstration to *Flight* of the little TK 4, caught out by a stalling speed as great as the Spitfire.

De Havilland's elder son, twenty-seven-year-old Geoffrey Raoul de Havilland, already experienced with the Albatross and the DH 93 Don, holder of 'A' and 'B' pilot's licences and 'A' and 'C' ground engineer's licences, became chief test pilot.

By this time Germany was testing a slick four-engined 25–30 seat equivalent of the Albatross – the Focke-Wulf Condor, which with 720 hp BMW radials could cruise at 215 mph despite nacelles which seemed of much greater drag than those for the Albatross's H engines. Thurston James, promoted to technical editor of *The Aeroplane*, reported that: 'Construction is a notable feat, because Herr Tank, its designer, and his collaborators, have hitherto built nothing bigger than the twin-motored Weihe, yet they successfully jump to a four-motored transport', and he gave top speed at 260 mph with a range of 1,180 miles. Whether a transport or not, it could have equivalent sea patrol use to the four-engined Sunderland, the prototype of which was ready for water trials on 14th October, and on the 16th Parker and Piper made two flights totalling forty-five minutes, but later the wings were to be swept $4^1/_2$ degrees further back and the step repositioned to cope with a more rearward C. G. centre of gravity due to changes in armament.

Long overdue, the latest British landplane, the Bristol 148 Army Co-operation low-winger to the same specification as the Lysander, was flown by Cyril Uwins on 15th October. Unlike recent low-wingers, its wings were untapered with heavily rounded tips adding to the impression of a sturdy, compact machine, and when I briefly flew it during Service trials I felt it was nicer to handle than the Lysander. Because of the clean design and retractable undercarriage it was 25 mph faster, though the 'Lizzie' scored with shorter take-off and landing run and gave a sense of safety in semi-stalled descents using considerable engine power.

One more tragedy marred October. Flt Lieut Alfred Montague 'Dasher' Blake, AFC, chief test pilot of Blackburns whom he had joined in 1927, now aged forty-eight, was found dead behind the closed doors of his garage where his car engine was running. No reason was revealed – but he was the oldest of the test pilots, though he did not look it. Stolid Flt Lieut H. Bailey, his assistant, took his place, Bennett-Baggs giving an occasional hand with production test flying.

Demonstrating the resilience of women came Jean Batten flying in from Australia, and landed to her predicted schedule at Croydon on 24th October, having beaten Harry Broadbent's solo time from Darwin, which she had left just before midnight on 18th October, by fourteen hours ten minutes. 'The usual battery of Pressmen and photographers were awaiting, and the public turn-out suggested the palmy days of record-breaking,' wrote a reporter. 'She taxied to the apron, climbed out, quietly cheerful as usual. Obediently she stood for the photographers and sound picture merchants, climbing on to the wing of her Gull, waving her bouquet of yellow chrysanthemums. She did not look as though she had been through nearly six days of very hard work with only a few hours sleep – but then a woman can do a lot of repair in very few minutes, even if she has to clear Customs at the same time!' Her elapsed time was five days eighteen hours fifteen minutes which was astonishingly close to the record of five days six and three quarter hours held by Waller and Cathcart-Jones with their racing Comet.

Except for thick weather she might have beaten them. Throughout the week, fog had delayed airline departures from Croydon, one machine taxi-ing around for half an hour to find the white fog line for take-off, and another lost for an hour during which a tractor sent to find it also got lost. So crude were ground communications that pilots had to be instructed by messenger, though in the air 'the new scheme of taking bearings on every message which the machine transmits is working very well, and control officers find it most useful for keeping track of the bearings of aeroplanes.' Flying was still a game of chance, whether for record-breakers or airlines.

Imperial Airways might well look at Le Bourget with cupidity, for in its new guise with an aerodrome enlarged to 1.15 square miles and an imposing main building 700 ft by 100 ft with a terrace on top for 4,000 people, it was magnificent compared with Croydon. Pierre Cot, the Air Minister, officially opening it on 12th November, with President Le Brun and members of the Government in attendance, said that it was the largest aerodrome in Europe and perhaps in the world though he knew that there were vaster projects such as Templehofer Feld which were well advanced, but France in her turn would do something even better.

Imperial Airways became the subject of attack in the Commons on 28th October by that pertinacious critic Robert Perkins, Conservative Member for Stroud and Vice-President of the British Airline Pilots Association (BALPA) on whose behalf he was speaking. 'I want to draw attention to the grave dissatisfaction among pilots employed by Imperial Airways and other airlines operating from this country. The Association has specific grievances. We have been refused the right to collective bargaining. There have been curious dismissals. Captain Wilson, pilot of the *City of Khartoum* which crashed two years ago in the Mediterranean, was dismissed and no reason given. I feel it was because he opened his mouth too wide when the Inquiry took place. Then when the Association wrote to Imperial Airways suggesting that the Budapest service should be suspended for the winter as it was not properly equipped, the two men who signed the letter, Captain Rogers who had been with the company since its inception in 1920 and Captain Lane-Burslem, our Chairman, were dismissed. That is victimization. Imperial Airways have done everything possible to break the Association, and refuse to recognize us or discuss with the pilots at all. Our fourth grievance is the question of wages; while directors' fees have been increased, they have cut pilots' wages. Then again, the majority of the pilots of the Association regard the equipment provided by Imperial Airways as not good enough. Machines with one or two exceptions are not equipped with apparatus to effect blind landings and few are equipped with any form of de-icing, and very few, if any, with a spare wireless set. I am told by experts that it is impossible to run a 100 per cent service unless your machine is equipped with those three things, and it is unfair that pilots attempt such a service and take risks when they have not got these elementary things. Finally, the machines on the London-Paris route are obsolete and certain others on routes in Europe are definitely unsuitable for winter service. We now appeal to the Government to call a conference at which this organization can meet the directors of the company. We have no alternative but to ask for an impartial inquiry into the whole position of the pilots engaged by Imperial Airways, and in fact into the whole organization.'

Mr G. Le M. Mander, Liberal Member for Wolverhampton, backed Mr Perkin, saying: 'In reply to a question yesterday the Under-Secretary refused to give any information on the ground that it was within the domestic capacity of Imperial Airways. But a matter of public policy is involved, a matter of principle, the issue of collective bargaining. It had been fought out many times in the industries of this country, and it has been won every time. Every employer, whether public or private, who stands against the principle of collective bargaining will be beaten in the long run. Neither the Government nor this House can escape responsibility, and we must press this issue until it is accepted by Imperial Airways.' In reply, the new Under-Secretary of State for Air referred to the Hambling Report as the basis of a relationship between Imperial Airways and the Government, whereby the Government should not exercise direct control over the activities of the company other than appointment of directors, and added that Imperial Airways, as a commercial undertaking had a right to discharge employees in what might be called the ordinary business procedure, but if there was any breach of a fair wages clause a procedure was open for investigation of complaints. He then said that after consulting the Government directors, he was satisfied that Imperial Airways were not opposed to collective negotiation and there had been no victimization either of membership of unions generally or any particular union. 'However the Hon Member asked for an impartial inquiry. I am unable to accede to that request, but if Members wish to bring forward specific cases and specific evidence they are of course at liberty to do so and I will give consideration to those submitted to me.'

Next day George Woods Humphrey, as managing director of Imperial Airways, replied to Mr Perkin's allegation by telling the press they were made under cover of parliamentary privilege and were damaging to British prestige and the company's interests at a critical time. He pointed out that the Airline Pilots Association had only been in existence four months and was not recognized by any air company, whereas the Guild of Air Pilots and Navigators had been operating many years and was regarded as the representative body. The company was not in the least averse to collective bargaining, and had done so with the Wireless Operators Association. As to remuneration, pilots on the Empire services received a minimum of £975 to £1,525 according to length of service, plus keep while out of England, and on European services graded from a miminum of £500 to £1,500 according to experience, licences, length of service, and type of aircraft flown. As to equipment, all aircraft had a Certificate of Airworthiness and were operated in accordance with the Air Navigation Direction under supervision of the AID. Aircraft on the London–Paris route were not obsolete but had been in use some years and were as airworthy as when built, and easier to fly and more comfortable than later and faster aircraft. He insisted there was no lack of foresight for in 1933 the company submitted a memorandum to the Government requesting a decision on policy after the expiry of existing agreements in 1937–9. 'Considerable time elapsed while the Government considered alternatives because of the need to obtain decisions from self-governing parts of the Empire. The Government's policy was announced in December 1934, but that September the Board had already given an ITP for 'Hannibal' replacements. Their delivery unfortunately is already a year overdue, and there is no immediate prospect of getting them because of the rearmament programme.' With a fine wave of the flag he added that criticism of lack of vision did no justice to the Board's steadfast purpose in concentrating on the Empire airmail scheme and converting it from a project to reality. 'It is unquestionably the greatest step forward in postal development since the introduction of the 1d post, and will give immense benefit to the Empire and promote its consolidation and unity. To carry out this programme, Imperial Airways needs the support of the Press and people of this country and of the Empire.'

These views were of course reiterated by Sir George Beharrell at the thirteenth AGM of Imperial Airways on 10th November, where profit was revealed as £164,734, resulting in a dividend of 7 per cent, and a bonus of 2 per cent for earlier shareholders than subscribers to the million shares issued in June. Before explaining the acccounts Sir George paid tribute to the late Sir Eric Geddes, chairman since the company's inauguration, and to Sir Samuel Instone, a founder member who had died the previous day. He reminded shareholders that at the AGM two years earlier Sir Eric announced the order for fast landplanes for the European services, but none had been delivered and the delay was a serious matter for the company. However there had been much progress in studying various lines of development and techniques for transatlantic work. Among these were larger long-range flying-boats for normal take-off, refuelling in the air in conjunction with Sir Alan Cobham's company, and means of assisted take-off to increase range, such as the Short Mayo Composite. Sir George regretted that the new Empire marine airport at Portsmouth was still meeting political and other obstacles, though the Portsmouth Municipal Council was promoting a Bill for authority to begin work. Meanwhile the inconvenience of the temporary base on Southampton Water had been reduced to a minimum by helpful co-operation from Vickers-Supermarine, the Southampton Harbour Board and the Southern Railway. Difficulties concerning new offices there, and the intended terminus in Buckingham Palace Road, London, had been overcome though the necessary Act of Parliament had caused delay. The London building was part of the long term plan for development of terminal facilities by Imperial Airways, the Southern Railway, and certain public authorities, and would have a platform with direct access for Empire air traffic to electric trains running to Langstone Harbour. Similar trains would run between Victoria and a large site near Eynsford, Kent, which was capable of development into an excellent airport, whereas Croydon had inherent drawbacks and transport to and from London could not be improved except at great cost. He concluded by reiterating the comments made by Woods Humphery, glossing over the salient criticism, and stressing that the Board had to be satisfied that it dealt with the real representatives of employees. But to those within the industry it was clear that the fight was against BALPA as a trade union, whereas GAPAN was a City Guild of solemn occasion and overdignified approach when it came to pilots' employment.

Authorized capital of Imperial Airways was now £5 million – yet Hawker Siddeley currently became even bigger, with capital increased to £6 million by the creation of two million redeemable cumulative preference shares of £1. Revenue for the previous year had been £783,437, resulting in an outstanding 32½ per cent dividend on ordinary shares together with a cash bonus of 10 per cent. No wonder that armament expansion was proving attractive to investors!

Despite earlier objection's by Lieut-Col Muirhead, an Air Inquiry Committee was authorized in the Commons on 17th November to investigate the charges of inefficiency made by Robert Perkins earlier in the month. Lord Cadman was appointed chairman, though in no way connected with aviation except as a petroleum supplier, and the members were Sir Warren Fisher, Permanent Secretary of the Treasury; Sir William Barrowclough Brown, Permanent Secretary to the Board of Trade; W. W. Burkett, an assistant Driector at the Air Ministry. Woods Humphery was furious; most of his pilots were delighted, and so were the independent air transport companies. Coincidentally, Capt Wilcockson was awarded the Johnston Trophy by the GAPAN in recognition of his Atlantic flights. There followed the resignation of Air Vice-Marshal Sir Tom Webb-Bowen who had been staff manager of Imperial Airways for the past four years: George Woods Humphery, the true architect of the British Empire's air communication system, quietly departed to the USA, and Sir Francis Shelmerdine left for Madeira.

As always Frederick Handley Page seemed able to cash in on such occasions. He invited the entire Parliamentary Air Committee to visit the greatly expanded Cricklewood Works in order to study his impressive production system for the Harrow. The *Daily Express* reported: 'Although all the Committee had been invited to see how the RAF's giant night bombers are produced, only twelve of the eighty members yesterday met big, bald F. Handley Page, Britain's "big bomber" pioneer. When Robert Perkins, whose fierce criticism of Handley Page's old type Imperial Airways airplanes led to the promise of an inquiry committee, flew in the bomber, he chose a position in the narrow domed gun turret at the tail and was bumped and shaken as the bomber taxied slowly through the mud.' So Handley Page had his revenge.

12

The four-year-old Comet *Grosvenor House*, scrapped as K 5084, re-built by Essex Aero Ltd as *The Orphan* of the Marseilles–Damascus–Paris race, and now named *The Burberry*, was in the news again. In the hands of Arthur Clouston, accompanied by Mrs Kirby Green, it disappeared through thickening mist above Croydon on the 14th November and re-appeared just after 3 pm on 20th November, having flown to Cape Town and back inside a week. On both the outward journey of forty-five hours two minutes and return of fifty-seven hours twenty-three minutes they had beaten Amy Mollison's record seventy-eight hours twenty-eight minutes of May 1936. Even C. G. Grey gave praise: 'The only thing that approaches it as a feat of individual skill is Lindbergh's finding Paris in the dark after flying from New York – but Clouston had to find his mark half-a-dozen times, not once. We have known him for a long time as a first-class pilot. The quaint things he used to do with his Aeronca, complete with dachshund, made him a name. The way he flew the little Hafner Giroplane marked him as a first-class handler of any aircraft and as a genuine aerial comedian. But now he has established himself as the finest British pilot of today.' A few might disagree but acknowledged his sheer grit.

Records were still attracting the Air Ministry. A speed version of the Spitfire with special short-life Merlin was being developed for this purpose. More within grasp was the world's distance record of 6,305 miles held by Russia, which some people contended was faked by using an intermediate refuelling point. What was required was an 8,000-mile ability, and there seemed a reasonable chance of this with a modified Vickers Wellesley powered with a high-compression Pegasus XXII, and fitted with a fuel jettisoning system. The fifth production machine, in Egypt on tropical trials, was returned to Farnborough as an engine test bed after strengthening the wing to a standard with which all Wellesleys in future would comply. The instigator was Roy Fedden, intent as always on establishing the superiority of his engines. By using a smaller supercharger suited to an operational height of 10,000 ft he expected to achieve very low cruising consumption yet have adequate power for take-off at an extreme overload of 7,900 lb above the normal gross weight of 11,100 lb. An RAF long-distance group was to be immediately formed under the command of ex-record holder Wg Cdr O. R. Gayford, and five special Wellesleys were allocated.

The Air Ministry also was concerned with the increasing number of accidents. An enlargement of the Accident Investigation Department was announced on 30th November, and Wg Cdr Vernon S. Brown, was appointed chief inspector in place of Major Percy Cooper, creator of the department in 1924, who now retired. Two inspectors, three assistant inspectors, and two investigating officers comprised Brown's staff. No job could be more thankless, and their conclusions could even be subject to libel action unless expressed as a witness at an inquest. Consequently much evidence and deduction was never made public. Certainly no criticism of lack of appropriate experience could be levelled at Vernon Brown, for he had learnt to fly in 1915, had been a test pilot, a CFI instructor, had taken the RAF's Henlow engineering course, and latterly was Major Cooper's deputy – yet, above all, it was his power of unbiased assessment which made him outstandingly suitable for this often tragic job.

Though not directly affecting aviation, a notable retirement at the beginning of December was that of the Farman brothers, Henri, Maurice, and Dick, all of British descent but born in France. Of the famous sixty-four-year-old Henri, the French historian, Charles Dollfus, said: 'He flew with calm and logic, and therefore had great influence on the beginning of aviation, for he inspired confidence in those who emulated him.' Ever since his first attempt at flying with a Voisin *Type Delagrange* Henri had continued to pilot the simpler of his firm's aircraft. When introduced to him at the Paris Aero Show in 1949 I found him to be a tall quiet man, clean-shaven and therefore unrecognizable from early photographs showing a beard. Trained as an artist, not as an engineer, his early Boxkites were flagrant derivations of the Voisin, using the same scantlings, but with more effective control as the result of his piloting experience. Later designs were intuitive and empirical, but he soon began to employ a structural engineer, as did Maurice who initially had no connection with Henri's work. Dick the oldest, became the business man, but he learnt to fly at fifty and continued until an accident ten years later. Dollfus said of the three: 'Their influence has been great because of the duration and strength of their firm rather than by technical novelties – equally by the great number of types constructed and their variety from the smallest to the largest French aeroplane; also by the strength and reliability of their machines, for there has hardly ever been a fatal accident to a pilot of the House of Farman except through the pilot's fault.' In future the name of Farman would be merged into that of the Société Nationale du Centre (SNCC).

There was no anonymity about the British shadow factory scheme. Progress had been tremendous. In August of the previous year the Rootes factory had been a field where now a splendid brick-built office fronted an extensive steel-structured building fully operational with automatics, capstans, milling machines, and the first batch of Mercury blower units was ready. Equally modern was the great Daimler building commenced that September and now in full swing with the manufacture of crankcases, oil sumps, air intakes, and similar items and the first set was currently completed on 25th September. Standard's factory, with a floor area of 230,000 sq ft was a show place with all the equipment a particular shade of blue and the men wearing white overalls; but it was also unique as the only factory with flow production on a conveyor system, and complete cylinder units were being made. Rovers had started to build their two-acre factory a year ago, and were in full operation with connecting rod assemblies and cam gear. At Speke Rootes were beginning to get their airframe factory in operation for the construction of Blenheims in an erecting shop a quarter of a mile long, 400 feet wide, and 40 feet high. Biggest and most imposing of all was the Austin factory, for its aircraft assembly shop was breathtakingly huge and there was a big separate engine component section for crankshaft and reduction gears and assembling and testing half the engines to be built by the shadow group – but as yet there was no sign of aircraft, though very elaborate jigs were being constructed for Battle wings and fuselages. To help speed matters Sqn Ldr Tom Harry England had retired from Aero Engines Ltd and joined the Austin Motor Co 'for duties not yet specified in connection with the Austin airframe factory at Longbridge.' All these great factories looked like being splendid white elephants if no war came, but the lavish expenditure, the fantastic extent of tooling and jigging, and the entire lack of financial risk were objects of envy to those resolute pioneers who had re-built the aircraft industry and lifted it to eminence despite the negative attitude of successive governments and the inadequacy of contracts which had prevented sufficient extension of their factories to meet the present needs. Without their strong sense of vocation there would have been no aircraft and engines for the shadow factories to build.

To encourage aircraft workers by extending a sense of Royal patronage, the King visisted three factories early in December, beginning with Vickers at Brooklands, where he was received by Sir Charles Craven, Sir Robert McLean, and Trevor Westbrook who showed him around the shops, and despite the impediment of his stutter the King spoke to a number of the staff and work people. From there he was driven round the famous race track to Hawker's assembly shops to be received by George Bulman and shown the Henley and Hurricane before continuing to the main factory at Kingston-on-Thames where Frank Spriggs deputized for Tom Sopwith who was abroad, and presented his colleagues F. I. Bennett, Sydney Camm, H. Chandler and R. W. Sutton before proceeding to the new experimental designs and basic Hawker production. Next day the King went to Handley Page Ltd at Cricklewood where he was met by S. R. Worley as chairman, with Handley Page beaming over his shoulder, and they conducted him through the shops where the last of the Harrows were being completed. A cup of tea in the canteen followed, during which the workmen dropped their jobs and lined up to see their monarch leave. 'Those in charge of his safety seemed momentarily anxious, but the King greeted the crowd with a friendly gesture and walked through them smiling.'

A few days later George Bulman gave a refreshing slant on aerodynamic safety when criticizing a strongly Farnborough-orientated lecture to the RAeS by Sqn Ldr H. P. Fraser who pressed the case for irreversible ailerons and a variable air brake: 'An inventor outside Farnborough who suggested such a thing as that horrible rat trap the irreversible control mechanism would be thrown into the deepest dungeons of DTD,' said George with his sad smile. 'The overshoot approach technique is old fashioned, and I suspect this obsession with forced landings is a Farnborough habit. The undershoot method is now in common use, and attempts to land large aeroplanes without motor to the point of touchdown should be forbidden. Take-off is the serious problem, and I think runways will have to be introduced. As for tricycle undercarts, I am all in favour.' This gave an opening to Hollis Williams who had been flying a Monospar fitted with tricycle chassis. 'It might be a throwback, but I regard it as the biggest thing which has happened for years,' he enthusiastically said. 'No other aid has so reduced the need for a high degree of skill. It brings fog landings within view, so the old type of undercarriage is already in the shadow of obsolescence.' When I flew this machine at Hanworth, accompanied by Hollis, there was no doubt that it removed all difficulty of landing and take-off but had the penalty of reduced payload because it was heavier than using a tailwheel. However Gordon England was hopeful that General Aircraft Ltd might secure an Air Ministry order for a new version of the twin-engined ST-18 Croydon with longer nose and retractable tricycle undercarriage. Currently he was negotiating with the British Aircraft Manufacturing Co Ltd for their factory, plant and assets, as the commercial failure of the Eagle and Double Eagle had led to a financial crash.

It was the crash of the homeward-bound Imperial Airways flying-boat *Cygnus* during take-off from Brindisi outer harbour on 4th December that became the first task of the new Accidents Investigation Department. Of the six crew and eight passengers, one of each was killed and the others injured including Air Marshal Sir John Salmond. The weather had been rough, with strong wind and high waves, and the boat touched twice after lifting clear, then hit nose down from twenty feet, stove in the fore end, and the inrush of water put her out of control.

So far the Empire flying-boats, with their higher wing loading and greater manoeuvring inertia, had been discouraging, for though by the end of the year twenty-two had been delivered, three were wrecked and lost. In the official Inquiry on *Cygnus*, Imperial Airways were censured for inadequate escape hatches, and all flying-boats thereafter had the roof emergency exits enlarged and push-out windows fitted. The crew was exonerated, but in attempting to take off quickly the pilot had fully lowered the flaps, and subsequent experiments by Lankester Parker revealed that though take-off was safe with rough water under these conditions, uncontrollable porpoising could develop on a calm surface – but tests with wave-impinged long rolling seas, such as those at Brindisi, were not attempted.

Of British-based Empire flying-boats, much of the maintenance was carried out by Imperial Airways in the main erecting shop of British Marine Aircraft at Hamble. As Henry Folland was making a success as managing director the Board now decided to change the company's name to Folland Aircraft Ltd, having written down the capital from £500,000 to £300,000. With the demise of BAC, that still hopeful pioneer George Handasyde had become director in charge of production.

More definite signals of success were flying from the New Luton factory of Percival Aircraft Ltd, for Edgar Percival had tested his new six-seat; twin-engined Q6 low-winger and found it extremely promising. 'He has built a flourishing business from very small beginnings,' enthusiastically reported C. G. Grey. 'The whole business has sprung from the sale of civil machines and has never had to depend on military orders. The factory is hard at work turning out Vega Gulls; we counted more than twenty in various stages of asembly. The first batch of five of the new Q monoplane is now going through, and another batch of ten will follow at once. It is an obvious Percival product to the last lick of paint, and is distinctly Vega-ish. There is something about the set of the wings which betrays its typical Percival design. A retractable undercart can be fitted if required, but the first machine has Gipsy Six motors and a fixed, trousered undercarriage. In this form the top speed is 195 mph and cruises at 175.'

Major V. W. Eyre, formerly a co-director of Tom Sopwith, was a director, W. A. Summers the works manager, and Arthur A. Bage the chief draughtsman. Recollecting his days with Percival Aircraft, Bage, who as a youngster had worked with Frederick Koolhoven, then with John North of Boulton & Paul before going to Glosters for six years and Handley Page for five, wrote: 'Joining Percival in 1934 was a gamble but a chance to get in on the ground floor, and it worked out right. The factory was at Gravesend and the design office in Grosvenor Place, London. When I joined, the design office consisted of Rowland Bound in charge of a junior draughtsman and a lady tracer; the general staff was a typist and an accountant. Bound left about six months after I joined and I took over his job. It was then that I realized he had been the technical brain behind EWD who was a brilliant pilot. From then on I had to do all the performance and stressing calculations as well as being responsible for design work and production drawings. I redesigned the Mew Gull for the 1935 race, and we went on improving it each year until it exceeded 250 mph, still with the same 200 hp Gipsy engine. I engaged two more draughtsmen and a stressman, and at the end of 1935 obtained Air Ministry design approval.' Nevertheless Edgar Percival states that he himself designed all the firm's aircraft – and in the sense of initiation, instruction, and authority he undoubtedly did, leaving his technicians to put ideas into effect and take responsibility for structural and aerodynamic integrity, and with his management ability, adequate financing, skilled flying, and air of *bonhomie*, had built up a sound and profitable business.

So had Richard Fairey. Presiding at the ninth AGM on 20th December he reported a profit of £248,178 enabling a dividend of 12½ per cent to be paid. 'This has been a year of continuous output,' he said. 'Preparations are being made for a new type to succeed present production. Relations with the Austin Motor Co Ltd which controls the shadow factory making the Fairey Battle are excellent. Our new research department, including a large and up-to-date wind-tunnel, and new chemical and physical laboratories, will be in operation next summer.' He was determined to avoid the pitfall of becoming too big and losing that dominant personal touch which had brought success, and in contrast to earlier opposition he now welcomed such Government-sponsored activities as the Austin factory at Birmingham from which he would reap profit without risk.

As Norman Macmillan said of him: 'He had a grim determination to make progress which gradually hardened his expression as success came and expanding burdens fell more heavily on him.' He could be intimidating and biting as he stared at his victim with pale-blue eyes, but he could also speak in moving terms, and many kind deeds went unrecorded. Of his work people he said: 'When a body of men and women a hundred strong have spent their lives in the same service in close company with each other working to the same ends, suffering the same triumphs and disappointments, it forms a bond between us hardly equalled by any other human experience.' Yet when his life-long business friend, Maurice Wright, 'obliged young John Fairey and his sister with an on-the-spot advance of about twenty-five pounds in francs when the kids were without money in Paris, Fairey sent for Maurice, accusing him of corrupting his son! He sacked young Peter Wright from the design office and refused to speak to Maurice for thirteen months, and Maurice was heartbroken. He even sacked his secretary overnight in Bermuda, with no compensation, no reward for service, no pension.' But Laurence Pritchard, who knew Fairey well, taking the broad picture pronounced him: 'A man greater than his achievements, great though they were. A sportsman, fisherman, helmsman, country squire, he was also a great president of the Royal Aeronautical Society, and in his heyday served it, guarded it, gave it his own great heart, and left it strengthened beyond measure.'

CHAPTER IV

1938: *The Cards are Stacked*

'If our friends think we can do them no good, or our enemies believe we can do them no harm, our condition is far from being prospered.'

Charles II (1663)

I

Low, overcast, stormy skies of early January brought a spate of airline accidents in Europe and the USA, but in Britain civil flying escaped major disasters, yet week by week the toll of Service fatalities mounted, provoking more questions in Parliament. Instrument flying experience in the RAF, despite use of Sigrist simulators, was limited, and particularly hazardous for new entrants. Thus in December a Blenheim, flown by a recently trained pilot, Sgt Sydney H. P. Smith, ran into a snow-storm above Leicestershire while on cross-country flight from Wyton, Huntingdon, to Ireland, and on attempting to fly blind he lost control, opened his roof and unfastened his belt ready to jump. By that time the machine was inverted and he fell out, but was able to open his parachute. His wireless operator Cpl Barnes, and crew LAC Dearn, were thrown in a dazed heap, and with amazing presence of mind Barnes scrambled into the pilot's seat, and though without pilot training, managed to get the nose down, dive through the overcast, and crash-land in a field without hurt. He was awarded the Air Force Medal for exceptional courage and devotion to duty. For young Smith there was a court martial, with the charge that: 'He was flying in cloud conditions higher than entitled to; and abandoned the aircraft without giving warning to the crew.' Squadron pilots knew that in no circumstances were Blenheims to be taken above clouds, but the Pilot Officer acting as meteorologist on this occasion was satisfied that an attempt should be made despite unfavourable weather, and if conditions worsened the machines were to return. The prosecutor agreed that 'until certain modifications have been made the Blenheim is not considered suitable for casual cloud flying.' Dr S. J. Peters MP, defending, said the case raised matters of considerable importance to the RAF. Smith's total instrument flying was eleven hours dual and two hours thirty-five minutes solo, so it was irresponsible to send a man of immature experience in an aeroplane which he had never flown in such weather conditions. There was no evidence that Smith got on his seat and jumped from the aeroplane, so he should not be accused of cowardice but of an error of judgment. The Judge Advocate General agreed that conditions were too adverse for any but the most experienced pilots, and promulgated a cautionary reduction to the ranks as a warning against overkeenness in getting young pilots to accumulate flying hours too quickly in machines which were difficult and complex after their simple Tiger Moth trainers.

Every interval of good weather was being used for RAF training, or testing prototype and production aircraft. On the frosty morning of 1st January Lankester Parker took the fully mounted Short-Mayo Composite for initial taxi-ing trials on the Medway, with Harold Piper sitting above at the locked controls of *Mercury*. Crouching in the blast beneath the tripod securing the seaplane was Bill Hambrook observing the release gear. First tests, powered only by the flying-boat's engines, showed that water manoeuvrability and behaviour were satisfactory, but gales intervened, during which the Composite rode safely at moorings, and not until the 19th was a high-speed run made when the Composite with combined powers became briefly airborne. Next day a full flight was made, but *Mercury*'s controls were kept locked as no separation was intended, though this would follow in February.

An alternative method of increasing the effective range of the Empire flying-boats was Flt Lieut Dick Atcherley's method of mid-air refuelling conducted with *Cambria* and the redundant twin-engined Armstrong Whitworth AW 23 bomber-transport loaned by the Air Ministry as a tanker. The patents were owned by a company named Interair Ltd financed by Leo d'Erlanger. The pipeline was on a reel beneath the pilot's cabin of the AW 23, and to connect the two machines a weighted line from the tanker grappled a line trailing from *Cambria*'s stern, which the latter's crew winched in, drawing the AW 23 line and its attached feed hose which was then connected to a pipeline leading to the *Cambria*'s tanks. Couplers were automatic and had quick-release devices. On completing refuelling the two machines merely pulled away and the connection was automatically broken, the tanker reeling in the pipeline.

Almost as important as refuelling was the ability to jettison the petrol load in an emergency – but this had its perils. The most experienced of America's airline pilots, Capt Edwin C. Musick, with a crew of six, were lost because of these hazards. On 11th January his four-engined Sikorsky S-42 *Samoan Clipper*, on the second scheduled trip of the trans-Pacific Honolulu-New Zealand service, developed an oil leak in one engine, and Musick radioed he was turning back. Two hours later he reported that petrol was being dumped preparatory to alighting at Pago Pago. Then silence. The burned out wreckage was found next day fourteen miles off shore by the American minesweeper *Avocet*. The stream of vaporizing petrol released in mid-air had caught fire and destroyed the machine. Few had greater experience of flying-boats than Musick, who joined Pan American on its formation in 1927, and that year was the first to fly the sea route to Havana, Cuba, and then captained the first flying-boats on the Caribbean and South American routes, building up his company's technique of ocean flying. In August 1935 he made the first survey flight from San Francisco to Honolulu, Wake, and Midway Islands in the then new S-42. Two months later he took delivery of the first of the big Martin flying-boats, and opened the first of the regular mail services to Hawaii and Manila. Everywhere his quiet personality and methodical manner inspired the utmost confidence. Nothing could be more ironic than the disaster which ended his life.

A fortnight later the new look in British airliners, the Armstrong Whitworth Ensign prototype, took the air. It had been ordered in September 1934 and a further eleven in May 1935, two more being added in January 1937. The intended delivery for the first two had been May 1936, with completion at one a month by March 1937, so the prototype was already twenty months late, but Armstrongs claimed this was largely due to many modifications required by Imperial Airways and not because it was built at Hamble instead of Coventry. The engines had first been run before Christmas, but there was difficulty with the fuel system, and not until Sunday, 23rd January, were modifications and final inspection completed. Next day, with that great aerial artist Charles Turner-Hughes at the controls accompanied by stocky Eric Greenwood to add strength if necessary, the impressively big, glittering machine taxied ponderously from the tarmac. Four ground runs were made at increasing speed, but the limited airfield length was insufficient for the customary brief airborne straight hop, so with engines fully opened a committed take-off was made, the machine climbing away across the Spithead waters towards the Isle of Wight. The rudder proved so heavy that it required the combined operation of both pilots, so speed was kept low and the undercarriage left extended during the brief fifteen-minute flight. Two days later the prototype was cautiously flown to Baginton for full contractor's trials.

The gales towards the end of January certainly showed that the Ensign, with its heavy wing loading, could be ground manoeuvred under its own power with safety and facility, whereas the long serving, lightly loaded Handley Page HP 42 biplanes after touching down had to await numerous ground crew to prevent them being blown over – yet pilots loved these machines because their low stalling speed made them the only large airliners which could be safely forced landed. However the British Airways new Lockheed Electras could also ignore the wind when taxi-ing, and their high cruising speed enabled them to keep close to schedules against big head winds. The Press misleadingly proclaimed their down-wind flights, such as the 215 miles from Croydon to Le Bourget in fifty minutes, as 'records'; but an HP 42 battling to Paris might take four hours, and the DH Dragons of local services also suffered big delays, for no schedule could be kept against winds often above 60 mph.

When the gales persisted into February, much publicity was made of a flight by one of the Hawker Hurricanes newly equipping No. 111 Squadron which was flown by the CO, Sqn Ldr J. W. Gillan, from Turnhouse near Edinburgh to Northolt in a tail wind, estimated at between 50 and 80 mph, which enabled the 327 miles to be covered in forty eight minutes, giving an average of

408.75 mph. Urged by the Air Ministry publicity section, newspapers hailed this as another great record proving the superiority of British machines. Few knew that a specially tuned Messerschmitt with greatly uprated 1,650 hp Daimler-Benz had achieved a world's record for a landplane of 379.39 mph.

Meanwhile the Aerodrome Owners Association opened their annual conference on 19th January, chaired by Sir Francis Shelmerdine at the FBI Assembly Hall, London. At the official banquet that night Lieut-Col Muirhead, toasting the Association, said that four years ago there were sixteen Municipal aerodromes, eight further sites had been bought and seven reserved; now there were thirty-six available, ten under construction and eight more in view, as well as about fifty privately owned. He warned that although the Maybury Committee had recommended the supply of essential services, the Government was faced with unprecedented demands, so civil requirements must give way to those of the RAF who themselves were seeking to acquire some of the established aerodromes.

First among the lectures next day, A. H. Wilson, Assistant DCA (Home), dealt with implementation of the Maybury recommendations, and stressed that 'the safety of civil radio-equipped aircraft engaged in cross-country flight, particularly regular commercial services which must operate in all conditions of weather, depends on adequate facilities for wireless communication between aircraft and ground organization. Because of progress during 1936 and 1937 there are now sixteen area stations and five short-range stations, in addition to Borough Hill Broadcast Station, all except Jersey and Sollas provided by the Air Ministry. With the exception of the extreme south-west, the whole country now has a network giving radio navigational assistance to aircraft operating on any route. There remains the problem of radio approach facilities at aerodromes. While radio navigational facilities can guide aircraft to the vicinity of any aerodrome, the final approach and landing in bad visibility can only be made with the aid of radio apparatus on the aerodrome. The rate at which the requisite radio and control organization can be developed depends on such matters as recruiting and training a large personnel and securing augmented accommodation for the expanded organization. It is not possible to say how soon all these facilities can be provided, but difficulties will not prejudice the programme of radio and control developments.'

Despite what C. G. Grey termed 'municipal melancholia,' more than a hundred conference delegates went by coach to Short Bros factory at Rochester on 21st January, and spent the morning looking round the Works and inspecting the Short-Mayo Composite which had made its first flight the previous day. On the production line were modified versions of the Empire boats with tankage for 1,500 miles, and engined with the more powerful sleeve-valve Perseus XIC, permitting 4,500 lb greater gross weight than the 40,500 lb of earlier machines. Ernest Jones, the CTO at Felixstowe, estimated that if powered with Hercules engines of 1,400 hp each, take-off could be achieved in sixty seconds with overload maximum of 54,000 lb – so Arthur Gouge was heading towards this target with an enlarged G-class design for which the Air

Ministry would subsidize the cost of three provided they were made available to the RAF if war came.

The corresponding team of Henry Knowler at Saunders-Roe were completing their R 2/33 four-engined A 33 fitted with a 90 ft parasol Monospar wing strut-braced from each outboard engine to a wing-like water sponson and the two inboard engines daringly mounted on the unsupported central portion of the wing to counter its airborne bending moment, leaving the outboard tapered tips cantilevered. They were equally busy with the R 1/36 – a smaller twin-engined, Hercules-powered, shoulder-wing flying-boat with a deep hull reminiscent of Short design, and construction was far advanced. Major Rennie at Blackburns similarly was designing to this specification, but proposed using twin 1,720 hp R-R Vultures and was endeavouring to eliminate the need for a deep hull by making the planing portion extendable on struts to obtain the requisite propeller clearance, stabilized by wingtip floats on long built-up struts which hydraulically folded upward to house at the wingtip in similar manner to the American Consolidated PBY flying-boat.

Saunders-Roe was involved in changes. Sqn Ldr C. J. W. Darwin resigned through disagreement with the Hon Clive Pearson, second son of Lord Cowdray, whose Whitehall Securities Corporation virtually owned the company. Pearson had ambitious ideas, but, as Lord Balfour of Inchrye later recorded, speculation with Spartan, Highland, Scottish, Hillman, and ultimately British Airways was: 'Great fun, but it must have cost the Pearson organization several hundred thousands.'

Sir Alliott Verdon-Roe, president and now sixty-one, played little active part these days, but was as inventive as ever at his delightful Regency house at Hamble, occasionally commuting by motor-boat to Cowes. Reorganization did not greatly affect the aircraft department, for Harry Broadsmith remained a director and general manager, but a new executive director was appointed to deal with financial and commercial matters. The last of the Mk II Londons was about to emerge and the shops were busily building Vickers-Supermarine Walrus amphibians. The famous boatbuilding business was transferred to a new company, Saunders Shipyard Ltd, the shares of which were owned by Saunders-Roe Ltd. Similarly the extensive plywood section became Saro Laminated Wood Products Ltd, which was previously their marketing company directed by Mrs Scroggs who became joint managing director of the plywood Works.

2

Charles Bruce-Gardner, who received a knighthood in the New Year's Honours, currently resigned as managing director of Securities Management Trust Ltd, and director of the Bankers' Industrial Development Co Ltd and Credit for Industry Ltd, to become chief executive of the Society of British Aircraft Constructors. His sponsor, Frederick Handley Page, impressed by

Bruce-Gardner's recent reorganization of Armstrong Whitworth Securities, explained on behalf of the SBAC: 'His previous efforts have been to resuscitate decadent industries. With the aircraft industry it is different, for it is living and full of virility. His duty is to co-ordinate all those live and competing forces so that the industry shall pull together to the best advantage.' The Press of course excitedly claimed that Sir Charles was there to clear up the industry's muddles as their own overlord.

No longer could the SBAC afford amateur status under the pressures of the expansion scheme. The new title of President replaced that of Chairman, currently held by Handley Page who now gained status as the first President under the new constitution. Not only was fifty-year-old Sir Charles an outstanding financier but also a trained engineer, and had been a director of companies engaged in iron, steel, coal, and similar industries, so SBAC members had every confidence in the new regime.

No aviation industrialist received high honour this year, but Air Marshal William G. S. Mitchell, Air Member for Personnel, gained a KBE, and there were the usual lesser honours and decorations. Some crucial appointments also were announced. That ex-RFC fighter pilot and early chief pilot of Handley Page Transport Ltd, forty-five-year-old Air Vice-Marshal W. Sholto-Douglas, received the newly created appointment of Assistant Chief of Air Staff. New Deputy Directorates of Operations were formed, with Grp Capt D. F. Stevenson as D-DO (Home) and Wg Cdr William A. Coryton D-DO (Overseas).

Britain was in the throes of consolidating operational strategies and tactics; Germany was gaining invaluable experience in the Spanish internecine war; France was reorganizing under a decree signed by the President on 21st January which extended powers of co-ordination and control of the Minister of National Defence and War over the Ministries of Marine and Air, and created the new posts of Chief of General Staff on National Defence and Secretary-General of National Defence, respectively allocated to General Gamelin and General Jocomet.

Former ex-Chief of General Staff, General Debeney, forecast in the Paris *Excelsior* that in a European conflict there would be insufficient disparity of material or quality of armies to make the chance of an overwhelming attack anything more than a gamble. He stressed the immense importance of dispersing industrial centres and people to the country rather than pack factories into thickly populated areas inevitably subject to bombing, and warned that it was elementary prudence to make preparations for a long war. Perhaps this led Sir Thomas Inskip contrarily to declare: 'When I read that Herr Hitler said at his New Year gathering that he hoped for a year of peace and looked forward to working for the peace and happiness of his German people, I think that was something to be accepted at its face value. I am prepared to accept it and build upon it a happier relation between Germany and this country.'

In Britain, the King was diligent in boosting morale with personal visits to factory and Forces alike. Towards the end of January, wearing the uniform of

Marshal of the RAF, he flew from Bircham Newton in his Envoy, piloted by Wg Cdr Fielden, to inspect the RAF College at Cranwell where in 1918 he had been a junior officer. Accompanied by Air Chief Marshal Sir Cyril Newall, CAS, Air Marshal Sir Charles Burnett, C-in-C Training Command, and Air Vice-Marshal Baldwin, AOC Cranwell, he inspected a guard of honour of flight cadets before touring the workshops, barracks, Church, flight shed and the College building which had been opened in 1934 by his brother when Prince of Wales. Cranwell was now a very big establishment, and in addition to the large staff of educational and RAF officers had 3,200 airmen, apprentices and boy entrants, and 144 flight cadets.

Not only was the RAF expanding in step with increased aircraft and aero engine production, but civil aviation was trying to follow the same course. To strengthen the Air Registration Board Dr Harold Roxbee Cox was appointed Chief Technical Officer, effective as from March. This smoothly brilliant, black-haired thirty-four-year-old ex-student of Imperial College of Science and Technology gained his first aircraft experience with the Austin Motor Company in 1918, and then worked under the late Lieut-Col Richmond on the R 101, transferring in 1929 to the RAE on structural investigations, but returned to Cardington as CTO in 1931. At most RAeS meetings he was a prominent participator, and was the youngest on the Council, but was regarded somewhat cautiously by older members as an oncoming young man who might presently attain high position. Responsibilities of the ARB were now extensive. Certification of civil aircraft, particularly for airline operations to meet ICAN regulations, was complex and dealt both with structural and operational requirements in more involved procedure than for military purposes.

The tremendous output of military aircraft necessarily required an even more complex organization, primarily dependent on the staunch AID now 2,200 strong, with resident inspectors at 175 locations and 1,600 approved firms supervised by visiting AID staff, aided by 6,000 approved firms' inspectors. As Air Chief Marshal Sir Cyril Newall said at the AID annual dinner on 18th February: 'The position has been made more difficult because all equipment development has to be done at a time of revolutionary change in aerodynamic design. The resultant complexity had thrown great strain on the AID.'

Handley Page was not wide of the mark when he said: 'There is no large body of men in the industry trained as engineers. Many, including members of the AID, must be regarded as amateurs. We want a lot of really good trained engineers. That is the subject of the SBAC scheme of scholarships.' Nevertheless, the industry had some bright youngsters as Handley Page well knew, for only a fortnight earlier at the annual RAeS Conversazione he had presented the year's prizes, causing a visitor to remark: 'It was quite an evening for the HP organization because 40 per cent of the prizes went to members of that firm!' In fact Lachmann received the Wakefield Gold Medal awarded annually for any invention tending towards safety in flying and two student members on his staff shared the Pilcher Memorial Prize; but other winners had no connection. For a paper on stability problems an outstanding young

Farnborough scientist, Alfred Pugsley, shared the Busk Memorial Prize with that strong personality Major B. C. Carter who specialized on airscrew blade vibration, and the quietly diffident twenty-three-year-old Arnold Hall, who had joined the RAE from Cambridge, made an auspicious start with the Arthur Yate Memorial Prize for a paper on turbulence and skin friction.

On 13th February a man died who was neither scientist, designer, nor administrator, yet had done more than most in his quiet determined way to make the general public air-minded through his own livelihood. He was forty-five-year-old Flt Lieut Percival Phillips, DFC, MM, who in attempting to fly from a small field at Gamlingay in Cambridgeshire, where he had visited friends, failed to clear a tree, crashed, and the machine burst into flames. A Cornishman, he owned a garage at St Austell, and after serving in the RFC and RAF formed the Cornwall Aviation Co and soon became a familiar lone-hand barnstorming figure with his scarlet Avro 504K which operated from the nearest suitable field of every town in England and carried thousand upon thousand passengers. Such was his reputation that both Cobham and Scott successively recruited him for their Aviation Day displays, and Phillips also owned Air Publicity Ltd as a banner-towing business. When a student I had flown with him from a little field at Sherborne, leaving abiding memory of evening sunlight on the golden buildings of that quiet Dorset town, and the song of the old biplane's bracing wires as we came sighing in to land. As one of his admirers said: 'He was a celebrity outside aviation, and within he was so old a friend of so many that he seemed an institution. We always looked on him as the one pilot sure of dying peacefully in bed. He was one of the kindest and most considerate of men, and all who worked with him loved him. He did a vast amount of good, and did nobody any harm.'

By contrast, Anthony Eden, as Foreign Secretary, was regarded as potentially harmful, for he disagreed strongly with the Prime Minister's foreign policy. Chamberlain was all for appeasement with Italy, and threatened to resign unless this was accepted by the Cabinet. Instead, Eden resigned, and so did his Under-Secretary, Lord Cranborne. Lord Halifax, a supporter of Chamberlain, was appointed in Eden's place.

Italy in 1934 had been instrumental in blocking German annexation of Austria but was now so preoccupied with Spain and Abyssinia that the Austrian-born Hitler, assured by his greatly enhanced military power, found it opportune to break the Treaty of Versailles forbidding union between Germany and Austria. Assuming Supreme Command of the Armed Forces, he despatched his troops on 12th March into that country, occupied Vienna, and took control of the Brenner Pass. The Luftwaffe played a prominent part with troop-carriers landing hundreds of infantrymen at Aspern Aerodrome, Vienna, 'just in case local Nazis needed support before the armoured cars fought their way through the battle of flowers,' reported *The Times*. Leaflets were dropped affirming Germany's peaceful intentions.

Instead of protest there was open support. The famous J. L. Garvin expressed in the *Observer* the prevailing sentiment: 'Germany after the Thirty

Years' War was broken into three hundred pieces. The process of reconstruction since then through three centuries has been long, slow, painful – an epic marked by repeated convulsions of arms. By the Versailles Treaty nearly twenty years ago the Germans of the Reich, of Austria, and of Bohemia, who form one continuous racial body in Central Europe, were sorted into three divisions more completely separated than ever in a thousand years. It was against nature, history, geography and economics. It could not last. When the Allies dethroned Vienna as the metropolis of the Danube; when they reduced Austria to an isolated fragment incapable by itself of an existence worthy of its past; when they did these things they made inevitable the ultimate *anschluss* between German and Austria which was achieved yesterday.'

In the Commons, Neville Chamberlain said: 'The hard fact is that nothing could have arrested this action by Germany unless we and others were prepared to use force to prevent it.' There was not a word from the League of Nations – but Winston Churchill solemnly warned that if Germany was allowed to go on conquering Eastern Europe 'the time will come when Germany will turn westwards.' Rashly, C. G. Grey commented: 'Mr Churchill's attitude seems that of many people in this country who cannot equal his intelligence and therefore have more excuse for being stupid. The people who govern Germany, whether Nazis or any other ruling class, are much too wise to think they could conquer and hold Western Europe. No German would be bothered with France as a province of the German Empire, or with England as an island colony.' That was the general attitude of wishful thinking.

3

The Report of the Cadman Committee on 8th March was described as 'the most sensible official document yet issued on civil aviation.' Its repercussions were to echo down the years. 'Those who appeared before the Committee say they have never met men whose normal interests are entirely outside aviation yet who showed such intelligence in finding out how things are misdone in the aviation industry. Lord Cadman's judicial attitude and evident determination to get at facts impressed witnesses strongly. Most were particularly struck by Sir Frederick Marquis, head of the great Lewis department store in Liverpool. The Cadman Committee has decided that more ought to be done in civil aviation than is ever likely in the next ten years!'

The Report was not easy to follow. The first thirteen pages of fifty-six numbered paragraphs gave the Government's observations and decisions on the recommendations, and the Report itself had thirty-one pages of 122 numbered paragraphs which were not correlated with those of the Government's. The basic premise was that association of military and civil flying in one Department of State was necessary to ensure the closest co-operation on questions of regulation and co-ordination of military and civil flying requirements at home and abroad. Accordingly the Government decided to up-grade the Department of Civil Aviation by creating the post of Permanent Under-Secretary of State, thus putting Sir Donald Banks as co-equal with the Secretaries of Admiralty, War Office, and other State Departments. The DGCA Sir Francis Shelmerdine, still retained right of direct access to ministers, and a Deputy-DGCA was to be appointed to aid with the day-to-day work and particularly finance. The Government also accepted the Committee's recommendation for the appointment of a Director of Civil Research and Production 'who will work in closest contact with research organizations, Air Ministry establishments, and the aircraft industry.'

It was emphasized that there was 'reason for more than apprehension' on British civil aviation. 'We view with extreme disquiet the position disclosed by our inquiry. There is not a medium-sized airliner of British construction comparable to the leading foreign types.' Accordingly the Committee suggested that an advisory panel should be formed of operators and constructors, which should not be restricted to the Air Ministry list of approved firms, to consider the probable airline requirements of two years ahead.

Railway companies were considered to be making useful contributions to civil air development and had provided capital and experience in a proper and constructive manner. Criticism that they attempted to stifle rival interest in air transport could be dealt with under the Railway (Air Transport) Act of 1929 which empowered the Secretary of State for Air to direct a public inquiry into any development which seemed prejudicially affected. In any case the Committee did not agree with subsidizing internal airlines, but until internal air transport extended, the State might make development grants towards capital expenditure on fully equipped aerodromes – though it was not mentioned that the RAF would certainly require them in war. The Committee stated that plans for diversion of Croydon airspace depended on development of an aerodrome at Fairlop by the City of London Corporation and on the decision of the Southern Railway about Lullingstone in which Imperial Airways were interested. 'Fairlop is neither fit for man nor devil,' fumed C. G. Grey.

Above all, the comments on Imperial Airways struck home. The Committee stated that personal contact between airline personnel and their employees must now be supplemented by collective representation of employees. As to the criticism that pilots' salaries had been cut, but dividends increased from 8 per cent to 9 and directors' fees doubled, the Committee pointed out that the original 1924 prospectus stated that any balance of profits, after payment of 10 per cent on paid up capital, was to be divided one-third to the Government towards repayment of the subsidy, one-third to reserve for development, and one-third in additional dividends to shareholders but the latter had in fact averaged only $4^1/_2$ per cent, and it was they who voted increased fees for the directors.

Dealing with the purchase of power boats from one of the directors, Hubert Scott-Paine, and the use of Dunlop products, of which company Sir George Beharrell was chairman, the Committee said that the aspersions against these directors were unwarranted. Several other matters were defended, such as the lack of Lorenz blind-flying equipment, because only German aerodromes were equipped with it. In answer to the charge that the course between London and Budapest was 'strewn with crashes' of Imperial machines, it was stated that no passengers were killed on their European services during 1935–37, although three crew died in the last crash. Nevertheless, an appendix showed that between 1934–37 sixteen passengers and seventeen crew were killed on other routes – but Deutsche Lufthansa, KLM, and Sabena had even more, while Air France had a massive forty-two crew killed, although only seven passengers.

As to the internal management of Imperial Airways and its relation with the Air Ministry and dealings with staff, the Committee stated: 'We are profoundly dissatisfied in regard to them. Although the carriage of air passengers in safety and comfort, and conveyance of mails and freight have been achieved with considerable efficiency, we cannot avoid the conclusion that management has been defective in other respects. Not only has it failed to co-operate fully with the Air Ministry, but it has been intolerant of suggestions and unyielding in negotiations. We consider that the responsibilities confronting the company can no longer be borne for practical purposes by a managing director. The chairman should be able to give his whole time to the direction of the business and should personally control the management of the company aided by one or more whole-time directors.'

The Government acted with promptitude. The 6th Earl Winterton, Chancellor of the Duchy of Lancaster, was appointed to the Cabinet as Deputy Secretary of State for Air responsible in the Commons for the Service side of the Air Ministry, and as an additional member of the Air Council. He had entered Parliament in 1904 'practically straight from the university, and for many years was regarded as an *enfant terrible*. As a Tory of Tories, his youthful exuberance made his intervention in elections highly entertaining and considerably effective.' In the 1914–18 War he served in Gallipoli, Egypt and Palestine, and latterly had been concerned with India – but he could only offer an astute mind on aircraft matters in lieu of aeronautical experience.

Another ex-soldier, William P. Hildred, was appointed Deputy-DCA, but similarly had no aeronautical experience. After the War he had entered the Treasury, became Finance Officer to the Empire Marketing Board, transferred in 1934 to the Ministry of Agriculture and Fisheries, and for the last two years had been Deputy General Manager for the Export Credits Guarantee Department – but at least that revealed world-wide experience of trade.

The new appointment of fifty-six-year-old Major Charles Stewart as Director of Civil Research and Production was welcomed not only by the British aircraft industry but others abroad, for since 1931 he had been the authoritative Superintendent of Technical Development (STD) at the RAE – not to be confused with the Superintendent of Scientific Research (SSR), the donnish, spectacled Herbert Stevens until recently CTO at Martlesham. Rugged Andrew Swan was promoted to SSR.

Although the Air Estimates were presented to the Commons on 15th March, the Debate on the Cadman Report was opened by Neville Chamberlain the following evening, and in the Lords next day. Labour's Clement Attlee moved an ultimately defeated amendment that 'The explanations and proposals of HM Government cannot be regarded as adequate to allay public concern,' alleging long-standing Ministerial neglect and gross inefficiency in the heavily subsidized Imperial Airways. Defence and cross-fire prolonged matters to a second session on the evening of 26th March, and when Lieut-Col Muirhead rose to close the Debate, Robert Perkins, originator of the Airways attack, congratulated him on the skill of his innings, particularly in letting difficult balls go past. The Cadman Committee had sown good seed, he said, and the result would depend on the action, not promises, of the Government. The changed management of Imperial Airways was good, but a changed outlook was also needed. The question of the pilots' collective bargaining was making progress. BALPA now represented 82 per cent of the pilots on regular airline work, and British Airways had come to a fair agreement with their pilots – yet the door of Imperial Airways remained slammed. When Lieut-Col Muirhead interpolated that collective bargaining recommendations would be made good, Mr Perkins said that a statement was vital to prevent more victimization, and he continued: 'The export market for civil aeroplanes is lost, yet there seems no great change in policy. Even if designers started immediately to equal the imminent new American types they would be two years late again, so why not buy from America and concentrate on a design suitable for far further ahead, otherwise KLM will probably spring the surprise of taking 100 passengers to India in twenty-four hours?'

Four days later, Imperial Airways issued a denial of the allegations against it, and emphasized that their managing director had been condemned by the Cadman Committee without the opportunity of saying a word against the criticism that he had been 'intolerant of suggestion and unyielding in negotiation, and took a commercial view of his responsibilities that was too narrow.' Ten weeks later the Commons were informed that: 'The Board of Imperial Airways have, with the concurrence of HM Government, invited Sir John Reith to become whole-time chairman of the company, and Sir John has accepted this invitation. The Board have also nominated our recent Deputy Secretary of the Ministry of Labour, Sir James Price, to be a director and thus will draw on his wide experience of labour and staff questions. The directors have expressed their strong refutation of the personal charges made against them, and are asking the new chairman to investigate the charges made against the management. As the appointment of a whole-time chairman carries with it the position of chief executive, the present managing director, Mr Woods Humphery, has placed his resignation in the hands of the Board, who have accepted it. To facilitate the task of the new chairman during the next few months, Mr Woods Humphery has offered to make available his knowledge and experience to the new chairman, which he has readily accepted.'

When dealing with Imperial Airways and its Westland Wessex, I had found Woods Humphery businesslike, terse, but completely fair. Great and experienced though he was, he had been overshadowed by the heavy personality of Sir Eric Geddes, who regarded every employee as a mere cog in the machinery, and Woods Humphery's job was to execute in an impersonal manner the orders of his leader with the full authority of his own technical and operational experience. As a result he had built up the longest, if not yet the fastest, airline in the world, and spanned the entire Empire. That he had to use antiquated aircraft was largely a matter of Government economics, comparable with starvation of the aircraft industry in earlier years.

His ascetic, darkly looming, piercing-eyed successor, Sir John Charles Walsham Reith, was the fifth son of the Rev. Dr George Reith, a Scot. Just as Woods Humphery had lifted Imperial Airways from an uncertain beginning to eminence, so had Sir John with the BBC where he ruled with a rod of iron. Few knew that before the war he and the then young Woods Humphery had worked at adjacent drawing boards in the marine office of Yarrow & Co on Clydeside. As one BBC man said: 'He will be a stickler for staff punctuality, though he may never meet them. He will want full details of school, university, previous experience, from every man working under him. He will turn up at staff dances, sports, dramatics, and unbend just enough to have every man and woman his friend.' According to newspaper stories, he was to get £10,000 a year instead of his £7,500 at the BBC. Everybody was crossing fingers, for he had no aircraft experience whatsoever. Nevertheless, there was general accord that under Sir John Reith there would be fair dealing, and when he took office on 4th July he cabled Woods Humphery: 'My first act on joining the great organization which owes so much to you is to send this signal of greeting and good will. I am glad you are standing by.'

4

Overwhelming the Cadman Report were the annual Air Estimates introduced by Lieut-Col Muirhead on 15th March, for what seemed an enormous sum of £103,500,000 gross – which was six times as large as in 1934, twice that of 1936, and 25 per cent greater than 1937. Expenditure on technical and warlike stores was £66,734,000 compared with £48,132,000 for the previous year. The Estimates contained provision not only for the current programme but the foundation of further expansion requiring a supplementary estimate. The programme targeted 1,750 aircraft by March 1939. Already there were 123 squadrons in the Metropolitan Air Force and twenty-six overseas. The FAA had the equivalent of twenty squadrons. Responsibilities of the former AOC-in-C Air Defence of Great Britain would be divided between an AOC-in-C Air Striking Force and an AOC-in-C Fighter and Army Co-operation Units, with a separate Maintenance Command for stores and repairs. Between April 1935 and March 1938 some 4,500 pilots and 40,000 airmen and boys had entered the Service, and 430 cadet pilots from the Dominions. It was estimated that manufacturing manpower had increased from 30,000 in 1935 to 90,000 at the beginning of 1938. 'The problem is to keep a balance between speed of production and being up to date,' said Muirhead. 'Developments and improvements have been so big and rapid and so fundamental that machines now turned out are only nominally new types.' Nevertheless gun supply was satisfactory, and power-operated turrets were a notable step forward. Production of radio sets was also very good. Swifter bomb-loading methods were being investigated, and because of the increasing importance of armament developments that department now had a separate Director. A new wind-tunnel was being set up at the RAE because 'at speeds of 500 mph and over, the question of compressibility of air assumes new importance which calls for practical experiment.' Fuel of 100-octane was being experimentally used and gave 10 to 15 per cent increased power compared with the standard 87-octane.

Turning to the contractual position, Muirhead said that 63 per cent of the prices had been settled, and 12 per cent further quotations had not yet been accepted – 'these two categories embrace 75 per cent of the contracts and cover half the expansion in aeroplanes ordered. In another 12 per cent of the contracts, batch costing has been resorted to instead of fixed prices; one important case has been settled by arbitration, and progress is being made in negotiation on all outstanding cases.'

Labour's one-time Under-Secretary of State for Air, Mr Montague, leapt to the assault by demanding a competent and full investigation into the military side of the Air Ministry, yet agreed that 'whatever may be thought about the ultimate virtue of force, most people in this country feel compelled to recognize the necessity for military preparations.' Bitterly he attacked the financial gains of industry, saying that Handley Page had a 50 per cent dividend and 100 per cent bonus shares; Hawker Siddeley made a gross profit of over £750,000; Vickers £243,000, Fairey £188,000; Hawker Aircraft £154,000; Armstrong Siddeley £133,000; Bristol £86,000. 'In time of war the nation must take over this industry,' he declared.

In a sweeping reply the Prime Minister insisted that representations of muddle and chaos were completely fantastic, nor was there any parallel between the Cadman Inquiry into civil aviation and an inquiry into what was and must always be the main function of the Air Minstry. There might be different opinions about the importance of civil aviation to the prestige of this country, but there could be no question about the vital nature of the Royal Air Force for the security of the nation. For that security the Government bore all responsibility and did not share it with any committee.

Other MPs had their say – sound comment, biased views, inaccuracies, prejudices, and even the humour of Lieut-Col Moore-Brabazon, who described his attitude towards the Air Ministry vote as an assenting negative in the form of the USA's 'Oh yeah!' However, Lieut-Cdr Fletcher, Socialist Member for Nuneaton, moved an amendment that the House was not satisfied with existing arrangements, and Mr Garrow Jones, Labour, seconded, challenging the Minister to deny that the front-line air strength of Germany was 50 per cent greater than that of the British. He particularly stressed the vulnerability of Britain's sea routes in the narrow waters, and emphasized that in the previous war we lost more than 3,000 ships 'which was about half our mercantile marine to-day.' He agreed that the entire British aircraft industry should be brought under Government control and that the Air Ministry needed a Minister really capable of doing the job. Sir Stafford Cripps, Socialist Member for Bristol, backed both MPs, but the amendment was defeated by 253 to 139 – which revealed an outstandingly large attendance for an Air debate.

The Air Estimates were followed by those of the Navy and included £5,718,000 for the expanding FAA now that it had been transferred from the Air Ministry to the Admiralty, though this excluded the purchase of aircraft and research and development which were still the responsibility of the Air Ministry. It was stated that *Ark Royal*, constructed by Cammell Laird for the 1934 programme, would be completed during the current summer; and under the 1936 programme, *Illustrious* and *Victorious* had been respectively laid down in April and May 1937 by Vickers-Armstrongs. Two further 23,000-ton aircraft-carriers, *Formidable* from Harland & Wolf, and *Indomitable* from Vickers-Armstrongs had their keels laid respectively in June and November 1937.

In the Commons on 24th March, Neville Chamberlain announced that still greater effort must be made to increase and accelerate the rearmament programme, especially that of the RAF and anti-aircraft defences as a priority in the nation's efforts. New factories would be erected, and increased output of anti-aircraft and other guns would be put in hand. The programme of air raid precautions would be speeded. As Sir Thomas Inskip stated earlier in the month: 'The balloon barrage for London is organized in ten squadrons. There will be four depots for storage and administration, of which three are under construction. Four-fifths of the balloons have been delivered and all the winches. New measures will increase our already powerful Coast defence, and the Fighter Force and anti-aircraft guns for the defence of London and other cities, besides other resources which cannot be mentioned, will be very effective against invaders.' As to merchant shipping he said there was little experience of air co-operation, and whether a convoy system would be used could only be decided by a proper authority when the occasion came.

As in war, so was private aviation a continuing story of tragedies, defeats, and successes. In the second week of March the enormously popular managing director of the Cinque Ports Flying Club, Bill Davis, brother of Brooklands' celebrated Duncan Davis, was killed with three passengers on taking off at Lympne with the Club's ST-4 Monospar. He and his vivid wife had been tremendously successful in fostering international sociability as an important function of amateur flying, and he had been a highly regarded instructor, but despite his piloting skill he failed to turn on both fuel cocks, the resulting failure of one engine swinging the machine out of control.

Flying the much trickier Comet, Arthur Clouston, accompanied by Victor Ricketts, a *Daily Express* reporter, achieved a triumph enthusiastically described by *The Aeroplane* as 'The world's greatest flight – so far!' They had flown from Gravesend to Blenheim, New Zealand, and back to Croydon between 8.20 pm on 15th March and 5.40 pm on 26th March. Sticking to his principles C. G. Grey said: 'The performance was reckless and foolish, yet it was worth doing, because it brought the furthermost Dominion of the Empire within a week of home. And that was magnificent. The whole thing was a triumph for Clouston's unusual ability as organizer and navigator and for the courage and endurance of both men. They showed how mails could be carried and how the farthermost Dominions could be reinforced today by air if things were properly organized. We hope these lessons will not be lost on those responsible for the Service and civil side of the Air Ministry.' Currently one of Clouston's jobs as an RAE civil test pilot was testing aerial wire-cutting devices by flying specially equipped experimental aircraft into balloon cables. Two decades earlier Geoffrey Hill, now immersed in his duties as a professor, made similar experiments there.

Three days before Clouston's attempt, Harry Broadbent had left Lympne in his Vega Gull hoping to beat Jean Batten's time to Australia by twenty hours, but on 16th March was reported missing until thirty-three hours later when Capt Hussey of Qantas, westward bound, spotted his machine on the small island of Flores 1,000 miles east of Batavia. Extreme fatigue had forced Broadbent to land, and his machine tipped on its nose, damaging the constant-speed propeller. Scrounging a lift in a Dutch Naval seaplane, he caught a Qantas flight to Sydney, had a new propeller air-freighted from England, and after an adventurous return by air and boat, managed to fit it, and marginally took off in no wind down a slope and over a cliff edge, and so to Sydney.

An accident of greater impediment occurred with the DH Albatross. The mailplane prototype E-2 had been delayed by system faults and modifications to the reverse-flow cooling, but it had proved to be the world's most economic aircraft of its type, achieving a gross ton-miles per gallon of thirty three, beating even the slick Comet; on 25th March, there was a minor disaster while young Geoffrey de Havilland was flying it, for both the hydraulic and emergency systems failed to extend the undercarriage and a belly landing became necessary. Skidding along the grass gave small risk of fire, but the damage meant five or six weeks of repair, bringing development to a halt because the second prototype E-3 was far from ready and the airliner versions even less so.

That same day George Bertram, promulgator of many 'Notices to Airmen', who since retirement as Deputy DCA in 1935 had been Air Ministry adviser, died aged sixty-two. Originally a mathematical lecturer and school inspector, he joined the Ministry of Munitions just before the Great War, and after the Armistice transferred to the Air Ministry as Secretary to Lord Weir's Committee on Civil Aviation. Although lacking all technical knowledge of aviation, this typical civil servant became Deputy to Sir Sefton Brancker in 1926. One who knew him well said: 'He was never exactly a red tape obstructionist, but was on occasion very correct and formal. Perhaps he was the right man to act as a shock absorber for the high speed methods of Sir Sefton Brancker. Nevertheless everybody liked him because he was essentially amiable.'

Though Bertram represented Civil Service caution, it was national inertia which currently was preventing any interest in air raid precautions (ARP).

However grave the politicians, the masses remained unmoved by threats of enemy bombers and clouds of poison gas. Winston Churchill's warnings might hit headlines, but only because the Press preferred to jeer at him. Certainly patriotic citizens of all ages, men and women, had been volunteering to help local authorities, and had attended courses on fire fighting, first aid, and the horrors of gas – but 'patriots are invariably a very small proportion of the whole, in time of peace anyhow, with the result that the Home Office has become anxious, and the Home Secretary broadcast an appeal for a million volunteers.' Even Chamberlain's muted reference to ARP in the Commons led one newspaper to explain that the Prime Minister was becoming an old man and had had a rather strenuous week. Though the Home Office became more specific in listing the principal ARP services, such as 'air raid wardens', 'auxiliary fire services', 'rescue parties', and issued plans for home-made air raid shelters, the public remained largely unconcerned, for these were prosperous times.

What newspapers featured week by week were many short paragraphs on strikes and industrial unrest – a recurrent characteristic in times of full employment. Though this was ascribed to professional Communist agitators, it was primarily due to cashing in on labour shortage. The industry must have manpower to meet its targets; men could therefore dictate terms. There was even sabotage, and on 30th March Earl Winterton stated: 'Since malicious damage at Fairey's Ringway factory last month there have been further instances at the Stockport factory to electric wires in four Battle aircraft. One of these aeroplanes was due for delivery to the RAF and the others for Belgium. A further case of severed wires in an aircraft for the RAF was reported by Supermarine at Southampton and similarly by Armstrong Whitworth at Baginton. Police investigation is proceeding. Manufacturers have been warned of the problem and all practicable precautions are being taken, but it is of the utmost difficulty to guarantee immunity against damage inflicted in a few seconds by an ill-disposed person apparently engaged on his normal work. I should add there is no evidence that such damage can be attributed to any agent of a foreign power.'

Germany was causing different concern. Now that Hitler's troops were on the Italian frontier at the Brenner Pass, Czechoslovakia was exposed to German advance from Austria. Propaganda infiltrated the world's Press alleging Czech ill treatment of Germans in the Sudetenland. On 24th March Chamberlain implied to the Commons that if Britain's ally, France, went to the defence of Czechoslovakia, then Britain might apply 'sanctions' on Germany in order to assist Czechoslovakia.

5

Combined Exercises were held by Britain in the western approaches to the Channel at the end of March. The Home Fleet's capital ships and destroyers commanded by the C-in-C were the attackers, and they had thirty-six bombers, twelve fighters, and seventeen reconnaissance aircraft. The defenders were under joint command of the C-in-C Coastal Command and the Rear-Admiral Submarines, their force comprising ten submarines of the Second, Fifth and Sixth Flotillas, three destroyers and two motor-torpedo boats, together with the ten RAF squadrons comprising twelve flying-boats, forty-five reconnaissance landplanes, and twenty-four torpedo-bombers. Lee-on-Solent was their HQ.

There was a bad start when one of the Stranraers, the fastest flying-boats in the RAF, its pilot blinded by patches of fog, crashed into the sea and was lost with the crew of five. Meanwhile the Fleet had been located laying a smoke screen which blended during the afternoon with cloud down to 200 feet, resulting in the recall of patrolling aircraft. The first air attacks were at dawn next day by Avro Ansons on sighting the convoy forty-five miles south of the Lizard, and a combined attack by Nos. 206 and 217 Squadrons followed. Only the *Nelson* brought its anti-aircraft guns into action before the squadrons dropped their bombs, and it was assessed that four of the thirty aircraft were hit, but Fleet casualties were theoretically high. Shortly after midday the Fleet was near enough for torpedo-bomber squadrons to participate, and their attacks continued throughout the afternoon until operations were called off at 6 pm. To those taking part it was all a medley of attacks without assessable pattern. As one air crew member wrote: 'Three small silver spots were our first sight of the enemy, and a few seconds later a hole in the clouds allowed sunshine to pour through and reveal a perfect view of nine destroyers in line-astern. The CO rocked his Anson in time-honoured way and put its nose down towards a small bank of clouds conveniently between us and the Fleet. Torpedo-bombers were disappearing seaward on their part of the business, and we flew up the line of ships with aircraft right and left breaking off to deliver further attacks. Aeroplanes and ships seemed to be coming and going at every angle. In brief clear intervals the ships could be seen altering course and generally doing precautionary air raid drill. Suddenly three searchlights showed that our machine was under heavy fire, so we withdrew and set course for home.' What it all meant would take days to assess, but undoubtedly the C-in-C Navy and AOC Air would be satisfied.

In future manoeuvres there would be Sunderlands to augment the Londons and Stranraers. The prototype with drastic modifications of sweep and step, and re-powered with Pegasus XXII engines of 1,010 hp, was re-tested by Lankester Parker on 7th March, followed by full load tests, and early in April Harold Piper flew it to Felixstowe for official trials. Already L 2158, first of the production batch of eleven, was in the last stages of completion, and on 21st April Lankester Parker made the initial flight.

In the Short's No. 3 Shop the 134 ft 4 in span G-class enlargement of the Empire flying-boat was beginning to take shape, though still more than a year from completion. Powered with four 1,380 hp Bristol Hercules IVC, the estimated speed was over 200 mph carrying forty passengers at a loaded weight of 73,500 lb. Yet the United States was still ahead. At Baltimore, Pan American's new 152 ft span Boeing Clipper, powered with four 1,500 hp Wright Cyclones, was almost complete. At an all-up weight of 82,500 lb it similarly had an estimated 200 mph, but would carry seventy-four passengers. Shorts, however, were on a new tack, for Gouge was designing a clean four-engined landplane airliner with almost identical mid-wing to the Stirling supporting a circular-section fuselage terminating in a tail somewhat like the DH Albatross's. Powered with turbo-supercharged high-altitude Hercules VIC engines it should have a top speed of 275 mph flying unpressurized at 10,000 ft, or 330 mph with pressurized cabin at 25,000 ft, carrying twenty-four passengers. Suitable means of pressure production and regulation had yet to be devised, but parallel preliminary work had been conducted in the USA on pressurization, as members of the RAeS heard on 21st April when they 'listened with their usual polite incredulity to Professor John E. Younger of the University of California describing recent developments of commercial high flying in the USA.' Experiments with an XC-35 version of the Lockheed Electra airliner had led to the design of the four-engined Boeing Stratoliner and Douglas DC-4 with pressure cabins, and initial flights of each were scheduled for the summer. Rubber gaskets sealed the doors and windows, and an aneroid capsule-operated spring-loaded discharge valve held a constant pressure from the exhaust-turbo centrifugal blowers which had the disadvantage that considerable power on the glide must be retained to keep sufficient pressure. In case of failure there was an automatic oxygen spray. The increase in structure weight was less than 1 per cent, but the air conditioning equipment weighed about 1,000 lb. Though a Farman with a pressure cabin had been flying in France, it was clear that the USA now had a very long lead.

Shorts received an order for three of their S 32 airliners, but Faireys did even better with their FC 1 tender for a 275 mph four-engined Taurus-powered, 30-passenger, low-winger fitted with tricycle undercarriage and similarly pressurized cabin with a differential of 6 lb/sq in, for fourteen were ordered by British Airways. Lobelle appeared to be taking a less prominent part these days and hankered to start his own business, so more and more responsibility was devolving on R. T. Youngman, head of the technical department and deviser of the high-lift flap of that name which Fairey patented.

British Airways were swiftly developing with multiplied services. Their summer timetable showed six a day to Paris except Sunday when there were five. A return ticket cost £6 6s for the ninety-minute journey in a Lockheed Electra. Pilots still lacked confidence in their high landing speed in poor visibility, and currently one of these machines overshot when landing in thick overcast clouds at Croydon and was damaged by smashing through the boundary fence, though without injury to passengers or crew. But the Electra had been an eye-opener. The Air Ministry was beginning to feel it could be used by the RAF. A Commission comprising tough Air Commodore Arthur Harris, Air Commodore (formerly Brig-Gen) Jimmy Weir, Augustus John's son Commander Casper John, Sqn Ldr 'Tiny' Horrex, and Freddie Rowarth of the

A & AEE, was dispatched to the USA to examine all types of American aircraft.

There was quick action to forestall the American purchase, but Earl Winterton reassuringly explained: 'The party of experts now visiting the USA and Canada are concerned solely with exploratory inquiries. They will investigate whether certain aircraft which might be suitable for RAF purposes, are available for early delivery. They will also examine the capacity and potentialities for production in Canada. It is the intention of the Government that all British firms suitable for aircraft production for the expanded programme shall be able to give the maximum output possible during the next two years. The bulk of large-scale orders has been placed, and further orders to secure maximum accelerated output will be given without delay.'

Blandly Mr Montague asked whether difficulties of labour, skill, and materials could be overcome by America, Canada, and Germany, yet in this country were insuperable?

'No Sir,' snapped Earl Winterton. 'I do not think my reply indicated anything of the sort.'

When Capt Harold Balfour asked: 'Is Scheme F for 1,750 aircraft by March 1939 now superseded by the expanded Scheme G; and when can we know what is the objective, in first-line aircraft and reserves, of the new scheme, and how long is that scheme planned to take?' Earl Winterton said that he could make no statement at this moment.

Three days later, on 29th April, the Air Ministry announced the formation of a Supplies Committee chaired by Earl Winterton, to place orders and arrange an extension of existing works or the creation of new ones. The members were Air Vice-Marshal W. L. Welsh, AMSO; Air Marshal Sir Wilfred Freeman, AMRD; Air Vice-Marshal R. E. C. Peirse, Deputy CAS; Mr A. H. Self, second Deputy Under-Secretary of State; Sir Charles Bruce-Gardner, representing the SBAC; Mr E. Bridges of the Treasury; and Mr H. Russell, secretary. 'The new plan was prepared by the Air Ministry after consultation with representatives of the aircraft industry. The composers hope it will provide increased output almost immediately. Provision has been made for the formation of extra squadrons and an increase in the strength of other squadrons, and for an increase in personnel of all ranks.'

Concentrated criticism of the Air Ministry followed in the Commons on 12th May when the vote of £1,490,000 for salaries of Air Ministry personnel was debated. Sir Hugh Seely moved a reduction of £100. He said: 'We are dealing with a Germany which has an enormous number of aeroplanes; I believe about 8,000, of which 3,500 are front-line. They can produce 400 to 500 a month, and within a year will have some 6,000 first line aeroplanes. Under the latest scheme we shall have 2,700 front-line machines in two years' time. When MPs inspected the Hurricane in June 1936 they were told 340 or more would be completed this year. That was two years ago. Can the Minister deny there are only twenty-eight in service? The Government has been asked since March 1936 when the Spitfire is coming into production. Can they deny there is only one in existence?' He continued in like vein on the superfluity of different types, of aircraft, gun turrets, engines – and proposed a Ministry of Supply; then he attacked the unbusinesslike methods of the Air Ministry causing huge overdrafts to firms because they were not paid on time. Further: 'The Minister cannot deny that training has fallen behind, that operational training is practically non-existent and cannot be pushed forward because of danger of wastage, that bombing accuracy is worse than before the expansion because of difficulties with machines and because essential ancillary goods have not turned up, that instrument flying and bad weather flying and night flying are worse than before. Armament training is behind, largely because of an almost complete lack of gun turrets.'

Both in the Commons and Lords, Earl Winterton had a rough passage 'simply because he has not been long enough in the job to know off-hand all the points in which he could have crushed or flayed his critics.' As he pointed out: 'Disappointment over production of aircraft is not always due to mistakes by the Air Ministry. There were those years of folly when the whole country demanded disarmament, the folly of Parliament and the country. Today there is a closer relationship between the Ministry and the industry. This is no sudden and panicky plan which has been proposed. It is a natural extension and acceleration of an already rapidly growing Air Force. I do not think that if the Archangel Gabriel stood at this box and produced an expansion scheme for the present Government it would satisfy the Rt Hon Member for Epping, Mr Churchill.'

In the Lords, Lord Weir reminded the House that he had once been responsible for aircraft supply in times of war, and said: 'As to the suggestion that the Air Ministry should have a production department to control the industry and teach it how to produce aircraft, I do not believe that at the present stage of airframe development anyone knows the best methods and means of production. Lessons are being learned every day in the factories, and nowhere else can they be learnt, much less taught. The duty of the Air Ministry is to make preparations for war by the erection of shadow factories. By the Cabinet's latest decision large additional orders have been placed so that each main contractor has a straightforward production programme. There is a healthy spirit of drive in every direction. I see nothing likely to prevent the realization of the new programme to this schedule, even after conservative allowances for technical delays and obtaining and effectively absorbing the new labour required and its applications to new classes of work.'

Endorsing these views, Lord Londonderry congratulated the Air Minister on planning this further expansion. He said he had personal experience of being a scapegoat but had no complaints, for if one person disappeared there were always others to take his place. If he could do anything to prevent Lord Swinton from becoming a scapegoat he would do it, for His Lordship was owed a debt of gratitude.

But Lord Swinton resigned, followed next day by Lord Weir and Colonel Muirhead. Despite eulogies by the Prime Minister, Lord Swinton was unmourned, indefatigable worker though he was, for many officials found themselves addressed as though they were public meetings; others felt that he kept too much of the detail to himself. In his place came the Rt Hon Sir Kingsley Wood, PC, solicitor son of a Wesleyan minister, plump and cherubic-faced behind round glasses. He had been very successful in all offices he held since his election to Parliament in December 1918; for he had been Parliamentary Secretary to the Ministry of Health, then Board of Education, and ultimately Postmaster General from 1931 to 1935. As his Under-Secretary, tall, quietly commanding, Capt Harold Balfour was appointed after resigning his directorship of British Airways Ltd.

Perhaps it was the perversity of the Labour Party which led its Mr Hugh Dalton to continue the attack on the Air Ministry a fortnight later by moving: 'That in the opinion of this House the growing public concern regarding the state of our air defences and administration of the department concerned calls for a complete and searching independent inquiry conducted with dispatch under conditions consistent with the national interest.' He barbed his speech with personal comments on RAF officers and officials of the Air Ministry, saying that except for Capt Balfour, no member of the Air Council had engineering qualifications, nor had ever done blind flying, and none could navigate an aeroplane or fire a gun in the air; the principal permanent official had been brought in from the Post Office, and the second from the Ministry of Agriculture, and they had no practical knowledge of air problems, nor among the higher echelon of the Air Ministry was there any up-to-date detailed knowledge of the vital problems. Of the Department of Supply and Organization, he said there was evidence from many quarters that it was in a mess, 'Its Director, Air Vice-Marshal Welsh, is not on top of the job. As to the AID no doubt Colonel Disney knows all about telephones and wireless equipment but nothing about aircraft production. His department is largely responsible for all the production muddles. In the Air Ministry there is a state of friction and jealousy between departments amounting almost to war.'

The Prime Minister replied that he would not deny there had been delays and disappointments in the programme which had to be altered and expanded from time to time. Three years of expansion coincided with one of those forward leaps which periodically takes place in applied science. The features were threefold: development of the all-metal monoplane, design of new engines of unprecedented efficiency, and the invention of the variable-pitch airscrew. The combination not only completely altered design but necessarily altered the strategy in making use of these newly developed machines. Evidently he had been well briefed when he discussed mass production, for he said: 'Fords have recently completed four million V-8 car engines under conditions effective for mass production – but such engines have about 1,700 parts, yet a modern bomber has 11,000 parts in the engine alone, and the airframe additionally has 70,000 parts for which 5,000 to 8,000 separate drawings are required. Although large orders have been placed, running to 700 or even 900 machines, the actual number of parts which can be simultaneously made are not comparable with the numbers we commonly associate with "mass production"'. That his knowledge of aviation was admiring rather than profound was clear from his commendation of 'the remarkable demonstration of the speed of the Hurricane which flew from Edinburgh to Northolt in forty-eight minutes', but though he did not claim this as the normal speed he said he was impressed that the pilot was able to find his way and steer a correct course by instruments alone, which 'shows that while we have developed these great speeds we have not neglected safety or adaptability.'

Winston Churchill added partial support to Dalton, asserting that the Air Ministry and War Office 'are absolutely incompetent to produce the great flow of weapons now required from British industry. However, though I have long pressed for an inquiry into the state of our Air programme, I am not sure that the new Minister would be helped if such an inquiry, searching and disagreeable as it would be, were proceeding day by day. As a very serious breakdown is now admitted and there is a fresh start, some at least of the arguments for an inquiry are removed.'

Conservative and Labour, somewhat apathetically, continued to fight it out. Finally, Sir Kingsley Wood answered the critics, saying that talk of serious breakdown at the Air Ministry was ridiculous, and he regretted Mr Dalton's reference to so distinguished an officer as Air Vice-Marshal Welsh who had done fine work and had high qualities, and he hoped the House would not accept the condemnation which Mr Dalton implied. Kingsley Wood agreed that special consideration must be given to Supply questions, and he proposed to enlist the help of leading industrial experts. Lord Nuffield had readily responded, and allocated the services of Oliver Bowden, his vice-chairman, and would help in the direction most needed, namely the provision of airframes. The Minister concluded by saying that he hoped the House would now realize that at this critical time an inquiry would hinder rather than help the Air Ministry. On putting the motion to the vote it was rejected with a Government majority of 185.

Listening to the Debate was Leslie Hore-Belisha, the new Secretary of State for War and a keen user of air transport. The Air Council's DH Rapide or the four-engined DH 86B were his usual mounts. On 14th April he flew to Malta, piloted by Sqn Ldr H. K. Goode, landing *en route* at Le Bourget where he lunched with M. le Chambre, the French Air Minister, then continued on a leisurely journey to Marseilles, next day to Naples, and on the third day reached Malta where he stayed four days and inspected units of the Defence Forces under Brig-Gen G. C. Stubbs. On 22nd April he left for Rome, visited Mussolini and dined with the latter's son-in-law, Count Ciano the Foreign Minister, then departed for England with an equally leisurely flight on 24th April. All in all an interesting 10-day holiday!

6

While one sector of aviation was absorbed in the problems of mass production of military aircraft, the flying and gliding clubs and private aircraft owners were happily disporting in ever increasing number. England on a sunlit day was still an attractive panorama except for scattered smoky industrial blots – and beyond its shores the whole world beckoned if one had wings. There was, for instance, Harry Broadbent 'gradually wearing out the air route between England and Australia.' Having reached Sydney he decided to attempt the eastbound record held by Jean Batten, for his Vega Gull with constant pitch propellor was somewhat faster than her fixed-pitch Gull and had a range of almost 3,000 miles. For this, his fifth flight between the two countries, he took off on 17th April, and in ten fast stages reached Lympne on the 22nd – a record of five days five hours twenty-one minutes that beat Jean's by thirteen hours, yet was the least exhausting trip he had made, cruising leisurely at 140 mph at 6,000 feet. A week later, Philip Wills made greater headline news by flying 206 miles in his glider from Heston to St Austell.

The annual aeronautically social event of the year, the RAeS Garden Party, was held on Sunday 8th May at Fairey's Great West Aerodrome. Mr and Mrs Handley Page, with their bevy of daughters, received the many distinguished guests. 'All the usual Pillars of the aircraft industry were there, bulked-out by people from the shadow factories and sub-contractors and sub-sub-contractors, and all the latest-joined makers of raw materials, parts and accessories lured into the aircraft industry by the RAF inflation panic,' sourly wrote C. G. Grey. Even *Flight* reported: 'It has become less and less a garden party and more and more a flying display on the grand scale. True, an official reception, a nicely served tea, and a theoretically restricted entrée are retained. However May proved so fickle that party frocks were at a discount; escorts seemed to prefer green Tyrolean pork-pies to toppers; and even the band had thrown up Sullivan in favour of Cole Porter. Not that the usual curtain-raiser to the flying season has suffered by this onslaught of informality. It is accepted that everybody comes to see the newest and best in British aviation, civil and military, to meet old friends, and to enjoy himself or herself generally.'

Chris Staniland had organized the flying, and the egregious William Courtenay, now a prospective MP, maintained 'an informative flow of loud-speaking.' Of the latest production aircraft allowed to participate there were the Blenheim, Lysander, Wellington, and tactfully the Fairey P 4/34 which though still on the 'Part Secret List' was officially permitted to perform in the hands of Staniland who looped and rolled with his customary elegance. Neither Hurricane nor Spitfire was available, but the Short Empire *Calpurnia* came floating across the aerodrome impressively low and then 'cut capers in the way of vertically banked turns which showed that even a hard working matron of the airways of the Empire could feel the Spring.' A number of smaller civil aircraft also flew, enlivened by a minor pile-up when Robert Kronfeld, now a naturalized Englishman, took off in a low turn with a coupé version of the single-seat Luton Buzzard and his wingtip touched the ground flinging the machine round with a thud. For romantics the Shuttleworth Sopwith Pup, a somewhat similar Hanroit biplane, and a 30 hp Anzani-powered Deperdussin cavorted prettily, and two Blériots were on display. Three sailplanes, flown respectively by Philip Wills, Dudley Hiscox, and Amy Johnson, were aero-towed and presently whispered around the aerodrome. Amy, who had changed back her name by deed poll, these days was living quietly in the Cotswolds, having failed to secure a job. Journalism and gliding were her current interest, and she had bought a Kirby Kite.

The day after the RAeS party, the King attired as Marshal of the RAF, flew in the Royal Airspeed Envoy to representative stations of each of his four Home Commands, accompanied by Air Chief Marshal Sir Cyril Newall, and piloted by Wg Cdr Fielden. At Northolt he was received by Air Chief Marshal Sir Hugh Dowding, AOC Fighter Command, Air Vice-Marshal Gossage and Grp Capt Orlebar, and inspected the Hurricanes of No. 111 Fighter Squadron commanded by the much publicized Sqn Ldr Gillan. At Harwell, one of the newest stations, he was received by Air Chief Marshal Sir Edgar Ludlow-Hewett, AOC Bomber Command, together, with Air Vice-Marshal Playfair and Wg Cdr L. G. Maxton, and inspected a representative Harrow, Blenheim, Wellesley and Whitley before watching five Hawker Hinds fly past followed by five Fairey Battles to emphasize the aerodynamic advance. Upavon was next, and Air Marshal Sir Charles Burnett, AOC Training Command, Air Vice-Marshal Pattinson and Grp Capt J. M. Robb, the CFS Commandant, received the King against a background of yellow-painted Harts, Tutors, Ansons, Furies, and Oxfords, and after lunch in the Officers' Mess, built in the far days when Capt Godfrey Paine RN commanded the Station with Major Hugh Trenchard as assistant, a brilliant display of aerobatics was given by two instructors flying Hawker Furies. Finally the King flew to Thorney Island where Air Marshal Sir Frederick Bowhill, AOC Coastal Command, Air Vice-Marshal H. M. Cave-Browne-Cave, and Grp Capt J. C. Russell awaited him, and he inspected lines of Swordfish, Shark, Osprey, Nimrod, Vildebeest and Anson. At the end of the day, from Buckingham Palace he dispatched a message to the Secretary of State for Air congratulating the RAF on the determined and successful way in which they were meeting the heavy demands made by the expansion of their Service.

Meanwhile the new Air Navigation (Financial Provisions) Bill resulting from the Cadman Committee recommendations was introduced by Sir Kingsley Wood as his first major task as Secretary of State for Air. The Government had increased the civil aviation subsidy from £1½ million to £3 million, and gave Imperial Airways monopoly of the Empire routes, with British Airways running all European routes except London-Paris which would be operated by an amalgamated company formed by Imperial and British Airways. Some £400,000 would be devoted to establishing daily services to all principal capitals of Europe instead of merely Paris, Brussels, Copenhagen. Stockholm and a summer service to Basle. Sir Kingsley admitted that no suitable British aircraft existed, but told the House of a promising medium all-metal airliner being built. This was the DH 95 Flamingo powered with two Bristol 890 hp sleeve valve Perseus XIIC engines. A shoulder-winger, it was the first over-all responsibility of tall, slightly stooping Ron Bishop who was now in unsmiling authority as chief designer. His task of devising a metal stressed-skin structure was unique in de Havilland history. Aided by the logical assessments of that mainstay of de Havillands, Charles Walker, a new technique of research, design, and manufacture had to be evolved, and it was to the credit of all that this was accomplished in very few months, resulting in a machine of 70 ft span and 17,000 lb all-up weight compared with 29,500 lb for the Albatross airliner. A distinctive feature was the 7 ft 4 in width of the cabin, seating eighteen passengers three abreast plus two by the entry vestibule. With a payload of 3,300 lb and three crew there was an estimated 600-mile still-air range and a maximum speed exceeding 200 mph – so if cost could be brought down by constructing it in reasonable numbers there seemed a chance of knocking the Lockheed 14, which carried six fewer passengers. For the moment it was a private venture, but Sir Kingsley Wood mentioned that 'the question of financial assistance for development of this machine is now being considered' and encouragingly added that the Ministry proposed to allocate £100,000 for air services within the British Isles so that companies suffering losses could be sustained on a commercial footing.

Harold Balfour, in his maiden speech as Under-Secretary, said that Government policy was to temper the strength of monopoly with a reasonable measure of competition. None could tell what would happen in twenty or fifty years but as policy had to be decided now, control of private enterprise with profits limited by the State was the soundest course to follow. C. G. Grey pontificated: 'Sir Kingsley Wood and Harold Balfour both acquitted themselves well, and one feels that when the former obtains a grip on his subject, things will begin to move. Much as one may long for an aeronautical Hitler to hustle British aviation, one feels that Sir Kingsley is likely to be as good a democratic controller as can be got.' When the House divided, the Bill was carried by 173 to 108, which, as one reporter wrote, 'means that the vast majority of Government supporters did not bother to come into the House, whereas all the Opposition took care to be there. This looks very like what would happen in another general election.'

Empire Air Day, in deluges of rain, followed on Saturday, 28th May. Mass formation flights above 170 cities and towns had been made by the RAF during the previous five days using ten routes twice daily to give the greatest possible number of people a view of them. Every station was set with appropriate aeroplanes in assorted patterns of old and new. Crowds poured into the aerodromes in their hundred thousands. At Cardington, barrage balloons were seen for the first time moored in the defensive pattern intended for urban areas. At Martlesham the display was worthy of Hendon tradition, watched by more than 15,000. At nearby Felixstowe people crowded in by excursion train and bus, and the Fairey Sea Fox, Short Singapore, Supermarine Stranraer, Saro London, Mayo Composite and Blackburn Perth were all on view, the latter arranged for the public to climb in, examine the interior, and go out by another exit.

Said one analytical reporter: 'British crowds are attracted to aerodromes because they are more or less forbidden territory. People get a kick from nosing round the sheds and feeling the wings of displayed aeroplanes and generally amusing themselves.' With a little diligence it was possible to discover how many squadrons had new equipment, such as No. 3 at Kenley with Hurricanes. None of the new aerodromes had runways, for the Air Ministry considered that the expense was not warranted.

Sir Charles Kingsley Wood, flying in the Air Council Rapide, beamingly toured the RAF Stations of Halton, newly established Odiham, Farnborough, Kenley, and Biggin Hill. Harold Balfour did likewise with Manston, the aerial centre of his Thanet constituency; North Weald; Mildenhall; Henlow; and Hatfield, where he inspected the Flamingo.

'Taking it all round, Empire Air Day may be reckoned a success in spite of the rain,' wrote an enthusiast. 'If so much enthusiasm was shown on such a day, what would it be on a really fine day? Next Empire Air Day there will be a few thousand bigger and better and brighter aeroplanes to show off to their owners the tax payers, and every station will become a major rather than a minor Hendon.' But increasing numbers, including the Prime Minister, wondered if there would ever be another Empire Air Day, and this was endorsed when the SBAC announced that no Trade flying display would be held this year because 'it is not in the public interest to show the latest types of military aircraft and engines, and the immense amount of work devolving upon the industry makes undesireable the diversion of time and energy to the organization of such a display.'

Empire Air Day also saw the demise of the Bristol Type 146 fighter, for it returned to Filton from Martlesham that day, flown by an RAF pilot, but collided with a sturdy set piece on the aerodrome and was damaged too expensively to justify repair for a machine which had no prospect of orders – nor had the untested Gloster F 5/34 and the two-year-old Vickers Venom, for they were rendered obsolete by the Spitfire.

The excitements of Empire Air Day faded into renewed contention on the subject of accepting refugees from Communist Spain. Expressing the typical outlook of that time, C. G. Grey wrote: 'We in this country have always prided ourselves on offering a free home for political refugees, but for years the lowest classes of Eastern Europe have been drifting in and settling. Already we have thousands of Spanish children supposed to be Basques, but are mostly lower classes of the Northern seaports. One would scarcely ask a foreign country to accept the sweepings of Cardiff or Liverpool docks as representative of the English people.'

Franco had inflicted heavy defeats on the Red Government who claimed that if only it had more armament there were enough enthusiastic volunteers to defeat the Nationalists. In the Commons, Miss Ellen Wilkinson asked whether any indication had been given of the number of Italian troops fighting for Franco. R. A. 'Rab' Butler, on behalf of the Prime Minister, replied that 39,000 were reported at the beginning of the Aragon offensive on 9th March. Bitterly Miss Wilkinson asked how many Italian troops to one Republican did

Franco consider necessary for victory? Meanwhile France was backing Red Spain, and was said to have recently dispatched 200 new aeroplanes together with train loads and road convoys of arms pouring in to Catalonia. Numerous British and American adventurers on alleged idealistic grounds were also taking part. An American who had flown for the Communist Front Populaire described his activities in the American paper *Sportsman Pilot*, and said that after eighty hours training he was assigned a Russian Chato biplane (I-15) and subsequently a Mosca monoplane (I-16 later known as Rata) armed with 0.30-guns in the nose and two 0.50s in the wings, and there was also the fast Katiska SB-2 twin-engined low-wing bomber based on the American Martin but vulnerable to attack by fighters because the upward field of fire was limited.

In the Commons, Clement Attlee asked whether Britain could take effective action to avoid 'acquiescence in the active generation of warfare', yet he proposed that the Spanish Government be permitted to import anti-aircraft guns 'not only to protect British ships but to protect their own women and children.' Harshly Chamberlain replied that HM Government was by no means acquiescing in breaches of international law. Changing tacks, Attlee said that besides representations to General Franco, Mr Chamberlain should do so to the Governments supplying aircraft, particularly Italy, but was told that Britain held Franco entirely responsible for the forces under his control and nothing could be done. Sir Henry Page-Croft, Conservative MP for Bournemouth, with a humerous thrust suggested that the wishes of the whole House perhaps could be met if Excess Profits Duty was immediately imposed on all engaged in the trade with Spain, whether shipping or industry.

Certainly major British aircraft manufacturers were far too involved with production problems to sell to Spain, and supplementary aircraft purchase from the USA for the RAF had become essential. British newspapers published an official statement: 'The Government has accepted certain recommendations from the Air Mission which recently visited the USA and Canada, and contracts are being negotiated for 200 aircraft for general reconnaissance and 200 aircraft for advanced training.' These were for a militarized version of the twin-engined Lockheed 14, later known as the Hudson, and the Harvard derived from North American NA-16 two-seat trainer similar to those ordered by Australia. The contracts were worth £5,400,000 and delivery was promised within a year. These would supplement the 160 Airspeed Oxfords, based on the Envoy, and the Rolls-Royce-powered Miles trainers. Of the Oxford, H. A. Taylor, Author of *Airspeed Aircraft*, records 'It was not, nor intended to be, an easy aeroplane to fly. It would, if not handled carefully, swing on take-off, and residual drift had to be kicked off accurately just before touchdown if it were not to try to career off the runway in a ground loop. If you could handle the Oxford with reasonable competence, you could tackle almost anything in the way of recalcitrant twins and multis. A wag described the Hudson, which had somewhat terrifying ground and near ground handling characteristics, as an excellent interim conversion type for those about to fly an Oxford!' Nor could the Miles Master 1 be considered perfect, but it was the only high-speed monoplane trainer available, and so received a very big order. What with this, and another big contract for the draughty but pleasant Magister two-seat light trainer developed from the Hawk, Phillips & Powis, and certainly Miles himself, were cresting to success.

Earlier expectation was pinned on the DH 93 Don, but de Havillands had not been very clever, nor was D.H. himself much enamoured of Air Ministry projects, so design had lagged and with the elaboration of the specification the structure weight became excessive 'until it was literally no good as a flying machine and shortly after deliveries began the type was wisely relegated to ground instruction duties.'

At Airspeed there was difficulty. Nevil Shute Norway, that keen but stuttering founder of the firm, could not accept a decision by his co-directors of the Swan Hunter Group to sub-contract the Oxford, as a matter of economics, to his *alma mater* de Havillands whom he had hoped to beat at their own game. So Norway resigned, and though not yet a celebrated novelist, decided to devote his time to writing. He said: 'It was a very good way; it enabled me to work only for myself and have a good deal of money and abundant leisure. I was forty years old. I sold my old ten-tonner and planned a new ship which I named *Runagate*, an old English word meaning an escaped prisoner.'

Hessell Tiltman continued as managing director of Airspeed with A. Townsley as general manager and director. Sub-contracts were duly placed with de Havillands and Percival for the Oxford. A fortnight later the managing director, George Wigham Richardson, who had succeeded Lord Grimthorpe in January 1937, announced that profit on the year's trading had reduced the trading deficit to less than £70,000.

Though Swan Hunter remained cautious of their aviation acquisition, the rival shipbuilders Harland & Wolf were vigorously active with the new Short & Harland Government-owned aircraft factory on Queen's Island, Belfast, where more than 2,000 were employed on construction of the fifty big Bristol Bombays which had been ordered late in 1936 but were still far from ready.

7

June opened with the International Air Race from Hatfield to the Isle of Man for any aeroplane faster than 100 mph. First prize was £125. Of thirty-six entries, most were British, though six were from Germany, two from Czechoslovakia, one from Latvia, and one from Switzerland – but the number of starters diminished to only sixteen, mostly the new keen youngsters.

First away was Ranald Porteous with the scarlet little Chilton low-winger reminiscent of the ultra-lights of the mid 1920s. Last was Alex Henshaw with his neat white Mew Gull of similar span and wing area but three and a half times the loading, and fitted with a windscreen and canopy so low that he sat on the floor. The sun shone brightly at the start, but by mid-morning the clouds dropped to 700 feet above the Isle of Man and soon rain was falling, but S. T. Low with his Gipsy-powered Comper Swift beat the field with an average of 167.5 mph, Henshaw second on handicap with 247½ mph, and Edwards third with 122.25, and the prizes were presented that night at the Palace ballroom. Sunday was a day of rest. Monday, dull and gusty, caused the postponement of racing until Tuesday when Low in the first race added another win, and the Manx Air Derby was won by John Rush flying the only biplane competitor, an ancient Hawker Tomtit. Thereafter the machines streamed across the sea for England and home.

In secret on 8th June, a small helicopter, the Weir W 5 with twin lateral rotors like the Focke-Achgelus, made its first short free flight at Dalrymple, Ayr, piloted by the chief engineer's son, R. A. Pullin – but with only 50 hp it was badly underpowered and there were additional problems for Pullin senior and Bennett to solve. But presently the main features were sufficiently proved for design of a scaled-up, two-seat version powered with a Gipsy Six, financed as previously by Lord Weir and his brother, Air Commodore J. G. Weir.

To make up for the absence of the RAF Display was the advertised reason for a great Air Display organized by the *Daily Express* at Gatwick on 25th June, though rival newspapers refused even to mention the event. Lord Beaverbrook avowed it would be the Display of the century. The crowd was huge, the long grandstand backed by a score of vast Union Jacks breezing from tall white poles. Notable foreign pilots had been contracted to perform: there was Al Williams, the outstanding American racing pilot with his petrol-advertising 940 hp Cyclone-powered Grumman Gulfhawk biplane; famous Marcel Doret of France with his Hispano-powered Dewoitine aerobatic monoplane with which he made long spins, and later flew a sailplane which he flick rolled and flew inverted, ending with a landing from a loop; his compatriot Réné Vincent jumped from 3,000 feet, releasing from one parachute to another, the fourth and last in French tricolours; Paul Förster was no less thrilling with the aerobatics and inversions of his Focke-Wulf Steiglitz a few feet above the turf. One of the Empire flying-boats heavily circled the airfield, causing a spectator to mutter that it was a year late! A noisy Lockheed Electra next screamed round, and the whispering four-engined Focke-Wulf Condor arrived on its first visit to England, followed by a Savoia S 83 and a Marcel Bloch 220 looking much like a Douglas DC-2. There was squadron drill by nine Hawker Furies; six Hurricanes made their *début*; three Furies aerobatted in tight formation. In sequence a Lysander, Hurricane, Battle, Blenheim, and Whitley duly disported though described as 'secret war planes.' Finally there was a set piece in Hendon style in which an oil refinery was bombed by nine Battles, two of which, streaming red smoke, were brought down by six Hawker Hurricanes, but as one scribe put it: 'By triumph of mind over matter the Battle pilots refrained from the Heaven-sent opportunity of writing *Buy the Daily Express* in smoke across the sky.'

Three days later Fred Miles's new Monarch three-seat cabin prototype, largely designed by his brother George, crashed in turbulent weather when

signed as the smallest racer that could accommodate a 140 hp Gipsy Major engine and fitted with retractable
lercarriage, slots and flaps, the TK 4 built by DH students first flew in July 1937 and had a top speed of
5 mph.

e mail seaplane Mercury was the 73-ft span upper component of the Short-Mayo Composite aircraft and was
vered with four Napier Rapier H air-cooled engines, but after successful airborne launched flights the war
ervened and Mercury was used as a seaplane trainer until scrapped in August 1941.

.ia was the lower component of the Composite and almost identical with the Empire flying-boat, but with different
ning bottom and fitted with centre-section cradle to carry Mercury though was later restored as a ten-seater which
s destroyed in an air-raid on 11th May 1941.

After making separate flights in the autumn of 1937, initial trials of the complete Mayo Composite began on 1st January 1938 and on 6th February the first airborne separation was uneventfully made, witnessed only by casual spectators near the mouth of the Medway.

High hopes were centred on the Armstrong Whitworth Ensign – the first metal-skinned airliner to be operated by Imperial Airways. Intended for the Empire routes, the prototype was flown on 24th January 1938, but only eleven ha been delivered by the outbreak of war.

The first DH Albatross airliner version was delivered to Imperial Airways at Croydon in October 1938 and became t flag ship of the F-class which carried twenty-two passengers and four crew. *Falcon* set a record on 2nd January 193 by flying a scheduled service from Croydon to Brussels in 48 minutes.

ilt as an Army Co-operation two-seater to the same specification as the Lysander, the Bristol 148 was sixteen nths too late for the competition but was found to be very pleasant to fly and the low wing no real drawback to view npared with a high-winger.

An outstanding airwoman of the 1930's was Jean Batten who made flights to Australia and back, England to Brazil, and England to New Zealand.

rly production Whitleys differed from later versions in having no wing dihedral. By the time production ceased in ne 1943 with the Mark V, 1,824 Whitleys had been built, though by spring 1942 they had been retired from mber Command sorties as the big four-engined bombers were operational by then.

The Bristol Type 146 single-seat fighter was on low priority because of Blenheim production, making its first flight o 11thFebruary 1938, long after the others of similar specification, but ended its career that year colliding with a displa set-piece when landing.

These were the long-range Vickers Wellesleys which on 5th November 1938 took off from Ismailia's 3,600-ft runway and headed for Australia, and one was successful in achieving the world's long distance record of 7,178 miles which remained unbeaten until 1946.

As successor to the Moth biplanes, Geoffrey de Havilland decided on a low-wing two-seater with open cockpit – the £575 DH 94 Moth Minor which made its first flight on 22nd June 1937. Within two years production had risen to eight a week, but the war caused its abandonment.

ıpplementing the aeroplane speed, altitude, and distance records held by Britain was the world land speed record :hieved by Capt George Eyston with this four-wheel, streamlined, Rolls-Royce-powered car.

o confirm the calculated aerodynamics of Short's proposed four-engined bomber a half-scale wooden mini version, owered with four Pobjoy engines was quickly built, flown in great secrecy at Rochester Airport on 19th September 938.

, closely guarded ecret, the Westland Vhirlwind led the way re-war as the first vin-engined single-seat ghter and was uniquely rmed with four cannon ı the nose, but though sed in the war the mitations of its small Rolls-Royce Peregrine ngines prevented the evelopment it ıerited.

With marked similarity to the Blenheim, the Bristol Type 149 Bolingbroke had a longer nose to accommodate a navigator, and 676 were built in Canada by the Fairchild Aircraft Company.

Another variant of the Blenheim was the Beaufort Type 152 which became Coastal Command's standard torpedo bomber powered with two 1,130 hp Bristol Taurus or the slightly more powerful Pratt & Whitney twin Wasps. First flight was on 15th October 1938 and the machine entered service a year later.

The long-range flight of three Wellesleys during their tour of Australia after the record flight.

uselage production of some of the 1,368 Lysanders built at Yeovil between September 1936 and January 1942, and :hers were made in Canada as target tugs.

he four-engined Saunders-Roe A 33 of 1938 was an ambitious attempt to make a monoplane-like four-engined ying-boat, but the sponson-braced structure proved disastrous.

'he Saunders-Roe Lerwick was the second of the new monoplane flying-boats ordered by the RAF, and gave the npression of a twin-engined Sunderland though the 81-ft span wings were more heavily tapered. In addition to three rototypes, twenty-six production machines were delivered to the RAF.

The Blackburn B-25 Roc was a two-seat fighter developed from the Blackburn Skua, but construction of the 136 production machines was sub-contracted to Boulton & Paul Aircraft Ltd, builders of the rival Defiant.

The Fairey Albacore naval reconnaissance biplane replacement of the Swordfish was used successfully by Nos 19 and 45 Squadrons.

De Havilland's first stressed skin, all-metal high-winger, the DH95 Flamingo, was a bid to counter the American twin-engined 12/15 seaters. First flight was on December 1938 and six months later it was in operation with Jersey Airways; but war put an end to production after sixteen had been built.

Massed production of Hawker Hurricanes at Langley Factory.

nitial production of Blenheims at Filton in the pre-war erecting shop. Before production switched to Mark IV in 1939 total of 1,552 Blenheim 1s had been built.

Early Spitfire production at Vickers Supermarine began with an initial contract of 310 for completion in March 1939. By October the order had been increased to 4,000 – and the Spitfire went on to become a legend.

Designed by Arthur Hagg, ex-chief designer of de Havillands, this Heston Racer was financed by Lord Nuffield to achieve the world's landplane speed record but its Sabre engine overheated on the first and only flight of five minutes and the Racer piled-up on landing.

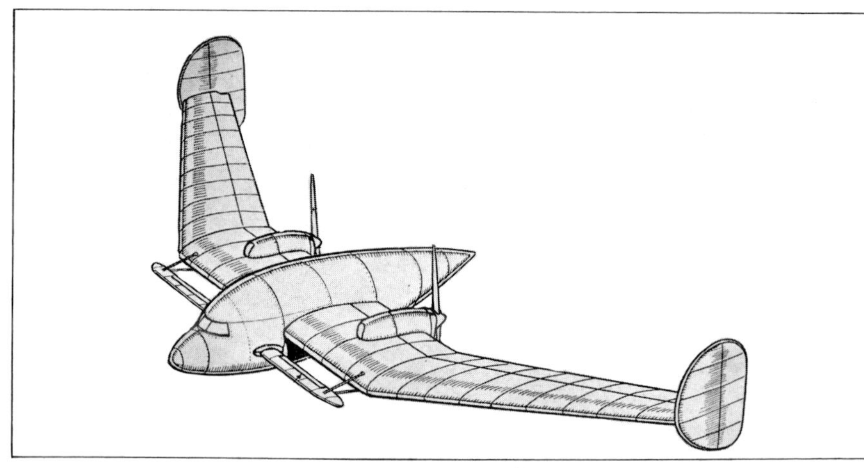

The Handley Page Manx twin-engined tail-less light plane was designed as a prototype for a 20,000 lb bomber, and featured a floating nose aerofoil to maintain trim with landing flaps depressed, but as Dr Lachmann, the chief designer, had been interned because of his nationality it was not flown until 1942.

The 103-ft span Sunderland, derived from the Empire flying-boats, takes off in all its maritime splendour.

'he prototype Short Stirling bomber came to a disastrous end on its first flight and had to be scrapped.

Success of the two-seat Bristol Fighter in the 1914–18 war led to the two-seat Boulton Paul Defiant of 1939 relying on a multi-gun turret for the rear cockpit, dispensing with forward fixed guns.

ast of the Sopwith
ove – later bought by
e Shuttleworth Trust
nd re-converted into its
arlier guise as a Great
Var Sopwith
up.

As a private venture to compete against the Hurricane, the Martin Baker company built the MB-2 light fighter of simple construction and easy maintenance facilities.

The torpedo-bomber Blackburn Botha had a shoulder-high wing for maximum downward view as it was designed for Coastal Command, but it lost out to the Beaufort and was directed to the RAF for North Sea patrols.

Largest of the pre-war airliners built for Imperial Airways the Armstrong Whitworth Ensign had a span of 123 ft and accommodated forty passengers in considerable comfort cruising twice as fast as the earlier Hannibals and Scyllas.

Golden Hind, launched on 17th June 1939, was the ultimate in pre-war airline flying-boats, spanning 30 ft more than ıe Empire class and designed to carry passengers from Ireland to Newfoundland without refuelling.

horts had shown the value of checking the aerodynamics of future designs by building a half-scale version, so aunders-Roe did the same with a Pobjoy-powered four-engined half-scale flying-boat in 1939.

)nce adopted, every ype of aeroplane had ıtensive development. 'hus the Hurricane ecame an even more ffective weapon armed vith four annon.

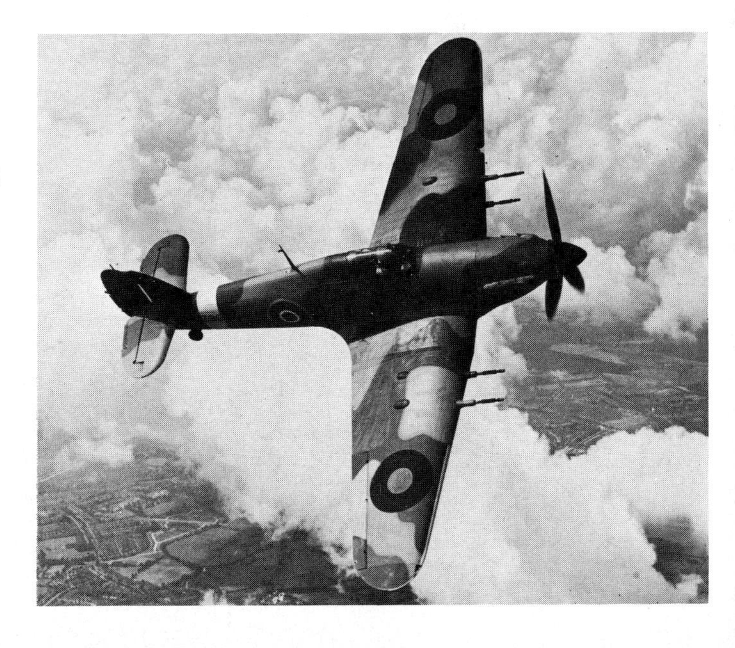

The original Bristol Beaufighter ready for test at Filton in July 1939 derived from the Blenheim as a heavily armed high performance night fighter with lethal power due to its highly secret AI radar.

Though similar in appearance to the Wellington, the Warwick was designed as a heavy bomber of slightly earlier specification than the Avro Manchester. The first prototype, flown on 13th August 1939, had Rolls-Royce Vultures, but the second had Bristol Centaurus as these were specified for production.

Competing with the Warwick was the Vulture-powered, impressive Avro Manchester of slightly smaller span flown on 25th July 1939 – but was one of the RAF's great disappointments, though it led to re-design as the immensely successful four-engined Lancaster of the ensuing war.

a) Robert Blackburn

b) Frank Sowter Barnwell

c) Hugh Oswald Short

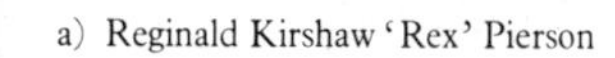

a) Reginald Kirshaw 'Rex' Pierson

b) Charles Richard Fairey

c) Reginald Joseph Mitchell

flying from Reading to Martlesham, and Wg Cdr Frederick William Stent, the company's forty-eight year-old test pilot, was fatally injured. No explanation was given, but probably a violent bump lifted him from his seat and the roof knocked him unconscious. Trained as an engineer, he had joined the RFC in 1914, took a post-graduate course in aeronautical engineering at Imperial College after the War, leading to combined engineering and piloting duties in Bomber Command, where he was OC No. 9 (Bomber) Squadron at Boscombe Down in January 1931, but retired from the RAF in January 1936 to join Phillips & Powis, and a few days later was lucky to escape by parachute during spinning trials of the Miles M 7 Nighthawk at Woodley. The revised version was therefore fitted with M 16 Mentor anti-spinning chines to the tailplane and given a rudder of considerably greater size, influencing George Miles in subsequent designs to depart from the customary shape associated with Miles light aircraft as typified by the Magister.

When the 1,000-mile King's Cup Race of twenty circuits between Hatfield and Barton Airport was run on 2nd July, Alex Henshaw, flying his slicked-up Mew Gull powered with a 200 hp Gipsy 'R', won the £800 first prize; Giles Guthrie was second with his 200 hp Gipsy Six II standard Gull; and Leslie T. H. Cliff, who had learnt to fly at Lympne at the same time as Ken Waller and was now an instructor, was third with his 130 hp Gipsy Major Miles Hawk Major. Edgar Percival who started scratch and was sixth on handicap had the second fastest time.

Several thousand people arrived at Hatfield in time to inspect the twenty three aircraft long before they started. There had been much polishing of the machines to reduce skin friction and many had been cleaned-up with fairings, flattened windscreens, venticular-section wheels, covered hinge joints for controls, and neater spats than standard. Specializing in this work was Jack Cross of Essex Aero Ltd of Gravesend. Only Kenneth Davies's Vega Gull of the British Parachute Co, the metal-skinned CW Cygnet, BA Double Eagle, Parnall Heck, and the ancient Comet G-ACSS entered by the now wealthy potentate 'Charlie' Devereux seemed untouched – but this last machine met disaster. Chris Staniland's first experience of it was ferrying from Gravesend with Capt Hopcraft on the morning before the race, and he admitted he was no enthusiast, so tried it again later in the evening, and then once more with Hopcraft, but it bounced semi-stalled, dropped its right wing, and skidded in a long curve until that wing hit the ground, broke off the tip, re-bounded onto the left wingtip and stopped. Clearly this DH machine did not like being flown by a Fairey pilot, but the feeling was mutual.

The start was uneventful. Throughout the long morning the competing aeroplanes re-appeared at unequal intervals, banking with a variety of techniques round the Hatfield pylon and disappeared again, eventually one or other landing for one of the regulation 40-minute stops to refuel pilot, passenger, and machine. As Frank Bradbrooke noted: 'The crowd showed signs of boredom. Perhaps they sighed for the counter-attractions of the Varsity match at Lord's, tennis at Wimbledon. or the river at Henley. Clouston, flying the BA Eagle, dropped out, some say because Mrs Clouston was not feeling well. Rounding sixty corners one after another at full speed in a long, bumpy, hot day in a closed machine for a lost cause, as they were out of the running, would be enough to try the stoutest.

'Most people settled down to tea while the race reached its climax. Sir Kingsley Wood arrived from the RAF Garden Party to present the King's Cup. Two laps from the end Henshaw flew into first place when he passed Cliff. From then on he widened the gap. Dead on estimated time his Mew Gull flew straight on across the aerodrome instead of turning at the pylon, circled to the blare of car hooters, and landed so fast that it disappeared over the brow at the far side of the aerodrome.

'Sir Kingsley Wood duly gave the Cup to Alex, with Lieut-Col Sir Francis Shelmerdine and Mr W. Lindsay Everard, chairman of the RAeC, in support, with proud "Pa" Henshaw and a few hundred others looking on. In fact Sir Kingsley gave the Cup to Henshaw four times over for the benefit of assorted Press and cameras. Even Capt Laurence Hope cannot boast that he had the King's Cup given to him so many times in so few minutes. In a private room, commandeered by Shell-Mex who had furnished the spirit of victory, the winner passed round the Cup full of champagne. One of his most cordial congratulators was Edgar Percival who must have had some consolation at least in having his Mew Gulls come home in first and second fastest time.'

Sir Kingsley Wood was also kept on his toes in the Commons. On 27th June Clement Attlee verbosely asked the Secretary of State whether he had any statement to make 'in regard to steps for unifying control of responsibility for development and production of aircraft and other materials.' Sir Kingsley then announced that Air Marshal Wilfrid Freeman would in future be responsible for production as well as development and research – a significant realization that huge production was useless unless experimental work resulted in a satisfactory machine. Everything in the way of re-equipping the RAF therefore hinged on Freeman. No burden, except that of Prime Minister, could have been greater.

To assist him he would have Air Vice-Marshal A. W. Tedder as Director-General of Research and Development (DGRD), and Ernest H. Lemon, chief mechanical engineer and director of the LMS Railway, was appointed Director-General of Production (DGP) though cautiously described by Major Bulman as 'no doubt highly efficient but hardly with background or breadth of experience calculated to win the confidence of the aircraft industry.'

These changes meant radical re-organization. That clear-thinking but modest artist-pilot Air Commodore Roderic Hill, brother of Geoffrey Hill, became DTD in succession to Air Commodore Verney, under whose care the new aerodynamically clean monoplane fighters and still unflown big bombers had been initiated. A quiet man, Verney had accomplished great work, well advised by the far-seeing young officers of DOR, most of whom later achieved high rank. The Directorate of Aeronautical Production became a separate department ranking equally with other senior departments, and its chief, Lieut-Col H. A. P. Disney, whose appointment had been considered ill-chosen by the industry, was diverted to the post of Director of Armament Production. In his place Lieut-Col H. W. S. Outram, who had built up the AID until it became the envy of the world, was appointed Director of Aircraft Production. Charles Walker was loaned by de Havillands to assist him and became Director of Production of Material – a vital post in getting supplies to manufacturers. Grp Capt John Sowrey from DOR was made Director of the AID, though he had no experience of such work other than as senior equipment staff officer ADGB 1930–1934. Both Major John Buchanan, as DDTD (Airframes) and Major Bulman, DDTD (Engines), were allocated new appointments, Buchanan as assistant to Lemon, and Bulman in charge of aero engine production.

Increased departments meant vastly increased staffs. Adastral House, Kingsway, though originally regarded as too big, had long proved inadequate. Offices were already dispersed in Ariel House, Audrey House, Bush House, Clement's Inn, Imperial House, Melbourne House, Princes House, Savoy Hill House, Victory House, and York House. Negotiations were in hand for a great block of imposing appearance in Berkely Square (later the American Embassay) where 1,700 officials could be accommodated.

Meanwhile the RAF had achieved a new triumph. The keynote of all British military exercises was defence rather than attack, so there were no attempts at simulating long-distance bombing except under the guise of long-distance reinforcement of Empire defence by air. To that end of the RAF's long-range development unit at Upper Heyford was gaining useful experience with its special Wellesleys which had a canopied rear cockpit and large streamlined external tanks at each semi-span from which fuel could be rapidly jettisoned in emergency. Their NACA cowled Pegasus XXII engines pioneered automatic boost and mixture control to eliminate the danger of weak mixture which might arise from faulty use of manual control, and they also introduced the Rotol constant-speed propeller. Though one machine had been lost without trace, five months of experimental flying in determining economic cruising conditions led to a 4,300-mile flight of the remaining four, under the command of Sqn Ldr R. Kellett, to the Persian Gulf and back to a landing at Ismailia, Egypt, 32 hours after taking off from Cranwell on 7th July. That gave a reassuring fillip to the practicability of long-range sorties.

A few days later the 21st birthday of the Gosport School of Special Flying was celebrated with a dinner at Brooklands arranged by Duncan Davis. Chief guest was the founder, the legendary, bearded, Colonel Smith-Barry who spectacularly arrived with his Puss Moth, greeted by many of his former instructor assistants.

RAF training was still based on his methods. Further facilities were vital. Grp Capt J. M. Robb was sent to Canada to discuss this requirement with the Canadian Defence Department, and agreement was made to enlarge the RCAF School at Camp Borden, Ontario, and build three other schools for *ab initio* Canadian and British RAF recruits, who would be established in Canada for a year, followed by a Short Service Commission, at the end of which the Canadians would go home as Reservists in the RCAF. In announcing this arrangement, Mr Mackenzie King, the Canadian Prime Minister, stated that long ago his Government had settled the constitutional principle that in Canadian territory there could be no military establishment unless owned, maintained, and controlled by the Canadian government as the hallmark of national sovereignty and the basis for friendly co-operation between the United Kingdom and Commonwealth.

There was pride in the cohesion of the Commonwealth, but also profit if Canada could engage in production of armaments and aircraft. To that end Marshal of the RAF Sir Edward Ellington arrived in Canada to join the British Air Mission and confer with Maj-Gen la Flêche, Deputy Minister of the Department of National Defence, and Air Vice-Marshal G. M. Croil of the RCAF, and then finalize arrangements for Canadian aircraft manufacturers to set up factories near Montreal and Toronto for the construction of bombers – for which purpose Frederick Handley Page, accompanied by his wife and daughters, had joined the Mission which included Henry Self, Sir Hardman Lever, and Lieut-Col Outram who had been visiting the Lockheed Aircraft Corp in California. Delays followed. There seemed a change of viewpoint, and it was suggested that the Canadian Government's hesitation was due to fear that if the Dominion engaged in the manufacture of munitions it would be impossible for her to declare neutrality should Great Britain engage in a European war. Supporters of the Canadian Government discounted this and claimed the real reason was that British manufacturers bitterly opposed orders being placed outside their own country. France was ready to jump in at the chance of Canadian manufacture, and already had offered private capital to erect a factory after discussing the proposal with William Tremblay, Provincial Minister of Labour in Quebec. Significantly Boeing Aircraft of Canada was converted to a wholly Canadian company registered as the Vancouver Aircraft Manufacturing Co.

Two months passed, and Ottawa then officially announced: 'As a result of exhaustive discussions, arrangements have been made by the United Kingdom Government in conjunction with the principal aircraft firms in Canada supported by prominent financial organizations, for placing contracts for bomber aircraft for the United Kingdom on a long term programme of purchases. A Central company is to be formed in association with these firms, which will operate two main establishments and contract directly with the United Kingdom Government. The United Kingdom representatives are also discussing arrangements for manufacture of other types of aircraft at other centres – for example Vancouver and Fort William.' The name of the new company had not yet been decided, but nine directors were appointed comprising the presidents of the Northern Electric Co, Canadian Car and Foundry Co, National Steel Car Corporation, Canadian Vickers, Fairchild Aircraft, Fleet Aircraft Ltd, Ottawa Car Manufacturing Co, Consolidated Paper Co, and the Canadian Bank of Commerce. Of these firms, only the Canadian Car and Foundry Co was regarded as sufficiently equipped to turn out complete aeroplanes rather than units, for they were making Grumman fighters at Fort

William. Inevitably the new consortium, soon registered as Canadian Associated Aircraft Ltd, was going to accept Handley Page's wordly advice to build Hampdens.

Sir Edward Ellington's visit took place only a fortnight after returning from Australia where he had been asked to advise on expansion of the RAAF. He found that squadrons were below strength, and particularly deficient in flight commanders and non-commissioned officers. Accidents were even worse that the RAF, and such was an Australian's independence that several fatalities occurred through disobedience to orders. Though Hawker Demon pile-ups had been ascribed to faulty maintenance and inherent defects of design, Sir Edward found no evidence of this; but he criticized the Wirraway NA 33 fighter, of which 100 had been ordered from the inadequate Commonwealth Aircraft Corp, and recommended that they should be regarded only as a temporary expedient as they were not fast enough. He urged there should be identical standard types of aeroplane in the RAF and Dominion Air Forces to simplify supply and spares provisioning, but was against the purchase of American machines. What the RAAF had were Hawker Demons, Westland Wapitis with wooden wings, Bulldog fighters, and an initial batch of Avro Ansons, DH Moths and Avro Trainers, together with old Supermarine Southamptons and ancient Seagulls. Air Vice-Marshal R. Williams, the CAS, had under his command some 3,000 men of all ranks, stationed at fourteen bases. As Major the Hon R. G. Casey, Commonwealth Treasurer, said: 'If Australia has to defend herself without the United Kingdom's help it will be a very poor outlook. Left to our own resources we could not hope to resist a determined aggressor.' Expenditure on air defence had increased in the past five years from some four million to the currently proposed £16½ million. Brigadier Street, the new Minister for Defence, was working on estimated total expenditure of £63 million by the end of the financial year 1940–41.

Whatever the glamour of flying there was no mercenary incentive to join Britain's RAF. A Squadron Leader's initial pay was only £1. 10*s*. 10*d*. per day, rising by a mere 3*s*. 4*d*. after eight years, though recently an Air Commodore's pay had been increased from £2. 18*s*. 4*d*. per day to £3 1*s*. 0*d*. Officers were retired on less than half pay after twenty-five years' Service, and even an Air Chief Marshal then received only £1,300 a year.

France, already with what was considered the largest air force in the world, proposed to increase personnel from 2,550 officers and 40,000 NCOs and men by 50 per cent, with more in 1939. Nevertheless General Vuillemin, Chief of the French Air Staff, who favoured an air pact between France and Germany, visited Field Marshal Göring for discussion of what was described as 'foreign policy'. To emphasize France's air strength a great *Fête de l'Air* was held at Villacoublay on 10th July, though hampered by low cloud and drizzle. Some of the latest French single-and multi-engined fighters and bomber equivalents of the Hampden were demonstrated, and by invitation of the French Government the RAF participated with formation flying of nine Hurricanes and three tied-together aerobatting Gladiators. 'The admirable President of the French

Republic, M Albert Lebrun, air-minded as always, attended the display, and cordially raised his bowler hat while the Armée de l'Air's band with exaggerated zeal repeated the last eight bars of God Save the King.' To further strengthen the ties of reciprocity and alliance, King George VI and the Queen visited France a week later, but travelled in the Admiralty's *Enchantress* escorted by units of the RAF. So concerned were the French for the safety of royalty that during their stay all private aeroplanes were forbidden to take the air.

While the King was still in France, exercises involving Britain's Coastal Defence culminated in a major combined operation of Navy, Army, and RAF from 20th to 23rd July, putting to test the defences of the Thames, the Medway, Harwich, Tyne, and Firth of Forth against sea and air attack, but concealing that it was designed to endorse the practicability of the secret line of radar stations from London to Scotland. The Blue attacking force had four battleships, an aircraft-carrier, four cruisers, fourteen destroyers, five submarines, and appropriate aircraft of the Home Fleet. Red defenders were limited to two cruisers and four destroyers but had the land fortresses of Forth, Tyne, Harwich, Thames, and Medway manned by Territorials and primarily relied on two squadrons of Saro London flying-boats, one with Supermarine Stranraers, and a mainstay of five reconnaissance squadrons of Ansons and one of Vildebeests backed by two squadrons of Gauntlets, one of Gladiators, and one of Hurricanes. Attack and counter-attack followed under difficult conditions of fog and low cloud. A special effort was made to interest the public in the progress of the exercises, and Pressmen were offered an air view of operations each day, though more seats were available than journalists willing to fly. Those who did were 'enormously impressed by a wizardry which brought an aeroplane across the open sea to arrive at the exact spot intended by the pilot.' To what extent that was assisted by radar plotting and radioed instructions remained a closed book. On the second day the entire Blue fleet had apparently been demolished, but it was allowed to continue the war in order to give aircraft and radar more practice. Significantly no attacking unit managed to get anywhere near the coast without being reported by defending aircraft despite continued poor visibility.

While the exercises were in progress the Short *Mercury*, piloted by Capt Donald Bennett, with radio operator A. J. Coster as crew, made a successful full-load separation from Capt Wilcockson's *Maia* at low altitude above Foynes on 21st July, and flew to Boucherville, Montreal, in twenty hours twenty minutes, having covered 2,930 miles against a head wind of 25 mph, and still had a marginal 80 gallons of fuel. After topping up, *Mercury* was flown to Port Washington, Long Island, on 25th July, and three days later, its duration limited as a seaplane, was flown back to Botwood, next day to the Azores, and so to Hythe via Lisbon on the 27th, refuelling at each stage. A fortnight later some of the satisfaction was diminished when a Lufthansa Focke-Wulf Condor landplane made a flight greater by 1,000 miles from Berlin to New York, returning nonstop two days later in slightly less time than *Mercury*'s outward flight. Nevertheless Bennett was convinced that *Mercury* could break the world's seaplane distance record, and the Air Ministry agreed to this some weeks later, but was more particularly interested in the RAF's preparations for an attempt on the world's landplane distance record by one of the four long-range Wellesleys which by now had returned to Upper Heyford.

8

On 23rd July Sir Kingsley Wood with Harold Balfour, Sir Francis Shelmerdine, and Sir Donald Banks, held a Press conference to announce the long-rumoured formation of the Civil Air Guard (CAG) whereby healthy applicants of either sex from eighteen to fifty would receive subsidized pilot training provided they undertook in a national emergency to serve in whatever aviation capacity they might be directed. Training would be at light aeroplane clubs, which would receive a doubled subsidy of £50 for every 'A' licence obtained. Ten hours subsequent annual flying per member would be subsidized at £2 per hour, and tuitional charges must not exceed 10*s*. per hour at week-ends and half that on days with less pressure. Clubs must undertake to operate a CAG section of at least twelve members. The Marquess of Londonderry was appointed Chief Commissioner, and the five other honorary commissioners were Mrs F. G. Miles representing the women, Lindsay Everard MP, who had done so much for private flying in Leicestershire, Major Alan Goodfellow of the Council of Light Aeroplane Clubs, and Robert Murray representing Scotland as a member of the Glasgow Transport Flying Club.

A deluge of membership applications followed. De Havillands saw this as the great opportunity to introduce the DH 94 Moth Minor as the logical successor to the Moth training biplanes, and put it into production at the knock-out low price of £575 for the open-cockpit version and £690 with coupé top enclosure. Not even Miles could do better. In the two years in which the Minor had been held in abeyance every pilot member of the sales, technical, and works staff had been allowed to fly it in both standard and experimental tricycle forms, but eventually the prototype crashed when young Geoffrey de Havilland and his assistant pilot, ex-apprentice John Cunningham, found recovery impossible during aft centre of gravity spinning trials. Both baled out, saw the empty monoplane plunge past as their parachutes opened, and were picked up by cars within a few minutes, then 'went to the Chequers at Cromer Hyde for lunch with Thom, Buckingham, Clarkson, Peter de Havilland, and Sharp, and the matter was hardly referred to.' The cure was the familiar tail chine fairing of the Tiger Moth.

Some twenty other companies, however ill-financed, hoped their light aeroplanes might be adopted by the CAG, and so did a dozen importers of Continental and American aircraft. Even General Aircraft felt they might back the desultory sales of twin-engined Monospars by taking constructional rights for the CW Cygnet all-metal low-winger as well as producing a 50 hp two-seat tricycle pusher monoplane selling at £500. Nick Comper was designing a somewhat similar single-seat pusher. Both were to discover that pushers had big trim differences between engine on and engine off.

However, there had been a shattering disaster with another hopeful ultra-light, for Bristol's world famous designer Capt Frank Sowter Barnwell was killed in a 25 ft span low-winger he had built for fun flying, powered with a two-cylinder inverted Scot Squirrel of a mere 28 hp. Nobody could claim he was a good pilot, but he was always keen to experience the airborne handling of each of his designs, though Cyril Uwins and the Bristol directors did their best to thwart him – which was why he designed this little machine. He made an initial short flight on 30th July at Whitchurch, newly dignified as Bristol Airport, and in the sunshine of 2nd August came again, spent a little time adjusting the engine, taxied to the far end, took off in 200 yards, climbed slowly to 70 feet only to dive suddenly into the ground and was instantly killed. Almost certainly he stalled, for the Inspector of Accidents reported that the engine was found in satisfactory condition, but he commented that Barnwell's log book showed that of his total of 485 hours since 1915 he had only flown thirty-one hours in the past five years.

He was universally mourned. I first met him in 1925 when I flew with Major Laurence Openshaw, the Westland pilot, to Filton, and as a student looked with awe upon this balding but legendary designer of the Bristol Fighter, this man with blue eyes scanning far horizons. Born in 1880, he became a schoolboy at Fettes, and from 1898 to 1904 was apprenticed with the Fairfield Shipbuilding Co at Govan where his father was partner, attended evening classes at the local Technical College and also in the winters of 1900 to 1905 at Glasgow University where he obtained his BSc in Naval Architecture. Already the idea of flying lured him and his brother Harold, and they made two gliders in 1905, then a fragile biplane aeroplane in 1908 fitted with an inadequate Peugeot motor-cycle engine. After a year in the USA, Frank joined his brother to form the Grampian Engineering and Motor Co Ltd at Stirling, and in 1909 they built a more rational but relatively huge 40 hp canard biplane of 48 ft span and great weight which hopped 80 yards before breaking in a heavy landing. Before the next design, a neat mid-wing monoplane, could be flown, Frank joined the British and Colonial Aeroplane Co in March 1911 as a draughtsman. His co-draughtsman Clifford Tinson recorded: 'He was ten years older than I at that time, but where some seniors are tempted to throw their weight about to emphasize their superior knowledge he was never like that, but was modest and unassuming, genuine and sincere, and entirely lacked arrogance. While not a man to suffer fools gladly, he was ever polite, scrupulously fair, and would never say anything unkind to rebuff or humiliate a person.

'He was a draughtsman of the old school, of the days before mass production, extremely neat and methodical in everything he did. Designs were prepared on cartridge paper in Indian ink, with centre lines drawn in red and dimension lines in blue, and the complete job was quite a work of art. He had a very high standard as a draughtsman and maintained it throughout his life. Because he was so painstaking he set an example which many followed, and among those who were on his staff it was never difficult to identify his disciples for that reason. He was indeed beloved and respected as a man and admired as a designer by all whose good fortune it was to work with him.'

Cyril Uwins, long the chief pilot and CFI of the School, wrote to me: 'He was a charming character of great personal integrity fully reflected in his approach to engineering problems. His aeronautical theory was in advance of the times, and my work benefited enormously from close association with him. A fanatical pilot, but regretfully not very skilled, his crashes were many. Pollard was his key man in our metal structural design including *Britain First*, the Blenheim and early subsequent types, though Frise and Russell were somewhat senior to him. Sir Stanley White, an extremely shy man, with little technical knowledge, relied on them all completely, but was always helpful in difficulty.'

Sir Stanley appointed the dour but able Leslie Frise, BSc, as Bristol's chief designer. In later years his one-time assistant, who eventually became chief designer as Sir Archibald Edward Russell, recounted: 'Frise now conceived the idea of designing a new fuselage for the recently developed twin-engined Beaufort reconnaissance torpedo-bomber to make it into a fighter with the more powerful Hercules engines. This proposal covering a single sheet of paper plus a drawing, was taken to Wg Cdr Saundby of DOR, as he was then, and approval was given on the spot to build four prototypes.' This was the two-seat Beaufighter with four cannon, intended as a stand-by in case the Westland Whirlwind cannon fighter designed by Petter and Davenport proved a failure, for an earlier Bristol design very like a Whirlwind with Hercules engines had been turned down.

Though Barnwell's crash closed an historic chapter, far too many others met their end in aircraft accidents. Only two days earlier America's Frank Monroe Hawks, the best known of all racing pilots, and who held the American trans-continental speed records, was killed when his small private aeroplane, on taking off from his private estate, struck electric power cables, crashed and burst into flames.

Concurrently a Dragon of Carol Air Ferries Ltd piled-up while attempting to land in bad visibility at Land's End aerodrome, the pilot killed and six passengers badly injured. Within days the Imperial Airways *Amalthea* of the Atalanta class crashed into a hillside near Kisumu, Kenya, a few minutes after taking off for Alexandria and all the crew were killed, though luckily there were no passengers.

In the Pacific the *Hawaiian Clipper*, piloted by veteran Capt Leo Terletzay, disappeared after signalling she was flying through rain and rough air 800 miles off-shore between Guam and Manila. Eventually an Army transport ship saw a great patch of oil and petrol some fifty miles from the last reported position, but no wreckage was discovered. Possibly the machine broke up in turbulence, for civil airliners, then as now, were designed to lower factors than military aircraft. Because of limited blind-flying experience, military as well as civil pilots still preferred to go under the weather rather than over, and surprisingly the majority of RAF aircraft still had no radio, nor did civil test pilots get such equipment until two years later. Flying remained a game of chance.

Within a week a Czechoslovak Savoia Marchetti airliner hit the top of the

Buchwalder Kopf near Strasbourg in dense fog and the crew of four and sixteen passengers were killed. A few days later the burned-out wreckage of a Mexican Douglas DC-2 airliner and the bodies of the crew and passengers were found near Vera Cruz on a mountainside into which it had crashed during blinding rain.

Experimental aircraft undoubtedly had their own perils. Guy de Chateaubrun, one of the best liked owner-pilots of France, was killed while flying the experimental tandem-winged Delanne 20 powered with a 185 hp Regnier engine and intended as a flying model of a new two-seat fighter. Apparently he was in a spin, and failed to recover, for, as I found on a later flight with a second model of this machine, the rudders mounted at the tips of the tailplane had too little lever-arm to be sufficiently powerful at slow speed.

Nevertheless the long fuselage of the Albatross gave its similar rudder disposition sufficient 'tail-volume' for directional stability, and as each rudder was in the combined slipstream of two engines a side they were fully effective but heavy to operate, necessitating full trimming tab with one engine out of action. By now the second Albatross, the unpainted E-3, arranged as a mail-carrier like the prototype G-AEVV at Martlesham, was on test at Hatfield. On 27th August Geoffrey de Havilland Junior taxied out for a series of take-off tests at maximum overload weight. 'The first landing went off well, though the undercarriage or tailwheel seemed to be making a great deal of noise,' said a witness. 'During the second landing the machine appeared to be flexing somewhat. After the third landing, as the Albatross slowed, a door in the middle of the fuselage flew open, there was a cracking sound, and the back of the fuselage broke just abaft the trailing edge, cocking up the nose and wing onto the trailing edge.' Devastated, the company at first tried to hush up the accident, but an amateur photograph in a newspaper next day showing the wrecked aeroplane prompted an official though not specific explanation.

The two halves of the fuselage were separated and reassembled in the jig after discovering that a vital wooden member had not been fitted owing to an error in workmanship and inspection. Rectified and re-built with 12½ lb of additional strengthening, the machine was ready to fly again within five weeks. Meanwhile the first of the five Albatross airliner versions, seating twenty-two passengers, was completed and flown, presently going to the A & AEE for flight certification on behalf of the ARB.

9

All new proposals involving big expenditure were being almost instantly settled by the Supplies Committee re-constituted as the powerful Air Council Committee of Supply (ACCS) which met twice a week, chaired by the striking and persuasive Sir Arthur Street, with Freeman as aeronautical expert, Sir Harold Howitt advising on contractual matters, Edward Bridges, 'a youngish, shy man with short shaggy, greying hair who had the schoolboy habit of chewing the end of his pencil' represented the Treasury with authority to give immediate financial approval, and Henry Self, 'a massive man in every sense' who was a Double First and brilliant administrator and had been called to the Bar but never practised, was to adjudicate legal aspects.

As a result of the Maybury and Cadman reports, civil aviation had become so completely divorced from the Air Ministry that an Order in Council on 23rd August authorized a Licensing Authority for home airlines, the members of which were A. M. Trustram Eve, KC, as chairman, Maj-Gen Sir Frederick Sykes, long forgotten as the first Controller-General of Civil Aviation, and Mr F. R. Davenport, a government official of whom the air transport industry knew nothing. To secure a licence, airlines must submit to the Authority all relevant information about the need and possibilities of proposed routes, details of the company's finances, and description of the proposed type of aircraft. Fortunately the licence only cost £10 per annum, but inflicted the unwanted chore that multi-returns of activities must be made. Immediately there was a rush of applications.

Summer also saw the sport of sailplaning expand. In a combination of boredom and excitement at Dunstable, the world's duration record was broken by Flt Lieut W. B. Murray and 16-year-old, spectacled John Sproule, who patrolled the ridge for twenty-one hours in their side-by-side two-seat Falcon III – but a few weeks later two Austrian pilots flew for almost forty-one hours, and later in the year two German pilots stayed aloft in their two-seat glider for two days two hours fifteen minutes.

The longest British sailplane flight was Philip Wills's 209 miles, but the world's record was 405 miles by a Russian pilot from Moscow to Saratov the previous year. The glider height record was held by a Lufthansa pilot with 23,190 feet using a Minimoa aero-towed to 3,000 feet. With a similar machine Philip Wills managed 10,800 feet, but as *The Aeroplane* commented: 'Great Britain does not offer the extremes of topography or weather giving the best chance for highest performance in motorless flying, but native design and construction is improving fast, together with the number and skill of soaring pilots.' Though cross-country flights totalled 740 miles in twenty-two journeys during Britain's National competition, the Germans at the Wasserkuppe summer meeting aggregated 47,000 miles from 1,300 launches and 2,750 flying hours were logged. To what purpose would those pilots be put?

Though Germany held the landplane air speed record and Italy the world's absolute air speed, Great Britain revived her prestige, and added to the laurels of the 2,300 hp Rolls-Royce 'R' engine, with a new world's automobile land speed record of 345.49 mph achieved by George Eyston in his *Thunderbolt* car. There followed great competition between him and John Cobb in the Napier Railton with twin 1,320 hp Napier Lion VIID engines at Bonneville Salt Flats, Utah, Cobb achieving 350.2 mph over the measured mile on 15th September and next day Eyston managed 357.5 mph.

A month earlier *Queen Mary* had regained the *Blue Riband* of the Atlantic from the *Normandie* with a record westward crossing in three days twenty hours forty-two minutes, but was likely to be surpassed by the world's largest liner, *Queen Elizabeth*, launched at Clydebank by the Queen on 27th September. The great shipbuilding firm of John Brown & Co who had built her now announced the acquisition from Petters Ltd of a controlling interest in Westland Aircraft Ltd. This represented the ultimate triumph of Sir Ernest over the efforts of Peter Acland and and Teddy Petter to oppose earlier proposals whereby the white-haired, elderly Petter twins hoped to recoup their long and recently profitable gamble of aircraft construction as a supplement to their main business of oil engine manufacturers which was concurrently sold to the Brush Electrical Engineering Co Ltd, and removed lock, stock and barrel to Loughborough leaving Westland with the entire manufacturing premises.

The John Brown interest had been aroused by thirty-two-year-old Eric Mensforth, MA, small and wiry in stature and of quick mind and intense energy, who was the elder son of Sir Holberry Mensforth, an outspoken director of John Brown who had risen through the shipbuilding ranks to authoritative eminence. Eric Mensforth gained a First in Mechanical Sciences Tripos at King's College, Cambridge in 1927, and then served an engineering apprenticeship with Mather & Platt Ltd in Manchester and continued with Bolchow Vaughan in Germany. After experience with steel production in the USA he became technical engineer to Dorman Long & Co Ltd and subsequently works manager of their Preston Works, but in 1937 joined Markham & Co Ltd, Sheffield, which John Brown had acquired the previous year, when he was made a local director. In that capacity he had been responsible for recent construction of an assembly shop at Westland bigger than the entire Petter oil engine factory.

Shortly afterwards, Sir Kingsley Wood had asked Lord Aberconway, the somwhat testy, autocratic lawyer-chairman of John Brown, to interest his Board in acquiring an aircraft business, then strengthen and expand it. Eric Mensforth, knowing the extensive Lysander contract, told his father that Westland might be suitable, though John Brown had been thinking of a flying-boat venture. The Board agreed. Lord Aberconway replaced Alan P. Goode, as chairman. Sir Ernest Petter left the Board. Dynamic Sir Holberry Mensforth and Stanley W. Rawson, the clever secretary of John Brown, tall, quiet, and with pebble lenses, became directors. At the first Board meeting Eric Mensforth and Peter Acland were appointed joint managing directors, and Teddy Petter was reappointed technical director.

The stage seemed set for shipbuilders to command the aircraft industry, for Vickers had Vickers (Aviation) Ltd and Supermarine; Swan Hunter & Wigham Richardson Ltd owned Airspeed Ltd; William Denny & Bros Ltd were associated with Blackburn Aircraft; and Harland & Wolf had Short & Harland Ltd. The latter were still struggling to build their fifty Bristol Bombays – but the main Short business at Rochester not only was steadily launching Sunderlands, in one of which Sir Wilfrid Freeman and Ernest Lemon flew, but was heavily involved in the design of the S 32 airliner and the four-engined Stirling of which the half-scale model S 31, designated M 4, was, as Chris Barnes records 'first flown by John Parker in great secrecy at Rochester Airport on 19th September after half-an-hour's taxi-ing trials on the 14th.' After eight further flights Parker flew it to Martlesham on 21st October. Handling was remarkably good, but the A & AEE pilots considered that the ground angle should be increased three degrees. Although the same lightness and quickness of response would not be expected in a machine twice the size and ten times the weight, these tests were encouraging confirmation of the general design of the full-scale bomber.

Despite the apparent strength of the shipbuilders, Hawker Siddeley were by far the biggest consortium, and their large new aircraft factory and aerodrome for Hawker Aircraft on 200 acres at Langley was nearing completion.

Hawkers had just suffered an unaccountable disaster. John Hindmarsh, Lucas's assistant and keen motor-racing driver, was killed when his production Hawker Hurricane emerged from low overcast cloud in an almost vertical dive and continued into the ground, smashing in to smithereens and burying the engine six feet deep. There was secret concern that the cause might be a shock-wave making the machine uncontrollable, for a nose-down change of trim in a TV dive had been noted with several prototypes.

A few days later cheery young Sqn Ldr M. J. Adam, who had regained the height record for Great Britain in the summer of 1937, was killed with his co-pilot and civil technician when his Vickers Wellesley crashed and caught fire half a mile from Farnborough aerodrome immediately after take-off.

Another well known pilot lost his life on 21st September – David Llewellyn, son of the President of the Royal Academy, who first came to fame with Mrs Wyndham's Parnall Heck by breaking the Cape to England record held by Amy Mollison. Since the death of Bill Davis he had been CFI at the Cinque Ports Flying Club. In that capacity he was flying with Lieut J. B. Kitson of The Blues in one of the Club's BA Swallows when it stalled and crashed close to Lympne aerodrome shortly after take-off.

9

An ominous announcement by the German Government prohibited all flying from 10th September over the Rhineland border except in special corridors. Ever since the spring Czechoslovakia had been the danger spot. Early in August Chamberlain had sent Lord Runciman, chairman of the great shipping firm of Walter Runciman & Co Ltd, to Prague as adviser on matters of conciliation. In mid-August German military manoeuvres were being ostentatiously displayed on the border. Attack seemed imminent. On 15th September Neville Chamberlain departed from Heston in a Lockheed Electra of British Airways flown by their chief pilot Capt C. N. Pelly, bound for Oberwiesenfeld Airport, Munich, *en route* to visit Hitler at Berchtesgaden for personal discussion, accompanied by Sir Horace Wilson, chief industrial adviser; and William Strang of the Central European Department of the Foreign Office. They were seen off by Lord Halifax and other officials including Dr Kort from the German

Embassy who complimented Chamberlain on his 'magnanimous initiative.' A more cynical commentator remarked: 'The trip was about as ordinary as it could be, for the principal passenger was an elderly and distinguished gentleman who had never flown before – yet no flight ever made has more thoroughly caught the imagination and attention of the world.' Chamberlain was convinced that what appeared a simple dispute between the Czechoslovak Government and its internal German minority could be resolved by wise concessions. Yet already Schuschnigg, his Government, and thousand-upon-thousand Austrians had been arrested. By now nobody in Britain had the illusion that we were not on the verge of war, yet only a few months earlier there was widespread disbelief that war could be possible.

At Munich, Chamberlain was met by von Ribbentrop as German Foreign Secretary, Sir Nevile Henderson, the British Ambassador in Berlin, and Herr von Dircksen, the German Ambassador in London. Crowds gathered to see him arrive, and Union Jacks mingled with German flags on the airport buildings. He continued by train to Berchtesgaden where, despite rain, more crowds lined the route to Hitler.

Next day he returned to England. On Sunday 18th September, Edouard Daladier, Prime Minister of France, and his Foreign Minister, flew to Croydon, and conferred at 10 Downing Street all day with Chamberlain, Lord Halifax, Sir John Simon and Sir Samuel Hoare, and it was announced that Daladier had agreed to support Britain's bid for Peace. In fact it was a plan for extensive cession of Sudetenland territory, forced upon General Syrovy, head of the new Czechoslovak Government, who said: 'We shall not attempt to throw the blame, but leave this to the judgment of history.' Most Britons felt that the Czechs had been disgracefully let down, and Chamberlain's popularity fell.

Forthwith this tall old man with an umbrella flew again to Germany on 22nd September, and met Hitler at Bad Godesberg on the banks of the Rhine. The Führer, believing he had Chamberlain on the run, dismissed partition boundaries agreed in London, and demanded immediate military occupation of a still greater area for which he refused international supervision of the takeover. A dispirited Prime Minister arrived back at Heston in the early afternoon of 24th September, and in a message to the public said: 'My first duty is to report to the British and French Governments. I will only say this. I trust all concerned will continue their efforts to solve the Czechoslovakian problem peacefully, because on that turns the peace of Europe in our time.'

The King signed a State of Emergency. The Navy was mobilized. Slit trenches were dug in London parks. Buildings were protected with sandbags. Trainloads of children with gas masks were evacuated to the country. London hospitals were made ready for casualties. The secret radar system was on 24-hour watch. Chamberlain wrote a last appeal to Hitler. In the Commons, amid 'scenes unexampled for enthusiasm and emotion,' Chamberlain read the Führer's reply inviting participation in a conference at Munich with Daladier and Mussolini. Bitterly C. G. Grey said: 'After all the talks during 1914–18 about a War to end War and self-determination of oppressed minorities, this is where weakness and disarmament have led us. Every country in Europe is mobilizing.'

That night, Chamberlain, whose Birmingham wood-screw business of Guest, Keen & Nettlefold had recently been joined by Tom Harry England disenchanted with Austins, broadcast to the nation: 'How horrible, fantastic, incredible, that we should be digging trenches and trying on gas masks because of a quarrel in a far-away country between people of whom we know nothing.'

For the third time he flew to Germany on 29th September, and at Munich met Mussolini; and Daladier and their entourages. Next day they conferred with Hitler 'the re-builder of the German Nation', and soon after midnight conceded all he wanted, but he accepted international supervision.

Before leaving Munich on 30th September Chamberlain notified the world's Press: 'This morning I had a talk with the Führer, and we both signed the following declaration:

> "We, the German Führer and Chancellor and the British Prime Minister, have had a further meeting today and are agreed in recognizing that the question of Anglo-German relations is of first importance for the two countries and for Europe. We regard the Agreement signed last night, and the Anglo-German Naval Agreement, as symbolic of the desire of our two peoples never to go to war with one another again. We are resolved that the method of consultation shall be the method adopted to deal with any other questions that may concern our two countries, and we are determined to continue our efforts to remove possible sources of difference and thus to contribute to assure the peace of Europe."'

That afternoon at Heston, as Chamberlain stepped from the cabin doorway of the shining Lockheed, he triumphantly waved a small piece of paper. 'We have plucked this nettle danger,' he said. He drove to Buckingham Palace to report to the King, and from the steps of 10 Downing Street waved to the cheering crowds, and declared 'I believe it is Peace for our time...'

Next day the Commons voted 366 for the Munich Agreements and 144 against. Thirty Conservatives disassociated themselves from Chamberlain, including Winston Churchill, Anthony Eden, Harold Macmillan, Duncan Sandys, and Robert Boothby. Clement Attlee vigorously asserted that 'A gallant and democratic people had been betrayed and handed over to a ruthless despotism.' Duff Cooper, First Lord of the Admiralty, resigned in protest. The *Star* said: 'A deep yearning for peace is shot through by dismay at the humiliation of surrender.'

The day before Chamberlain's first hopeful flight to Germany a new rigid airship, of much the same length as the ill-fated *Hindenburg* but a fifth-greater capacity, was launched at Friedrichshafen and named *Graf Zeppelin II* (LZ 130), the original having been scrapped. Piloted by Dr Hugo Eckener and Capt von Schiller, with thirty-five crew and seventy-four passengers, an eight-hour trial flight was made. Because the USA refused to export helium, the airship used hydrogen, but special precautions had been taken to avoid fire risk and to condense the water content of exhaust gases as additional ballast.

The limelight briefly switched to this event and that. Sir Kingsley Wood MP, was appointed Honorary Air Commodore of No. 901 (County of London) (Balloon) Squadron. Harold Balfour completed a 12,000-mile air tour of RAF stations overseas. The Royal Airship Works at Waddington was renamed the Balloon Development Establishment. There were the inevitable RAF fatal accidents – six in a week. In a Japanese air raid on Nanking 500 people were killed. Charles Lindbergh was invested with the Order of the German Eagle by Field Marshal Göring. Roy Fedden was elected a Fellow of the Royal Society of Arts. Igor Sikorsky visited England to offer his big flying-boats as rivals to Short's. But such things were among the trivia of the sweeping saga of aviation: significant or meaningless, normal or tragic, profit or loss.

At Vickers-Armstrong Ltd the clash between Commander Sir Charles Craven and Sir Robert McLean became unedurable and Sir Robert left. Alex Dunbar accepted Sir Charles's invitation to take charge of the aviation interests, for much earlier he had been at Weybridge for almost seven years, and subsequently served in every branch of the Vickers Group. He was a director and general manager of the English Electric Steel Corporation as well as director of other companies including Firth Vickers Stainless Steel. As co-directors at Vickers he had Sqn Cdr James Bird, Frederick Yapp, Major Hew Kilner, and John Reid Young as accountant, though all were fated to retire in December. Pierson and Wallis, after three years, were still trying to finalize their big B 1/35 Warwick bomber due to changes of mind by the Air Ministry on the types of engine and their delayed development.

Choice of the small Rolls-Royce Peregrine had similarly dictated the characteristics of Westland's striking four-cannon twin-engined F 37/35, presently known as the Whirlwind, which had been taken by lorry to Boscombe Down for first flight above a countryside giving better chance of a forced landing than around Yeovil. As Petter's personal assistant, Dennis Edkins, recollected: 'This aircraft, potentially a winner, had three things against it: an engine that never came up to expectation even though made by Rolls-Royce in their heyday; insufficient effort and understanding by the works management of what it takes to get an aircraft from development into production; too many technical innovations such as a one-piece high-lift Fowler flap from aileron to aileron which also controlled radiator cooling, ducted radiators of light alloy with horizontal tubes and reverse-flow, plenum chamber inlets, exhaust carried dangerously through the petrol tank, pure magnesium monocoque structure, butt-jointed longitudinal plating, high strength steel castings, torsion taken on the engine cowls, sealed inset balanced nose for controls, offset rudder hinge – all adding up to longer development time than necessary.

'Petter certainly had a good grip on development process of such a highly complex product and well understood the necessity of applying resources to ensure timely development cycles. One of my jobs was to extract and plot much data on rates of expenditure on tooling and hardware, and these were used in fairly frequent memoranda to the managing director to the effect that "we obviously couldn't expect to have such and such a component ready on the date quoted because the effort expended isn't enough as proved by our calculations." He was also good at structuring an organization, and well aware of the need to give this full attention, but was less than successful in picking some of his staff.' Nevertheless it was his senior designers and draughtsmen who carried the job through, for despite the streak of original genius, Petter was still relatively inexperienced when he started the design.

Smoothly finished in dark grey, the prototype L6844 made its first flight in the calm early afternoon on 11th October. Short taxi-ing runs had previously been made at Yeovil where the idiosyncrasies of the Exactor throttles for the oppositely rotating engines were experimented with. Normally a slight hydraulic pressure could be felt when the levers were pressed forward, but at times a large arc of movement made no difference to the rpm, the cure being to re-prime the system by pulling back the throttle and pumping it back and forth – but there remained the potential danger that an engine might not respond quickly in emergency.

Boscombe Down, like all RAF stations, still had a grass surface. Preliminary runs were made on the modestly long east to west length, but a stop was necessary because the engines were steaming. The eventual third run was too ambitious, and on throttling it seemed the machine might overrun because of the aerodrome slope, so power was fully opened and a take-off made with a marginal hundred yards to spare before reaching the fence. This alarmed the Westland and RAF spectators. As always with an initial flight, only tentative investigation of characteristics was made after reaching a parachute-safe height, but in the course of half-an-hour it was obvious there were various imperfections of control. Thus the side-hinged rudder was ineffective over the first five degrees and too heavy at big angles; there was a directional oscillation with the rudder free; the tip slats extended with an unacceptable jerk; as speed increased there was a suspicious nose-down tendency; in tight turns the machine shuddered, suggesting interference between fin and tailplane despite the latter's high position to overcome downwash when the flaps were lowered. As for the engines, performance was satisfactory except that they were running too hot – and I still did not like the smoke and flames exiting at the trailing edge. Inevitably in later days there came a moment when an exhaust pipe burned through as predicted – but amazingly the tank of fuel did not burst into flames; instead the incandescent gases burned through the aileron push rod, and instantly the machine went out of control with a vicious starboard roll as the Frise aileron on that side flew up to full lock. Fortunately it was possible to hold the other aileron at equivalent lock and skid level, then fly home with both ailerons upturned, using rudder alone.

Four days after the Whirlwind's flight, the Bristol Type 152 Beaufort twin-engined reconnaissance-torpedo-bomber development of the Blenheim was given its first flight by Uwins at Filton, eleven weeks after its famous designer's death. As far back as September 1936, seventy had been ordered, and this was the first production machine. The special low-drag Taurus nacelles with vertical flanking air exits proved too heat retentive, and soon normal gill cowling had to be fitted, but still the engines were too hot. As with most aircraft, one or other undercarriage set dropped first, but in this case produced a directional swing due to asymmetrical drag from the Blenheim type door across the twin legs, later overcome by using sideway opening doors. Turbulence from the blunt dorsal gun turret termination of the raised fuselage also presented a problem, and the sequence of test-flying defects and modifications, particularly engine and oil cooling, pointed to far longer development time than envisaged. Meanwhile the shops were not only busy with these machines but also the Bolingbroke, now known as the Blenheim Mk IV which was similarly in production by A. V. Roe and Rootes at their respective shadow factories after a preliminary run of Blenheim Is.

10

Though the breathing space given by Chamberlain's 'Peace at any price' was urgently being used to increase production both in Britain and Germany, there was no hint of war's alarms at the great five-day Lilienthal-Gesellschaft International Conference opened at Berlin by General Milch on 11th October. 'This year's meeting was more pleasant than ever because everybody was so relieved at the passing of that absurd crisis which should never have been allowed to become an international incident,' wrote one visitor. 'I cannot see any of these technicians of many nations who got together so closely in Berlin, willingly turning their great abilities to make war on one another.' Certainly there were overtones of politics in Dr Baemker's speech as President of the meeting, for he emphasized the transformation of the German Fatherland during the past five years, and declared that Berlin as the capital of the Reich, Nürnberg the home of National Socialist rallies, München the heart of the Movement, and Vienna 'capital of the Ostmark whose glorious past stood unforgotten' would be given a new future by Adolf Hitler, creator of the Third Reich. There was music and dancing, and Sir Nevile Henderson presented the RAeS Gold Medal to Dr Eckener. Sessions of lectures followed from outstanding German scientists and designers, and by Igor Sikorsky on 'Flying-boats', Roy Fedden speaking in German on 'Bristol Aero Engines', D. W. Tomlinson of Transcontinental and Western Airways on 'High Flying Air Transport', Jimmy Ellor of Rolls-Royce on 'Supercharging', W. G. A. 'Bill' Perring of the RAE on 'Propeller Research', and Harry M. Garner of the M & AEE on 'Hydrodynamics in Relation to Seaplanes'. Every German aircraft and aero engine firm, each manufacturer of raw materials or parts and accessories, sent picked men to the Conference; but British attendance was only sixteen including the lecturers, the Air Attaché and his assistant, and the irrepressible 'Charlie' Devereux of High Duty Alloys saving the rest of the huge British aircraft industry from non-representation.

'Apparently the impact of this super Conference on the sensibilities of the RAeS has made them self-conscious about their miserable housing,' wrote C. G. Grey. 'For the Society to have just three rooms on the third floor in a side street off Piccadilly is an insult to British aviation. Whenever they arrange for a lecture they have to hire a hall in which its members can attend the discourses which have made the Society world famous. The Council has now fixed on No. 4 Hamilton Place in Park Lane as its new abode – but once again it has no lecture hall, so they must build one.' Thanks to the secretarial zeal of Laurence Pritchard the RAeS was prospering. In addition to its handful of eminent Fellows, there were 400 Associate Fellows, 30 Graduates and 500 Students.

Whereas RAeS lectures were usually too scientifically intense for newspaper publicity, those of the Royal United Service Institution often created considerable attention, such as the lecture on air raid precautions by the Deputy Under-Secretary of State, Home Office, on 26th October when he warned that almost nothing had been done, and an attack might come with appalling suddeness. 'One of the problems is maintenance of essential production during air raid conditions. The whole industrial population cannot go to ground if the war effort is to be maintained.' Of more immediate concern was the difficulty of securing the right type of recruit for ARP work, most of whom were artisans, whereas 'those who might have brought personal leadership have not come forward.'

In the Commons, Sir Samuel Hoare said one million volunteers were urgently required, and he blamed local authorities for failure to take action. If war came, a further half-million paid whole-time workers would be needed for ARP, and the policy would be worked out in detail by Sir John Anderson, who was appointed Lord Privy Seal and Minister of Civilian Defence on 1st November. Under questioning, Hore-Belisha, Secretary of State for War, admitted deficiencies in anti-aircraft organization and equipment, but said that at the accelerated rate of production almost the whole requirement of 3.7-in guns would be met by the middle of next year; the 4.5-in guns were in production, and we had acquired manufacturing rights for suitable guns against low-flying attacks, but he did not reveal they were Bofors, many of which would be imported. In the Upper House, Lord Strabolgi moved that an independent committee of inquiry be set up to examine the state of national defences with particular reference to air raid precautions, but having aired the disquieting state of defence to his fellow peers amid mixed agreement and disagreement, he withdrew his motion. Meanwhile the Labour Party's new manifesto bore the heading 'A Supreme National Effort for Peace' and described the defence of the country as 'criminally neglected'.

Such constant self-criticism must have been encouraging to the Germans – but at least Great Britain's prestige was enhanced with the RAF's achievement of the world's distance record by three Wellesley monoplanes, two of which, respectively piloted by Sqn Ldr Kellett and Flt Lieut A. N. Combe, taking off from Ismailia on 5th November, flew 7,157.7 miles nonstop to Darwin despite adverse winds which caused the third machine, flown by Flt Lieut

H. A. V. Hogan, to land for a top-up at Kœpang, Timor, after surpassing the USSR's record. Ismailia had been chosen because of its longer and drier take-off run, and allowed a great circle course, chiefly over British or Allied territory without crossing the Himalayas. As *The Aeroplane* said: 'No praise can be too high for the nine young men who undertook this journey. To sit for two days and two nights cramped in an aeroplane, flying much of the time over open sea, calls for a large supply of that three o'clock morning courage which Napoleon Bonoparte admired so much. Whether we are justified in asking such young men to risk their lives hanging on to a single propeller and aero motor can be argued. Everybody knows that Pegasus engines with Rotol propellers with run many times fifty hours nonstop, and nobody expects a Wellesley to be shaken to pieces by such a smooth-running motor, yet the best of motors stop sometimes, and when an aeroplane has only one the risk seems unjustifiable.'

Joseph Stalin, dictator of Russia, viewing the British long-distance triumph with a critical eye after the failure of his own team to achieve more than 4,000 miles, said on 27th October when his men returned from the Far East: 'The lives of our brave and sometimes rash air heroes are dearer to us than any record, however great and noisy.'

Though the Wellesley was the aerodynamic and structural guide for its twin-engined relative there were problems with the production type of Wellington which was a complete re-design of the prototype and incorporated structural features of the larger and more powerful Warwick. An early undercarriage failure and subsequent installation of Pegasus VII engines delayed tests for over two months, and then dives showed that the machine became increasingly nose-heavy with speed, and Pierson found himself at loggerheads with the aerodynamic experts of the RAE who proposed an inset-hinge balance instead of servo-tabs. Only when he reverted to horn-balanced elevators did the behaviour begin to improve, but it led to the usual sequence of small changes to horn balance and trim tabs, and there remained a big nose-up change of trim when flaps were lowered.

Though Wellingtons were delayed, the first production Spitfire had flown in July; the next went to Rolls-Royce, and the third to No. 19 Squadron at Duxford on 4th August; others following at weekly intervals.

Already Nos. 3, 56 and 111 Squadrons had been fully equipped with Hurricanes. Amazingly, in view of Sir Kingsley Wood's urgent plea for fighters, Hawkers were able to accept contracts for the Hurricane from Belgium, Yugoslavia, and Rumania. To meet the huge production requirement, a new shadow factory was being built at Hucclecote for Hurricane construction by the Gloster Aircraft Co, and negotiations were advanced for manufacture in Canada.

Tom Sopwith, his hair now greying, remained the enigma of the company, for he was seen but little, and like a wise leader dealt only with major questions, leaving the various businesses of the Hawker Siddeley complex to his fellow directors. His early close friend, Major V. W. 'Bill' Eyre had retired, and sharp-visaged Fred Sigrist, no longer fit and active, now spent long periods in the Bahamas. The mainspring was Frank Spriggs, his early role as office boy unsuspected by those who saw only a keen-minded, immaculately dressed, talented managing director of the Group – though it was Philip Hill, better known for his Covent Garden deals and Beecham-Eno flotation and activities with Woolworth and Rank, who had been chiefly responsible in forming the Hawker Siddeley combine. Pairing with Spriggs was H. K. Jones, 'once a very humble and hard working boy who cleaned gear boxes in the old firm of Panhard & Levasseur where Sigrist was then a leading engine hand,' and he now was general manager, the pair making a perfect team. F. I. Bennett, one of the originators of the Hawker Engineering Co, kept a smaller orbit, but still had great responsibility as works manager of Hawker Aircraft Ltd.

Hawker Siddeley's announcement of a $32^1/_2$ per cent dividend and cash bonus of 10 per cent did not obscure the imperative need of the Group to secure more funds to finance work in progress, so £5 million in short term first debenture stock was to be created. At the AGM, Sopwith said: 'We have been called upon to concentrate to the utmost on production of military aircraft and this has been a contributing factor to the loss sustained by one of our subsidiaries (Armstrong Whitworth) with a civil contract entered into with Imperial Airways prior to this Group's existence. In view of the unsettled world political situation it is unwise to attempt any forecast of the future, but it seems reasonable to anticipate a fair commercial return for the work and responsibility we are undertaking.' Meanwhile total assets were valued at almost £10 million, with liabilities of nearly £$3^1/_2$ million.

There was a reference by Sir John Reith at the fourteenth AGM of Imperial Airways on 14th November to the delayed Ensigns. Significantly he said: 'Our policy has always been to buy only British aircraft and engines, but if delays in delivery of new machines continues the Board may be forced to apply for permission to adopt some variation.' Nevertheless a gross profit of £296,824 had been earned for 1937–38, affording a 7 per cent dividend. Touching on political criticism he said that the company, mainly through Woods Humphery, had done magificently. The staff had grown from 1,200 in 1933 to 3,600, and worked in thirty countries, many climates, many languages, and twenty-five currencies. There had been great difficulties, but faults had been magnified and achievements belittled.

Within a fortnight Imperial Airways was in more trouble, for *Calpurnia* was wrecked on Lake Habbaniyah, Iraq, when trying to alight during a sand-storm. There were no paying passenters, but its Captain, Ernest Henry 'Tich' Attwood, and First Officer Alexander N. Spottiswoode, the flight clerk and radio operator, were killed and three others injured.

II

The bombs were still falling in Spain. Madrid, Valencia, and many another city was feeling the impact of Franco. On 28th October his brother, Lieut-Col

Ramon Franco, once the hero of Spain and currently commander of a Nationalist air base, was killed in a seaplane accident in the Mediterranean.

In Germany, General Felmy submitted to Göring a memorandum of proposed air operations against England. Thirteen squadrons were to be used against naval targets and to mine enemy waters. Thirty squadrons of Heinkel He 177 long-range bombers were to provide a strategic air force against England, and fifteen medium bomber squadrons and twelve dive-bomber squadrons against France. Twenty thousand aircraft were to be available by 1942. As a warning growl the British Air Ministry invited the Press and public to Southampton on 25th October to witness the transport *Nevassa* depart with a thousand RAF personnel aboard, the largest contingent ever dispatched to overseas stations in the peace-time trooping programme.

Armament expansion was gripping the USA. President Roosevelt stated that considerations of national defence forced him to reconsider the next American budget, with 'a major shift in allotment of funds towards rapid and heavy development of aircraft.' Bernard Baruch, his confidant on finance and European politics, had impressed him that Germany was threatening the USA in South America which he described as 'the last Continental outpost of free governments'. Some felt that all Baruch saw was a better opportunity of supplying armaments to Britain and France if the US aircraft industry was rapidly expanded. Currently, on 9th November, Theodore P. Wright, Director of Engineering of the Curtiss Wright Corporation spoke to a crowded meeting of the RAeS at the Institution of Mechanical Engineers on 'American Methods of Aircraft Production' – from which it was clear that they were leading the way with lofting full-scale layouts, use of forgings, castings and extrusions, and scientifically regulated proper and logical flow of materials, tools, and finished parts in precalculated time sequence through every stage of construction. In many ways it was an eye-opener to the British aircraft industry.

Whether for manufacturer or air transport operator the tempo was increasing. *Frobisher*, first of the five DH Albatross airliners and able to carry 22/23 passengers in stages up to 1,000 miles 'by day and by night', was ready to enter service on the London-Paris route. Flying qualities were stated to be particularly good. With either pair of engines out of action the Albatross 'with the aid of the powerful rudder bias will turn against the dead motors, and will also take-off and climb to 14,000 feet at full load on three motors.' H. A. 'Tony' Taylor who tried the controls sitting alongside young Geoffrey de Havilland, wrote in *Flight*: 'The Albatross seems very stable on all axes yet is remarkably manoeuvrable for its size. Admittedly this was being demonstrated by the test pilot who has done more flying with it than any other – but the cloud base was so low that steep or climbing turns demanded a confidence in the machine which was proof in itself. When I took over, my performance consisted of gentle turns and a little adjustment of trimming devices. A movement of an inch or so on the fore-and-aft trim wheel is enough to make the difference between a gentle climb or level flight. In steep turns no retrimming is necessary even when banking quite tightly; obviously the elevator control is sufficiently light, yet there is not a trace of instability in the fore-and-aft plane, and she flies happily on ailerons alone, or for that matter on rudder alone, though this is almost rigidly heavy. Except at very low speeds, directional control is not needed, and with modern speeds it is hardly to be expected that a pair of five-foot rudders will be easy to move. If attempts are made to lighten the control with servo-tabs there is always risk that the rudders will try to take charge during more violent corrections at low speeds. Undoubtedly the most interesting feature is the way in which the Albatross takes off more or less by itself; only incidental rudder corrections are necessary, and then only during the early part of the run. During take-off the slotted flaps were moved to the 30 degree position by a little four-position switch, the size of which is somewhat disproportionate to its importance. Once in the air the large switch to start the electric undercarriage motor is moved and the warning light beside it flickers from green to red. When reasonable height has been attained the throttles are brought back to cruise and the constant speed airscrew lever adjusted to hold revs at 2,200 when each engine gives 325 hp at the optimum altitude of 11,000 feet. During the final part of the approach and in landing the effect of the slotted flaps, compared with split flaps, seems marked, though approaching at about 80 mph is comfortingly steep without being too steep. The effect of being perched high in the nose, even of a very big machine, makes the process of approach and landing seem easy, and is much simpler than from some queer abode near the tail with a range of view measured in a few segregated degrees. Our landing seemed a bit of a wheeler – but that is how all aeroplanes should logically be put down. The undercarriage is there for use. When every transport aeroplane is fitted with a tricycle the fact will be appreciated.'

Flying-boats were also having their initiation. That autumn both the Saunders-Roe four-engined A 33 strutted high-winger and the S 36 Lerwick twin-engined shoulder-winger were test flown. Frank Courtney, star of earlier days, still with rimless spectacles, had reappeared from the USA and was given the task of flying the 90 ft span A 33 which had been built to the same specification as the Short Sunderland. On 10th October, after some days of engine adjustment, it was launched on the sunlit, unruffled surface of the Medina for taxi-ing trials which seemed promising enough to continue two days later under similar conditions. On the 14th, at a loading of 10,000 lb below its designed maximum of 41,500 lb, Courtney took it to the Solent where the sea was smoothest, and made the initial flight after a somewhat sticky take-off due to porpoising and wave impact on the wing-like sponsons. The combination of so many unconventional features necessitated cautious step-by-step testing, and the radically novel wing bracing, or what appeared to be lack of it, caused an expectant distrust. Courtney made several further flights, with the usual consequent minor adjustments, and then tried in somewhat heavier sea conditions, but under the impact of Solent waves on the broad big sponsons the machine began to porpoise, and within seconds dropped with a heavy impact, jarring the structure so hard that the starboard inner section of the monospar failed and the wing twisted so considerably that a propeller blade pierced the hull and sponson. A stand-by motor-boat rushed to the scene and towed the mangled flying-boat back to East Cowes slipway – but now that the fears of the sceptics had been confirmed the Air Ministry decided the damgae was too great to warrant reconstruction.

A month later the first of the Lerwicks to specification R 1/36 was ready, and gave the appearance of a scaled-down twin-engined version of the Short Empire boats. There were problems from the beginning. Despite the deep hull it was a wet boat, and while lifting onto its step the bow wave sometimes impinged on the propellers. Always there was a heavy swing to starboard which could not be fully corrected by the rudder and necessitated momentary throttling of the opposite engine. A tendency to porpoise was evident, resulting in months of work on step modifications. When Courtney tried single engine behaviour, the rudder was impossibly heavy and could not turn the machine against the other engine. Worst of all, take-off was so long that the service load had to be limited to 8,300 lb, and even when medium supercharged Hercules IIs were ultimately fitted to give greater power low down there was little improvement, and speed was disappointing, for it was 15 mph below the estimated 230.

12

Mustapha Kemal, known as Atatürk, creator of the new Turkey and builder of the Turkish Air Force, died on 10th November. Recently the Turkish Air Force had ordered Lysanders, convinced by their jovial 18-stone test pilot who watched the Westland prototype start its run apparently 150 yards in front of a cottage on the aerodrome boundary – though actually the machine was ruddered slightly to one side to give the impression of climbing over the cottage with a margin to spare. Taxi-ing the Lysander to exactly the same spot the Turk took off directly at the cottage. The ambulance man started the engine. The Westland party blanched. But he made it – just!

Rarely had firms a demonstrator machine of their own, but one was borrowed from the Air Ministry by 'contract loan', the risk of replacement being the manufacturer's. Thus a French pilot trying a Lysander at Yeovil after watching a demonstration which included a fully-slotted descent, was foxed by the steep slope of the north side of the aerodrome while endeavouring a similar slow landing and hit with a thump that snapped one undercarriage leg completely off at the built-in 'weak point' just beyond the outrigger strut attachment point, and then to everyone's horror, opened the engine and flew round before making another attempt inevitably ending in a nose-over.

In France, strikes delayed the opening of the annual Paris Aeronautical Salon until 25th November. While far outnumbered by the exhibits of twenty-eight French firms, Britain had a Blenheim, Hurricane, and Spitfire as representatives of the RAF's rearmament, and though de Havilland had abandoned such international occasions, Phillips & Powis were there with a Miles Monarch and Faireys displayed a Tipsy B two-seat trainer both on their British and Belgian stands. Newspapers made a headline display of Jeffrey Quill's 'record' fifty minutes with a Spitfire from Croydon to Le Bourget, and Sir Kingsley Wood accepted the invitation of the French Air Minister, M. Guy la Chambre, for an official visit. Accompanied by Air Vice-Marshal Welsh, he flew to Paris on 3rd December, and before returning next day they visited the Experimental Establishment of the Armée de l'Air at Villacoublay where the latest Service aircraft were demonstrated. Sir Kingsley told French reporters: 'I return to England strengthened in intention to achieve our common end – namely, to see that our defences are such that we may even more effectively exert our influence for peace.' As C. G. Grey ironically commented: 'Apparently this did not include presenting designs of the Spitfire to the French National-Socialist aircraft industry.'

That industry, under the great domed glass roof of the Salon exhibited its typical characteristics of opulence and stability, although at the SNC de Moteurs and SNCA du Sud-Est most of the workers were on strike, and SNCA de l'Ouest and SNCA du Midi had closed and dismissed the strikers. The Salon's opening ceremony had become a great formal occasion, with an impressive build-up for the arrival of President Lebrun 'who climbs rapidly the steps. Hastens through the doors. Embraces high notabilities. Circumambulates the aisles at a minimum speed of 30 kilometres, but arrests momentarily to greet chiefs of foreign exhibitors, for above all a President must be a diplomat.' For once international jealousies had been overcome, and the three British military aircraft were on one and the same Air Ministry stand, although tucked with Gallic discrimination in the south-east corner rather than a central position. Equally discreetly the German stand was inconspicuously placed, but its centre-piece, the twin-engined Dornier Do 17 slim-fuselaged bomber certainly attracted British attention. The Heinkels and Messerschmitts and Focke-Wulfs were ominously missing.

So to mid December and a trade agreement between the USA and Great Britain affording prospect of food and arms supply if there was war. British aircraft production for the year was mounting towards the target of 3,000. New types were about to be flown. First was the Taurus-powered Fairey Albacore three-seat TSR dive-bomber replacement of the Swordfish. Ungainly and an anacronism, it had none of the earlier Fairey elegance and could be mistaken for the 1934 Armstrong Whitworth AW 19 GP biplane. Not unexpectedly it handled similarly to the 1925 Fairey IIIF but the massive rudder was extremely heavy, and because of greater inertia the manoeuvrability was less than its grandparent. The steeply sloping nose and high position of the pilot in front of the wing gave a much better forward view than the Battle, and the relatively light wing loading gave ease of landing far greater than the more modern Blackburn Skua of which the first production machine was being performance-checked at Martlesham and the second having armament trials, but the third had been damaged at Worthy Down in a landing accident the previous month, and the next piled-up on 16th December during full-load trials when alighting on *Courageous*. However, pilots were soon to get accustomed to this type, and eventually it had its brief moment of glory as the first British aircraft to shoot down a German machine in the following year.

Currently the Skua's Fleet fighter development, the similary powered B-25 Roc employing a dorsal four-gun Boulton Paul power-driven turret in a wider fuselage, had its first flight in the hands of broken-nosed, bristly-moustached Flt Lieut Hugh Wilson, standing in for Bill Bailey who was aboard *Courageous*. Inevitably the Roc had the Skua's characteristics worsened by the discontinuity made by the gun turret and greater all-up weight. Subsequently Martlesham criticized the high landing speed, instability in turns, and general trickiness. As Stainforth, currently a Wing Commander, found when he flew L 3057 in order to write the official Pilot's Notes, it was nose-heavy, and a touch on the brakes turned it over – with the result that he refused to fly it again. Service pilots had to put up with these characteristics, for the Roc was yet another example of contractual production prior to first flight, but excusable because the contract was made after limited experience with the Skua.

Blackburns were too committed to produce the Roc, so the order for 136 had been sub-contracted to Boulton Paul who were still struggling with production of the Defiant RAF equivalent. Yet another Blackburn pre-flight contract was for their twin Perseus-powered B-26 Botha TSR which had internal torpedo stowage instead of the semi-exposed torpedo of the Bristol Beaufort to the same M 10/36 specification. Construction had been neck-and-neck between the two machines, and though the Beaufort marginally won with its initial flight on 15th October, the Botha shoulder-winger of a foot greater span and all-up weight of 17,600 lb, compared with 21,230 lb, had its first flight by Flt Lieut Bailey when the factory at Brough returned after the Christmas holiday. There were the usual minor problems, and elevator control was barely adequate. Speed proved a little less than the more functional-looking Beaufort, which was more adequately powered with two 1,130 hp Taurus. However 442 Bothas had been ordered in December 1936, and production was well under way both at Brough and the Blackburn-Denny factory at Dumbarton – but it was a machine with which none would be enthralled.

On the same day that the Botha flew, Geoffrey de Havilland Junior, accompanied by George Gibbins, commenced trials with the DH 95 twin-engined all-metal Flamingo shoulder-winger of 70-ft span. So confident was young Geoffrey that immediately on becoming airborne he retracted the Dowty undercarriage – and at the conclusion of the flight reported particularly pleasant handling characteristics. The Flamingo was a triumph for Ron Bishop, its thirty-four-year-old designer. Cruising speed was 10 mph greater than the latest Lockheed 14H which had 750 hp P& W Hornets, so it seemed the Flamingo might prove highly competitive in the world market.

In any case the Lockheed 14 was not particularly liked. There had been too many accidents, and since the fatality to Capt E. G. L. Robinson with a British Airways Lockheed 14 in November, when he crashed on the rocky Somerset coast during a test flight, there had been mounting criticism that the machine viciously dropped a wing in landing or when turning at low speeds. Nevertheless assembly of the Hudson military version was to proceed at Liverpool Airport at Speke where a bevy of American aircraft workers had already arrived. Meanwhile Lockheed were rebuilding a Lockheed 14 with longer fuselage and bigger chord Fowler flaps to be known as the Lodestar, powered with twin 1,200 hp Wasps or Cyclones giving over 272 mph and a cruise of about 250 mph.

From the USA came a reminder of earlier days when Harold Bolas was the dynamic genius of Parnalls but had never quite made the grade. Crouch-Bolas (Aircraft Appliances) Ltd was registered in Britain as a private company with a capital of £125,000 to acquire the rights from Crouch-Bolas Inc of their system of using the slipstream from large propellers to increase lift. Sponsors of the English company were William Watkins and Bernard 'Bones' Brady of the Aircraft Exchange and Mart, and apparently Short Bros were interested in the application for flying-boats.

Shorts were prospering, and had a trading profit of £211,580. 6*s*. 1*d*. All profits, said Oswald Short, had been ploughed back into the business, and from 1916 to 1935 no shareholder had received a dividend – but a splendid aircraft factory had resulted which was now a national asset. He vigorously denied that a ring of aircraft constructors exacted high profits from the Government, for there was a strict investigation of costs and allowable profit was both fixed and very low. Nevertheless the Board was now able to recommend a dividend of 30 per cent plus a bonus of 12½ per cent. He also confirmed the directoral appointment of that terror of the industry, Sqn Ldr Hugh Walter McKenna, late of the A & AEE Martlesham, but made no mention of Arthur Gouge, for Oswald secretly regarded him as a mere employee rather than a great engineering practitioner.

There were always MPs ever ready to pursue the antagonisms between labour forces and employers. Questions were asked in the House on what was alleged to be a sell-out by Petters Ltd when the entire Yeovil factory was acquired by John Brown & Co, for this included the big foundry which had become redundant when Brush Electric bought the Petter engine business and transferred it to Loughborough, leaving a large number of unemployed foundry workers at Yeovil. Sir Thomas Inskip cautiously replied that they had a chance of work with Westland, though not in connection with cast iron as there was other ample foundry capacity available in the country. Certainly Sir Ernest had made no provision for them; but as soon as the problem became obvious, Acland, prompted by Eric Mensforth, agreed that every effort should be made to re-train them, and John Fearn was so successful in organizing this that ultimately there were no redundancies and good relations were cemented with their trade union.

Meanwhile the Whirlwind had completed preliminary trials after much trouble with cooling and considerable experimental work with the rudder which eventually was given a 'hollow-ground' tadpole section. The Air Ministry was anxious to have a preliminary assessment despite residual defects, so the machine was flown on the last day of December to the RAE at Farnborough instead of the A & AEE at Martlesham, but in any event was considered so promising that Westland was told that an ITP for 200 would be placed in the New Year. But what would that New Year bring? Not only the British aircraft industry but the whole world was on tenterhooks.

CHAPTER V

1939: *The Winds of War*

'Might not a bomb no bigger than an orange be found to possess a secret power to destroy a whole block of buildings – nay to concentrate a force of a thousand tons of cordite and blast a township at a stroke?'

Winston Churchill, 1925

I

Since Chamberlain's temporizing at Munich the initial relief at escaping from war had changed to more sombre realization that what he had gained was a margin of time in which to re-arm with the utmost speed, for in the past year Hitler had brought under absolute rule ten million Austrian and Sudetenland workers and soldiers, and the balance of power in Europe was in his favour. Moreover the much suspected Ribbentrop had gone to Warsaw on an alleged diplomatic mission which could only suggest that the German intention was turning in the direction of Poland. Accordingly Sir Kingsley Wood emphasized: 'The National Government is determined in the New Year to continue every effort to eradicate all differences which would make for war, supported, I believe, by the majority of our fellow countrymen and women. We believe that the strength of Great Britain is a decisive factor in the preservation of peace. We are therefore continuing to press on with all speed in rearmament of the country.'

Canada and Australia were equally intent on implementing their Forces. The Royal Canadian Air Force had only a few squadrons, but the current Budget was £2,350,000 to acquire ten squadrons of Auxiliaries with an authorized strength of 1,450 men. Meanwhile Australia's Prime Minister, J. A. Lyons, was arranging for an Air Mission from the RAF, Air Ministry, and British aircraft industry, to visit Australia with a view to 'developing Australian industry to a stage in which it could produce all Australian defence and aviation requirements, and also those of New Zealand and a proportion of those at Singapore.' Within a few days it was announced that the Mission would consist of Sir Hardman Lever who had headed the mission to Canada, Colonel Sir Donald Banks, and Air Marshal Sir Arthur Longmore. The prospect of aircraft and engine production by Canada and Australia was welcome – yet the idea of purchasing from the USA was still received with a growl, particularly as the Lockheed 14 was increasingly regarded as a tricky machine which whipped into a stall, and recently the French Service Technique stated it was not safe unless pilots were specially trained on it; nor was the Harvard free of similar trouble, and although fixed slots had been tried as a cure, the wing ordinates were being modified.

Possibly criticism of the Lockheed 14 was subtly sustained by de Havillands, who since October had been trying to enter the circle of military contractors building metal aircraft. On an issued capital of £800,000 turnover had increased from £1,800,000 to £2,600,000, and profit was a hopeful £152,813. Eighteen years earlier the embryonic company's first year had produced £2,387 profit on a capital of £18,000. If any firm had vindicated the policy of making civil aviation 'fly by itself' it was de Havillands. Yet as Major Bulman wrote: 'One never understood the disfavour of the de Havilland firm in the Ministry. Perhaps it started with Geoffrey himself, that great, shy, dedicated and most gallant Englishman, who imbued his ever devoted staff with his own detachment and self-reliance, and a sense of non-conformity and reluctance to bow to official dictates.' The Air Ministry remained suspicious of past attitudes, and because de Havillands essentially were a wood-working firm, decided that the business should continue with its Tiger Moths, Queen Bees, Moth Minors and Airspeed Oxford trainers. There was, however, a ray of hope when Sir Wilfrid Freeman requested the prototype Flamingo to be sent to the A & AEE for evaluation, and this led the de Havilland Board to continue with a small pre-production batch, encouraged by an order for two from Jersey Airways Ltd and interest displayed by the Egyptian Government – but it seemed unlikely that the firm could compete *de novo* with already extensively established production of Electras and Hudsons. However Sir Wilfrid remained interested in an elegant wooden bomber with which de Havillands were wooing him. Meanwhile the shops were full. Frank Hearle, the general manager, was currently appointed managing director, and L. C. Lee Murray became general manager of the aircraft factory at Hatfield, and John J. Parkes went as general manager to the Airscrew Division at the original Edgware factory which had been retained for this purpose with a sub-division at Bolton.

The de Havilland constant-speed 'hydromatic' propellers had received full approval and four were being fitted to *Cabot*, first of the improved S 30 Empire flying-boats, all of which had flight refuelling gear and were cleared for an airborne refuelled weight of 7,000 lb more than take-off weight of 46,000 lb. The North Atlantic was the aim. To that end Imperial Airways created a special division headed by Capt A. S. Wilcockson who had made the first flights over the proposed route in 1937, and Captains Donald C. T. Bennett and Jack Kelly-Rogers were his senior pilots.

A corresponding landplane transatlantic service was being formulated. Design was well advanced of both the Short and Fairey four-engined monoplanes to specification 14/38. At the Fairey AGM it was optimistically stated: 'The firm has, as usual, taken this order off the drawing board and expects the type to go into production long before the appearance of the prototype. The directors are hopeful that the results will mean building up a permanent civil branch of the firm's business.' Already the output of Fairey's four factories exceeded the original estimates, resulting in a profit of £395,125 and tax free dividend of 12½ per cent with bonus of 2½ per cent.

All aircraft manufacturers had made sweeping changes in establishing production flow systems now that money at last was available. The Hayes Works of Faireys had been increased in size by 25 per cent. At A. V. Roe & Co the energetic Roy Dobson now employed some 5,000 men, and when his new £1 million factory was ready he expected to double the employees within twelve months. The Bristol Aeroplane Co had extended both its engine and airframe factories and the buildings now covered over 2,000,000 sq ft making it the largest single aircraft manufacturing unit in the world. Every firm had put into operation long-considered plans for extension.

This was no less necessary at Croydon Airport, and the Under-Secretary for Air stated that it would have to be shut down for major reconstruction – but its area could not be stretched because of surrounding houses, and problems of interference hampered the Lorenz beam and approach flight paths. Extra hangars and main buildings were urgently needed, but the latest large civil aircraft were too big for complete enclosure, so nose hangars were essential to shelter wings and engines, leaving tails in the open. Spring was the target for the new fog-line runway, but because landing speeds continued to increase, a new airport in open country would soon be essential.

Langstone Harbour as a marine base was still hanging fire but at Southampton, Imperial Airways were extending the docking arrangements for flying-boats, two of which could now be dealt with simultaneously instead of leaving one moored until the previous one had cleared. Passengers recently even had the experience of salvage when *Calypso*, piloted by Capt Brown, was forced down with engine trouble in the open sea north of the Channel Islands, and the crew of six and two passengers were taken aboard a tug which towed the flying-boat into Cherbourg harbour. That added to the almost universal belief that it was safer to fly overseas in a flying-boat rather than a land aeroplane, but the continuing crashes of airliners of all nations, of which a Swissair Douglas DC-2 was the latest, still made flying seem a dangerous venture.

Imperial Airways was suffering many cancelled flights because of icing conditions and impossible weather and hardly merited its name, for apart from the depleted fleet of Empire flying-boats, only six old HP 42s and a few odd DH 86s were currently flying. All the Ensigns were having constant-speed propellers fitted to their Tiger IX engines and safety modifications incorporated in the elevator and aileron control circuits which the airline stated 'would make the machines less tiring on long journeys by making the controls lighter and perhaps improving stability.' There was a general feeling that the Ensigns were underpowered.

Meanwhile the Frobishers were grounded because the undercarriage of one had collapsed when the machine slid sideways against the edge of the tarmac apron at Croydon while taxi-ing in a strong side wind, but after modification all were reinstated on 20th January. Three days later the Empire flying-boat *Cavalier*, on the New York-Bermuda run, was forced to alight after two engines stopped because of carburettor icing and the remaining two began to fail. A few minutes later came the radio message 'Sinking', and within fifteen minutes of touch-down *Cavalier* broke in two. Only after ten hours in the sea were ten of her thirteen people rescued, two having died from exposure, and one drowned. A long drawn inquiry was conducted by the Chief Inspector of Accidents in which criticism was made of Exactor hydraulic controls which were favoured by several designers because they required no mechanical joints and therefore were easy to thread by devious route to the engines. The Inspector found the hot air intake shutter inadequate because gaps around it caused cold air to offset its effect, and ruled that a carburettor air temperature indicator was essential and there must be adequate manual control of oil temperature or by thermostat or viscosity regulator. The Inquiry criticized Imperial Airways' failure to instruct passengers on emergency drill; none had been told they must put on their safety belts before alighting, nor even that they were sitting on them or how to put them on, and it was recommended that a raft or life-boat be provided with emergency rations and pyrotechnic signals.

The rival British Airways, under the experienced business management of Major J. R. McCrindle, with Capt Alan Campbell-Orde as operational manager, spurred by knowledge that it would presently amalgamate with Imperial Airways by Government edict, was purchasing further Lockheed 14s; but there had been surprise when arrangements to take the first air mail from London to Lisbon on 2nd January had suddenly been cancelled. Nine days later it was revealed that General Franco had refused permission for the airline to cross the north-west corner of Spain because he had not yet been afforded diplomatic recognition by the British Government. There was no doubt that Franco had all but won the war. On 26th January Barcelona, the Republican capital, was captured by his forces. Madrid inevitably would go.

In Britain the tempo was steadily increasing. Air Marshal Sir William Mitchell, Air Member for Personnel, was appointed AOC-in-C Middle East with effect from 1st April, and was succeeded by Air Vice-Marshal C. F. A. Portal, previously Director of Organization. Three new Directorates were created, and Grp Capt K. C. Buss, formerly Deputy Director of Operations and Intelligence, became Director of Intelligence; Grp Capt John C. Slessor, formerly Deputy Director of Plans became Director; and bulky Grp Capt R. H. N. S. Saundby, formerly Deputy Director of Operational Requirements, became Director.

The Air Ministry announced that an Auxiliary Air Force Reserve was to be formed to train ex-airmen of the Auxiliary Air Force for return to duty with their squadrons in time of war, and this was matched by Admiralty orders announcing the institution of a Fleet Air Arm Supplementary Party for service at the bases of Lee-on-Solent, Ford, Worthy Down, and Donibristle, for duties as guards, boats' crews, aircraft hands, parachute packers, and so on.

There were also Cabinet changes. The Rt Hon Sir Thomas Inskip, Minister for Co-ordination of Defence, became Secretary of State for Dominion Affairs in place of Ramsay MacDonald's son, and Admiral of the Fleet Lord Chatfield took over Sir Thomas Inskip's position. The Earl of Munster was appointed Parliamentary Under-Secretary for War in place of Lord Strathcona who had resigned. There was also Lord Winterton 'who was slightly connected with aviation for a few months, and becomes Paymaster General. In his day he was a chronic rebel in the Commons, but seems to have lost his kick without having gained much in authority.'

On 24th January Capt Harold Balfour, speaking to the Royal Empire Society announced that members of the Civil Air Guard would be individually classified in categories related to their prospective value in war. Already 6,000 had joined, and sixty flying clubs were participating with 330 machines for instruction. About 1,400 members had gained 'A' licences, and the rest were being taught. A few days later Sir Kingsley Wood, indefatigably visiting every aircraft firm, further delineated the scheme when he opened an extension of the Phillips & Powis factory where he was received by the chairman of Rolls-Royce, A. F. Sidgreaves, and subsequently he flew in a Miles Monarch piloted by managing director Fred Miles. That week Sir Kingsley also visited the secret premises of Parnall Aircraft Ltd at Tolworth, Surrey, accompanied by Air Marshal Welsh, Air Marshal Sir Cyril Newall, Sir Charles Bruce-Gardner of the SBAC, and Sir Edward Campbell, to see an exhibiton of gun turrets built to the designs of Capt 'Archie' Fraser Nash who demonstrated their accuracy by fastening a pencil to the muzzle of a .303 gun in one of them and writing 'Kingsley Wood' in two-inch letters on a piece of paper.

On the last day of the Parliamentary recess Sir Kingsley Wood flew from Biggin Hill to Portsmouth Airport where he inspected the Airspeed factory, now employing 1,800 men, saw the new Alvis Leonides radial engine under test in an Envoy, and watched George Errington demonstrate the Oxford. From there he went to Hamble to see Air Service Training Ltd, and visited the nearby works of Folland Aircraft Ltd where he said he was impressed by the efficient way in which the company had got down to sub-contracts. He then called on Cunliffe-Owen Aircraft Ltd at Eastleigh and inspected the much publicized Burnelli twin-engined 'flying wing' which had been built in the USA where it found no market, so the possibility of the Air Ministry ordering it as a bomber was remote. Nevertheless Sir Hugo Cunliffe-Owen saw that his financial venture in purchasing its manufacturing rights might open the door to sub-contracting from Supermarine and other major companies engaged on the expansion programme. Then travelling by car from Eastleigh Airport, Sir Kingsley went to the Supermarine Works at Woolston where he saw hundreds of Spitfire fuselages under construction and witnessed the somewhat ponderous but effective aerobatics of a Walrus flying-boat before returning to the Supermarine assembly shops at Eastleigh where Jeffrey Quill and George Pickering impressively flashed round the aerodrome, slow rolled and looped, with two Spitfires.

2

On the opening day of the new session, Sir John Simon, Chancellor of the Exchequer, explained the Government's compensation scheme for loss of life, injury, and damage to property, in the event of war. Many dismissed it as just another exercise to keep 'the man in the street' on his toes.

Three days later Hitler announced a general reorganization of the Luftwaffe. His communiqué stated: 'The measure is intended to strengthen the preparedness of the Luftwaffe, increase its striking force, and arrange for further growth in personnel and material. This entails re-constituting the Luftwaffe in three groups. Air Fleet No. 1 (East) will be commanded by General Kesselring at Berlin. Air Fleet No. 2 (North) will be commanded by General Felmy at Brunswick. Air Fleet No. 3 (West) will be commanded by General Sperrle at Munich.' Thus Berlin would handle direct danger from Russia; Brunswick would guard against attack from Scandinavia; Munich would cope with hostilities from France and Britain. In addition General von Ritter was appointed Air Force General with the Navy, and General Bogatsch became Air Force General to the Army, indicating separate Naval and Army co-operative branches of the Luftwaffe. General der Flieger Erhard Milch, Secretary of State for the Air Ministry, was appointed Generaloberst and Inspector General of the Luftwaffe. General Ernst Udet, hitherto Chief of the Technical Bureau, became Generalflugzeugmeister, equivalent to combined British Air Members of Development and Production. General Rüdel became Head of the new Air Weapons Committee, and General Kühl was appointed Chief of Training. General Stumpff was appointed C-in-C the Air Defence Force.

The British Press in their several moods of sensationalism, snide comment, or judicial analysis, duly reported these significant appointments – but it meant nothing to the average Britisher, though there was a stir of response when Admiral of the Fleet Lord Chatfield broadcast from the BBC on 6th February: 'I believe in England being strong so that the large part of the world which looks to us for protection may feel they can rely on us. Only if we are well armed are we free to act as we consider right without risk of being misinterpreted; nevertheless it would be of little use unless we possess that spirit which made us great. Let us therefore have courage and confidence in ourselves, knowing that we retain those ancient qualities of our race without which mere arms would avail nothing.'

But there was also action. At Bristol, six weeks' work on the Beaufighter project, based on existing components of the Beaufort torpedo-bomber, had already resulted in re-design of the front fuselage and general re-arrangement, targeting the first flight at only six months ahead. Concurrently the Air Ministry confirmed 200 Whirlwinds for Westland. The English Electric Company at Preston, builders of electrical rolling stock and bus bodies, was given a contract for 225 Hampdens at its reintroduction to the aircraft industry after thirteen years. A. V. Roe received contracts for 1,500 Ansons. Orders for Lysanders,

Spitfires and Hurricanes were increased. More Hudsons and Harvards were bought from the USA. Specifications for replacement of even the latest prototypes were issued, such as S 24/37 of 6th January in lieu of the Albacore, for which Blackburn, Fairey, Hawker, Supermarine and Westland tendered, and contracts were won by Supermarine for a variable-incidence design of Joe Smith's and by Fairey for a shoulder-wing monoplane later known as the Barracuda. At Armstrong Whitworth, John Lloyd had the difficult task of completing the design of the twin-engined Bristol Type 155 Albemarle medium bomber to specification P/a 9/38, which he had undertaken because of Bristol commitments. Although graced with Bristol elegance, simplicity of detail was the criterion as it was proposed to sub-contract extensively to firms without much aircraft experience – though it was interesting that Westland was given an initial contract, soon cancelled, to build its one-piece wing centre-section which had a steel tube girder spar running across the fuselage between the engine bays. Vickers were exploring specification B 1/39 for a geodetic four-engined 115-ft span bomber, and in the shops their B 1/35 scaled-up Wellington named the Warwick was nearing completion, but although its Rolls-Royce engines had just been received there was sudden panic about the strength of the geodetic construction so all work stopped pending official clearance. Even further advanced at Glosters was George Carter's twin-engined F 9/37 single-seat fighter with two nose cannons and four machine guns, for its Taurus engines with three-bladed Rotol variable pitch propellers had just been installed – but he too was busy with initial low-wing layouts of a big B 1/39 bomber. Of existing types all parent firms were involved with later variants. Early in February Jeffrey Quill flew the first Spitfire to be fitted with two 20-mm cannon, and as a safety measure it also had four .303 guns.

Such intense preoccupation for war was in no way apparent at the crowded Aerodrome Owners Conference which Sir Kingsley Wood opened at the Central Hall, Westminster, on 1st February, where for the next two days the exhibition of a huge range of aerodrome equipment attracted hundreds and lectures were fully attended. So, for that matter, was a lecture on military *Air Co-operation* given by Grp Capt A. J. Capel to the Royal United Services Institution where he made it clear that command would be vested in the GOC-in-Chief who would decide when RAF support was needed, but on his staff would be an Air Officer Commanding who worked out the scale of the operation in relation to the target. It soon became obvious that tactics were still being considered in terms of the 1914–18 War. At Old Sarum the School of Army Co-operation in the time-honoured way viewed relief models from an enclosed gantry where pupils directed gun-fire by clock code system just as I had seen as a youth at Reading in the Great War.

But if there was still an old-world outlook, world records remained a lure. Lord Nuffield decided to finance a 32 ft span low-wing racing landplane designed in wood by Arthur Hagg for Heston Aircraft Ltd, and powered with a 2,300 hp Sabre loaned by Napiers in the hope of restoring lost prestige. A speed of 480 mph was estimated with a maximum safe duration of fifteen minutes.

Neither concept nor construction gave any sign of de Havilland practice despite Hagg's long connection with that firm.

There was also young Alex Henshaw who for some time had been preparing a record flight to the Cape and back using the Percival Mew Gull with which he won the King's Cup, but modified for 'grand tourism' by Jack Cross of Essex Aero Ltd with tankage of 87 gallons, sufficient for almost nine hours enabling 2,000 miles to be covered in still air. Alex left Gravesend Airport at 3.30 am on 5th February under a full moon, but minutes after he took-off the ground became covered in fog. Most of Europe was under cloud, yet in a gruelling series of stages he reached Cape Town in thirty-nine and a half hours for the 6,030 miles, beating Clouston's time by five and a half hours. After a day's rest he began his return at 10.18 am on 8th February, and in five long laps of turbulent weather and blind flying landed back exhausted at Gravesend in the early hours of the 10th, having equalled his outward time almost to the minute and beating Clouston's previous home record by seventeen hours twenty-six minutes. He was lifted from the cockpit in a state of collapse.

In avuncular fashion C. G. Grey said: 'Nobody admires his courage more than I do. That does not prevent me from thinking that he is a silly young ass ever to have started. Yet I can understand the spirit in which it was done. It is the same kind of youthful idiocy which in my days led people to cycle from Land's End to John o' Groats or London to York or 24 hours non-stop on road or track. Flying to the Cape and back in a high-speed aeroplane is certainly more dangerous but no crazier. I am told that Alex is now talking about a trip to Australia. I hope he will not waste himself and father's money, not to mention his father's nerves, on doing it in the Mew Gull.'

That machine's very light elevators made it lively to fly, so the slightest 'pump handling' with finger and thumb inflicted noticeable 'g', and though the ailerons were stiff but pleasant, the rudder was equally touchy but hardly needed as the machine was directionally steady. When Bradbrooke tried the Mew Gull he said he could not imagine how Henshaw managed to endure five hours of continuous blind flying in tropical storms. Certainly blood all over the cockpit was witness to the difficulty.

Some months earlier the khaki-painted first Lockheed Hudson and first North American Harvard had reached England. Accidents with the Lockheed 14 were continuing and it was being treated somewhat gingerly, yet the Hudson was a splendid machine as the A & AEE began to discover. The Harvard arrived with fixed wingtip slots, and Martlesham made tests with them and without, but decided that handling was nicer without the slots except slow speeds – when there was a tendency to drop a wing in the manner long reported. Robert P. Alston, one of the scientific officers of the RAE, was sent to Martlesham to investigate. On 17th February, the machine, piloted by Sqn Ldr Robert Cazalet with Alston as observer, spun into the ground and both were killed. A witness stated that it began to spin at about 3,000 ft, and the pilot streamed the tail parachute customarily used to help recovery of experimental machines. The spin stopped, but the parachute apparently broke away, and the

Harvard went into another spin and crashed. By the time of this accident Harvards were pouring into the country, and it was estimated that the first 200 would be completed by June. There would also be scores of Miles Masters, for the first of the production line was about to have its 715 hp Kestrel XXX engine installed.

From the statement on Defence presented by the Prime Minister to Parliament on 15th February it was clear from the highest financial allocation ever made for the three Services that the Government was convinced of the inevitability of war. In 1937 funds had totalled £262 million, and in 1938 were £388 million, but for the year beginning 1st April 1939 would be £523 million. Parliament was asked to vote £66,561,000 for the Air Services after deducting appropriations in aid and other receipts from the gross of £220,626,700.

A month earlier President Roosevelt sent to Congress his estimate 'of the minimum immediate requirements of the USA for National Defence' – a sum of £105 million, of which £40 million was to be spent before 30th June, 1940. Of this the US Army would get £90 million, of which £60 million was to buy 'several types of airplanes' providing 'a minimum increase of 3,000 airplanes, though orders placed on a large scale should reduce the unit cost and increase the numbers cited by about 25 per cent.' The President said that £10 million should be made available immediately 'to correct the present lag in aircraft production due to idle plants.'

French Estimates revealed a total of £60 million compared with £55 million the previous year, and the official intention was to build an average of 500 aeroplanes a month – believed to be about the rate of German production. Currently the French bought from the USA 200 Harvards, 100 Douglas DB-7 Boston bombers, 115 Martin Type 167 bombers, 200 Curtiss P-36 single-seat fighters, and 20 Chance Vought dive-bombers for the French aircraft-carrier *Béarn*. That would bring their current total to about 1,000 American aircraft and it was hoped to achieve 2,500 by the beginning of 1940. Inevitably there was French industrial criticism of these purchases, and a forced landing by two Curtiss P-36s on a cross-country flight in which one was slightly damaged was quoted to show that they were unreliable. One of these machines from l'Armée de l'Air had been sent to Boscombe Down for assessment by the Air Ministry and I took the opportunity of flying it. There could not have been a more pleasant single-seat fighter, for the controls were light and perfectly matched and stability was such that the machine would continue, hands off, at any attitude from climb to dive. Aerobatics were a delight, but it was about 15 mph slower than the Hurricane. A British order for 100 followed for a version known as the Mohawk with 1,200 hp Wright Cyclone instead of the P & W twin Wasp – but somehow these machines lost the perfection of the French version and were heavier to handle.

The Defence Debate opened on 20th February, and Sir John Simon spoke at length, duly opposed by Hugh Dalton on behalf of Labour who moved that the limit of borrowing powers be reduced, criticized the organization for Co-ordination of Defence, and questioned the wisdom of appointing a sailor to co-ordinate the three Services. In sequence the various protagonists caught the Speaker's eye and launched their criticisms and occasional laudatory comments. In the following week Simon strongly supported the British aircraft industry, denied it was making inordinate profits, and moved the second reading of the Defence Loans Bill. Mr Pethick-Lawrence, that one-time supporter of suffragettes, continued Dalton's theme by moving an amendment regretting the necessity for an unprecedented Defence programme, and urged that the Bill ought to be preceded by more effective co-ordination of the Services, organization of supply, and elimination of excessive profits. Sir Archibald Sinclair defined the case for a Ministry of Supply which would enable the Government to suppress profiteering and eliminate departmental battles for priorities. Sir Kingsley Wood explained that if reasonable prices could not be obtained, they would be fixed by arbitration, but in any case there was accumulated and accumulating knowledge of the cost and performance of each firm's previous contracts. 'Finally the Air Ministry has still another safeguard in having comparison with world prices. I made enquiry to-day and found that now we are fully in our stride, our costs are definitely lower than for corresponding types, say, in America.' He acknowledged that: 'The country is fortunate in having a strong core of aircraft manufacturing firms who have done so much in the rapid expansion of the industry by their special knowledge. These firms have rendered great service in training labour and giving advice to newcomers to aircraft manufacturer. Many of these firms have passed through difficult times in past years. Many had to write down their capital to low levels because of that, but fortunately they have in a number of cases been able to rebuild their organizations which are now capable of meeting the country's defence needs on a considerable scale.'

But what worried the industry was getting the Air Ministry to ensure adequate backing for the supply of raw materials of every possible kind, ready availability of machinery and hand tools, and sufficient time for an extended clerical staff unaccustomed to aviation to order the wide-ranging necessities of every contract. In the shadow factories the motor-car men were finding that aero engine construction had special problems, and those preparing to build airframes were discovering that the techniques involved accuracies and difficulties beyond their experience in car body-making and general engineering, although the aircraft firms were rendering every possible assistance. Nor were matters made easier when the huge shadow factory which Lord Nuffield had built at Birmingham was switched from jigging for Battles to Spitfires as it was already realized that the Battle was outmoded and the Austin shadow factory at Longbridge could meet the limited demand.

Inevitably the Estimates were accepted, for almost every MP now realized the necessity of defence. The new provisions increased the Metropolitan Air Force from the current 1,750 aircraft to 2,370 during the year, plus an overseas strength of 500 and completion of the expansion of the FAA. Even the non-aggression politicians realized the benefit of having the balloon barrage scheme extended to forty-seven squadrons under a separate Command for administration and training, though operational control remained that of AOC Fighter Command.

On 27th February the Prime Minister announced to the Commons: 'His Majesty's Government have given very careful consideration to the position in Spain, and to the action they should take in the light of all information at their disposal. In these circumstances they have decided to inform General Franco of their decision to recognize his Government and the Government of Spain, and formal action has been taken in this sense to-day. His Majesty's Government have noted with satisfaction the public statement of General Franco concerning the determination of himself and his Government to secure the traditional independence of Spain, and to take proceedings only in the case of those against whom criminal charges are laid.' That same day the French gave *de jure* recognition of Nationalist Spain. Sir Maurice Drummond Peterson was appointed British Ambassador, and eighty-three-year-old Marshal Pétain was appointed by France. As *The Aeroplane* commented: 'There is no such thing as a successful rebellion. If it succeeds it becomes a government.'

There was general relief at the cessation of hostilities. For the Germans it had been a good practice war, although politically the Nazi and Fascist participation had been their determined opposition to the Communist threat in Europe as justification for active support of Franco. The lessons had been invaluable. They had evolved a new concept of air warfare with their *blitzkrieg* Junkers Ju 87A dive-bombers which became an effective terror weapon operating with pin-point accuracy. The Heinkel He 111 twin-engined bombers, equivalent to the Blenheim, had been equally proved. Despite lurid stories of civilian massacre and bloodshed published by the more sensational British newspapers throughout the war, unbiased observers reported that objectives had always been military targets, and though such places as Barcelona had been much damaged in the port area, which was a legitimate military target, practically none had been done to the city itself, nor to Valencia or Madrid or any other great city of Spain, and even the bombing at Guernica was ascribed to civilians running in panic and mixing with the retreating troops.

But over Europe the clouds were gathering. On 1st March, in celebration of German Air Force Day and the fourth anniversary of his appointment as C-in-C, Field Marshal Göring said: 'Fear of the German Air Force, the mightiest in the world, has prevented the war agitators from barring the way of peace-loving statesmen to our Führer and to a just understanding – but in no field, either of airmen, anti-aircraft, Intelligence, troops, or civil aid protection, can we afford to rest. A glance over the frontiers shows that great States which regard themselves as protectors of civilization are in the grip of unexampled armament fever. It is for us not only to maintain but increase the advantage we undoubtedly have in the air and which even the foreign world admits. We must produce aeroplanes in numbers and quality which seem unimagineable, but which in the Reich of Adolf Hitler is possible. We must make ourselves independent of raw materials which the foreign world in no circumstances supplies to us in adequate quantities. We must sharpen the sword which serves the security of the German nation. But however many aeroplanes we manufacture, it is the men who decide the issue. The word "fear" has been struck from the German dictionary. We know our own worth; we do not underestimate that of the opponent, but derive from it only the will to increase our own worth. The Air Force demands another gigantic and powerful effort. We should think only of securing finally for the Air Force an advantage which can never be overtaken, happen what will.'

Within a week it was reported that forty German divisions were mobilizing – yet Chamberlain was telling journalists that 'Europe is settling down to a period of tranquillity'. This was followed by German troops marching into Prague after Hitler launched an ultimatum, and on 14th March he announced that Czechoslovakia had become a German Protectorate. Full occupation by German troops began next day. Two days later, in a speech to his Birmingham constitutents, Chamberlain reproached Hitler for a breach of faith. 'Is this in fact a step in the direction to dominate the world by force?' he asked in vain. Even Russia was disturbed, and on 18th March proposed a conference of British, French, Polish, Rumanian, Turkish and Russian governments to concert means of resisting aggression against any one of these countries.

Lord Trenchard, on the day after the fall of Czechoslovakia, opened the Defence Debate in the Lords and asked the Government whether adequate steps were being taken to ensure vigorous counter-attack against any hostile force rather than devote too large a proportion of the country's resources to defence. 'All the critics seem to think of is nothing but defence, and yet, as far as I can see, they disagree with the Prime Minister's great policy of avoiding war. Their attitude is incomprehensible. How could all these deep dug-outs, anti-aircraft defences, and fighters have prevented the Sudeten Germans from being incorporated in Germany? The only thing to stop war-makers is the knowledge that if they attack they will be hit twice as hard as they have hit.' He concluded his long speech by asking Lord Chatfield to strike a note of assurance that if war came Britain would use its air power to bring about an overwhelming victory and put up with the loss of life that was inevitable.

It was a logical assumption that Hitler, on the basis of espionage and published information, might decide to strike before the latest types of fighters and bombers under construction were ready for production in series. The Spanish War showed that he might be able to inflict a telling blow despite the cancellation of his strategic bombers. Yet in the Commons, R. A. 'Rab' Butler, replying on behalf of the Prime Minister, perhaps with reassurance in mind, was content to quote that a resolution of the League of Nations Assembly of forty-nine States was promulgating rules of war by which 'intentional bombing of civilian population is illegal; objectives aimed at from the air must be legitimate military objectives and must be identifiable; an attack on such objectives must be carried out in such a way that civilian populations in the neighbourhood are not bombed through negligence.' Were we really playing a cricket match?

On 28th March Madrid surrendered to General Franco's forces. Next day

Anglo-French Staff conversations began in London. Two days later there were British and French guarantees of aid for Poland in the event of an attack on her independence.

3

Though the absolute world's speed record had been held by Italy's Warrant-Officer Agello since 1934 when his Macchi-Castoldi racing seaplane achieved 440.67 mph, the Germans now increased the landplane record from the 379.625 mph made by a Messerschmitt 109R (actually Me 209) to a world record 463,943 mph on 30th March by Flugkapitän Hans Dieterle with a Heinkel He 112U (actually He 100) powered by a Daimler-Benz boosted to about 2,000 hp. The Heston-Nuffield-Napier-Hagg monoplane was still merely a design, and by the time of scheduled completion a year hence its speed was likely to be insufficient unless the engine could be up-rated.

For military aircraft that all important factor of speed was compromised by practical considerations of armament, for turrets were not conducive to streamlining. As with the de Havilland proposals for an unarmed twin-engined bomber, so Volkert at Handley Page was thinking on similar lines, backed by Lachmann. In a memorandum to Handley-Page, Volkert stated: 'It is quite obvious that the 400 mph long-range heavy bomber can at last become a practical reality,' and gave convincing comparison between the HP 56 twin Vulture-powered bomber he was designing and a similar machine with no defensive armament and consequently a smaller, lighter, ideally streamlined airframe, which could carry twice the bomb load, cruise at 300 mph, and required only half the crew. He deprecated the tendency of armed bombers to need heavier and yet heavier defences to fight interceptors and warned that ultimately this could result in a negligible bomb load. 'Adherence to precision bombing calls for far greater elaboration and perfection of equipment and far more skill than dropping bombs at random. In time of war the most efficient bomber is one carrying out the greatest destruction with the smallest crew and simplest equipment. The supply of highly trained pilots, navigators, gunners and bomb aimers necessary to operate efficiently the excessively complex equipment of to-day is bound to be limited and incapable of rapid increase or replacement. It would be far safer to fly in widely-spread formation of separate units which obtain position by short wave radio from two "pathfinder" machines, duplicated against emergency, who would send all instructions for bombing and navigation.'

There was no immediate acceptance of that novel proposal, but unknowingly he had added to the case presented by de Havillands for a simple and therefore cheaper bomber whose defense was speed.

'To invent something you can't turn on a tap, you have to feel inspired,' said Volkert. 'Inspiration comes in many ways, but rarely after reading an Air Ministry specification. Some I know seek inspiration from music – perhaps the piano accordion,' he added with a dig at Handley Page who played that instrument. 'Others apparently get ideas from the Aero Show in Paris – where of course there are other shows equally inspiring! If you miss some one at the Aero Show during the day, some one you specially want such as an important government official or one of your directors, you are bound to run across him later at the Folies Bergères or Bal Taborin. The ability to invent is not of course the sole monopoly of the design office. There are other examples – such as any firm's estimate of its own production capacity.'

His close colleague Lachmann was inventing in a different way. Some months earlier he and Cordes had flown to Yeovil to discuss Westland's experience of the Pterodactyl. Though the speed potential of swept-back wings was still not evident in Britain there were several prospects for tail-less machines, but Petter argued that they offered no benefit because the wingtip controllers at negative angle nullified the tip lift, so bigger wing area was necessary. Lachmann questioned various factors of control and stability, then proceeded undeterred at Cricklewood with the formulation of a twin pusher-engined tail-less machine which had a nose-mounted small auxiliary aerofoil to counteract 'large changes of longitudinal trim inherent in tail-less designs, particularly those with flaps,' but in fact the flapless Pterodactyl had shown no trim defect other than from its curable torsional weakness. In Germany, Alexander Lippisch was currently flying a tail-less machine with similar nose winglet.

Handley Page's planning that March for Short & Harland Hampden production coincided with completion of the first Belfast-built Bristol Bombay and the start of its Service career – yet the last of the equivalent Handley Page Harrows designed to the same specification had been completed in December 1937 and was now the RAF's major bomber equipping Nos. 37, 75, 115, 214, and 215 Squadrons, No. 271 Home Transport, and No. 93 for Night Defence. Pilots regarded it with affection, for it was easy and safe to fly. 'Why sweat at something difficult like the Blenheim when you only need to sit here and drive?' one of them said to me.

The three Harrows which had been withdrawn from reserve and modified by Flight Refuelling Ltd as tankers carrying 900 gallons of transferable fuel were now ready, and two were shipped to Hattie's Camp, Newfoundland (Gander), and the third was being sent to Hamble so that Geoffrey Tyson could become accustomed to the refuelling system in conjunction with the Empire boat *Cabot* flown by Lankester Parker. Concurrently Pan American was preparing their big Boeing flying-boats for the co-operative North Atlantic service.

Most of the Empire class boats had been delivered and although remarkably satisfactory, eight lost in two years was a heavy toll; additionally there were a number of major accidents such as *Capella* wrecked but salvageable on 12th March at Batavia. Two days later *Corsair* created enormous problems by getting mysteriously lost, running short of fuel, and forced landing on the isolated River Dangu in the Belgian Congo, stoving in the hull and sinking in shallow water. A team led by twenty-eight-year-old Hugh Gordon, an ex-apprentice assistant pilot of Shorts, flew to Juba, then went by lorry along unmade tracks to Faraje

150 miles away, and, guided by local inhabitants, completed the last long miles on foot, recruited a party of local tribesmen and managed to beach the great flying-boat. There amid tropic humidity, flies and mosquitoes, they proceeded with repairs – but how to get *Corsair* out again would present a problem.

On the same day that *Corsair* nearly met its fate, the King and Queen visited Short Bros at Rochester. There they saw the three fuselages being framed for the big S 32 airliners and the completed hulls and 134-ft span wings of the three transatlantic *Golden Hind* class, both types almost giants compared with the RAF Sunderlands and Empire boats for Australia that enabled the Royal party to contrast the comforts of civil boats with the spartan Service versions. In pouring rain, Lankester Parker demonstrated a Sunderland, taking off from his usual small 800-yard clear length of the winding Medway. Subsequently at Rochester Airport the Queen was enthralled with the S 31 miniature replica of the lofty, almost completed, four-engined Stirling prototype, and when Harold Piper flew the 'little bomber' she was so delighted that a repeat performance was requested later, to the dismay of the ground crew who knew how difficult the Pobjoy engines were to start. The machine was also still slightly unstable but a larger tailplane proved acceptable on 12th April, and the Stirling's was then made of similar proportions.

In the USA the Boeing B-17 equivalent of the Stirling had been flying satisfactorily for some time, and latterly its civil airline development, the 307 Stratoliner, with a pressure cabin accomodating thirty-three passengers and crew of five, was under test. KLM were interested in it for the Batavia route, and sent their technical director Pieter Guilonard and A. G. von Baumhauer, chief engineer of the Netherlands Technical Air Staff, to assess it; but on 18th March, while they were flying as passengers, it broke in the air and all were killed. Almost certainly the cabin explosively failed – a problem which Hollis Williams was hoping to solve with the experimental circular-section pressure cabin under construction by General Aircraft Ltd for the adapted airframe of a Monospar ST-25.

Five days before the Stratoliner catastrophe the Stirling was rolled out, and Lankester Parker made several taxi-ing runs, getting used to his high perch, and on one occasion became airborne for 50 yards. He found no difficulty, for in effect this bomber was an Empire boat on wheels, yet looking up at its massive shape, everyone, including Oswald Short, was impressed.

Next day came the first flight by Parker, with Sqn Ldr Eric Moreton, on loan from the RAF to help with Sunderland testing, as co-pilot, and George Cotton behind them to pump down the huge articulated undercarriage as it would take all of ten minutes to operate it should there be electrical power problems. The onlookers saw a stately take-off and watched the big, camouflaged machine in the distance as the pilots explored its behaviour. After twenty minutes, the Stirling came in to land, gently touched down but suddenly swung violently round and the stalky landing gear collapsed, tipping the machine with a thundering scrunch on wingtip and nose, so badly damaging the structure that the prototype was a write-off. The simple cause was a seized brake, for there was no experience on which to base their design as aircraft became heavier, resulting in teething difficulties that could delay flying for weeks.

In limited degree this had occurred with the Whirlwind at Yeovil, but the hydraulic Exactor controls, as with the Empire boats, were a bigger problem. Even the fuel cocks of the Westland fighter were operated hydraulically using the same pressure accumulator as the brakes, and a wary eye had to be kept on pressure, particularly if an engine failed and hand-pump operation was necessary. The hydraulically-operated large Fowler flap was linked with the radiator shutters which closed except for a small gap when the flaps were in flight position and progressively opened with lowering flaps until a gated position at which the shutters were fully open. Nevertheless plumes of steam often occurred during taxi-ing, sometimes with temperatures at danger level necessitating a stop to cool down.

All this complicated the urgent problem of getting the Whirlwind's controls right. Alterations of servo-tab gear ratios were made, changes to the sealed leading edge of the ailerons, adjustments of slat venting to prevent a jerk on opening, and the addition of a slight horn balance to the off-set rudder. There was also a break-down of airflow causing severe shuddering on a tight turn, leading to the discovery that the tailplane junction with the fin was the culprit. Flights were conducted with various concave fillets but only worsened matters, yet presently an acorn-shaped fillet made the machine acceptable though still with slight buffeting in a hard turn. Only when all this was completed could control examination at higher speeds up to maximum dive be investigated, and then an increasing nose-down trim change could be felt. Nobody was aware that this was symptomatic of Mach trouble. At 445 mph it seemed the machine would go out of control if dived faster. Petter would not believe this, and brought Clouston in from the RAE for an independent opinion. Though eventually he dived only to 430 mph he described the Whirlwind as 'a tricky demon of speed' – which it was not, as many a pilot later confirmed.

To change from flying this latest example of aeronautical art to the FAA Swordfish, of which there were now twelve squadrons, was liking stepping back to the Dark Ages. Because it was easy to fly and appeared to look after its pilot, it was deservedly popular – a kindly old biplane with whistling wires and draughty cockpit. There was no complexity of cockpit drill; one need only ensure that fuel was on and the trim wheel set, then on opening up it ran a mere hundred yards and gently rumbled into the air. Ailerons and elevators were moderately light and the rudder somewhat heavy, but all had an immediate grip, with the result that the Swordfish handled firmly, and approaches to a carrier deck at very slow speed felt completely safe. There was also the reassuring feature that even when the deck heaved up suddenly, the massive undercarriage could take a tremendous bump. As Terence Horsley recorded: 'A Swordfish will manoeuvre in a vertical plane as easily as it will straight and level, and to bring it out of a dive does not require a strong-man act on the stick, helped maybe by a few desperately hurried turns on the trimming wheel. Instead, gentle pressure pulls it out quickly and safely, and even though it comes down from

10,000 feet, the air speed indicator never rises much beyond 200 knots.' The technique was therefore to get almost directly above the target, then wing over in a dive until it showed over the centre-section, and keep going until within 200 feet of the water.

Latest of aircraft-carriers was *HMS Illustrious* of 23,000 tons launched by Lady Henderson, wife of Admiral Sir Reginald Henderson, on 5th April at Vickers-Armstrongs Ltd, Barrow-in-Furness. Her sister ship *Victorious* was being built at their Tyne yard. Like the subsequent *Formidable* and *Indomitable* they followed the general design of *Ark Royal* which was described to the Institution of Navel Architects by Sir Stanley Goodall, Director of Naval Construction, as the first aircraft-carrier to be wholly designed as such, and on a water-line length of 721 ft had a heavily armoured flying deck of 800 ft with two catapults at the fore, and three two-storey lifts to the hangars. Her defence was sixteen $4^1/_2$-inch dual-purpose guns and four pom-poms. With 103,000 horsepower available on three shafts her maximum speed was $31^3/_4$ knots. For the first time electric welding had been used extensively, resulting in a weight saving of 500 tons – but in aeronautical respects the Admiralty remained remarkably conservative and refused to consider twin-engined aircraft.

The Navy Estimates for £11,800,000 proved double those of the previous year. Mr G. H. Shakespeare, Financial Secretary to the Admiralty, declared that the Fleet Air Arm was the most backward branch in the Service because of the previous system of dual authority, but the Navy would now take over all first-line aircraft, immediate reserves, and aircraft for specialized Naval training. In 1937 only fifty officers and forty-eight men had been entered for training, but in the year just ending the intake was 507 officers and 206 men. There were now some 1,300 Naval maintenance personnel in training of whom 700 had been recruited from civil life. The Admiralty would take over the RAF Stations at Worthy Down, Donibristle, Ford, and Lee-on-Solent, and possibly Eastleigh and Lympne. Aircraft contracts for the six new carriers had been placed and production was in full swing, with delivery of several types starting during the year. The first squadron of Skuas would be embarked in the *Ark Royal* on her return from the Mediterranean. Mr Shakespeare boldly added that by 1942 the Fleet Air Arm would not suffer in comparison with any other navy in the world. By then the personnel would be 10,000 and all aircraft of the latest type.

Not merely territorial waters but Pacific defence was concerning both the USA and Great Britain. On 13th April *The Times* stated: 'Air Marshal Sir Arthur Longmore, a member of the British Mission to Australia, is remaining in New Zealand to represent the Air Ministry at the Pacific Defence Conference between British, Australian and New Zealand representatives.' Longmore had been born in Sydney, and the Mission was headed by Sir Donald Banks. Australia had already agreed that a Defence Loan of between £15 and £20 million would be raised in London largely to finance British purchases, and another would be raised later in New York to tap the American market. The Mission had been advising on the establishment of factories costing some £70 million to provide armaments and build Bristol Beauforts – affording work for 6,000 to 9,000 men. In New Zealand a factory would be erected near Wellington by de Havillands, and Hugh Buckingham had left England to organize this new undertaking where trainers and possibly small transport machines would be built. But it was not merely British possessions in the Pacific that were of concern. The USA was well aware that Pearl Harbor was vulnerable, and that the Japanese, flushed with victory over the Chinese, might try to grab the East Indies or even the Philippines – yet, considering the might of the American Navy, it was thought improbable.

4

On 1st April the Royal Air Force came of age. Soon it would be armed to the teeth.

The King's message to the RAF said: 'I have known the Royal Air Force from its earliest days, and am proud that its spirit remains unchanged – a spirit that has enabled it to surmount many difficulties and rise, true to its motto, to even greater heights of achievement.' Among many congratulations in similar vein, Air Chief Marshal Sir Cyril Newall warned the Third Arm: 'The task which in these critical times we may have to perform is greater than it has ever been before; but the success of the vast expansion we have undertaken is evidence of untiring efficiency and unshakable determination. I am convinced that if the RAF should be called upon once more to defend our Country and Empire it will maintain that reputation for high-hearted courage and devotion to duty which it has so fully earned.'

Crucial to the future of Poland, there had been deep suspicion of Herr Ribbentrop's intent when he visited Warsaw earlier in the year. Ostensibly Germany objected to the corridor across East Prussia granted by the Versailles Treaty and demanded instead a Polish corridor from Slovakia through Ruthenia and some eighty miles of Polish territory to the Ukraine. On 22nd March Memel had been ceded by Lithuania to Germany. Poland appealed for help. On 29th April the British Cabinet agreed that Anglo-French military conversations be resumed and a British Expeditionary Force be prepared of nineteen infantry and two cavalry divisions, of which four divisions were to be ready by September 1939 and six by January 1940, and the Army was to be raised to thirty-two divisions of six Regulars and twenty-six Territorials. Two days later, Chamberlain gave assurances to Poland that Britain would come to her defence, having conveniently overlooked that six months earlier that country had joined in the pillage of Czechoslovakia and grabbed back the Bohemian coalfields which had formerly belonged to them. Meanwhile British newspapers showed photographs of Mr Hudson of the Department of Overseas Trade *en route* to Russia with his arm affectionately round the neck of the broadly smiling Ambassador Ivan Maisky.

Despite war's alarm, the Government intended during the session to set up a public corporation to take over Imperial Airways and British Airways. The shade of Edward Hillman had been triumphing ever since the Cadman Report recommended that British Airways should establish services to all European capitals, but it was difficult to assess the share valuation for £333,000 was lost in the last two years and the balance of assets over liabilities was a mere £58,000. Nevertheless British Airways had built up an excellent organization at Heston for inspection and overhaul of their three Junker-Ju 52s, six Lockheed Electras, and recently acquired Locheed 14s, and its manager, C. T. S. Capel, had specialized departments for electrics, hydraulics, instrument testing, engine overhaul, and hydromatic airscrews. Air Ministry approval for major repair and alteration had been obtained, and currently permanent wingtip slots were being fitted to the 14s to overcome the much-criticized wing-dropping tendency. The original Handley Page patent for slots expired a year earlier, so slit slots could now be used without payment of royalties, thus explaining an enthusiastic change of attitude of many designers towards the slotted wing.

Forty-year-old Alan Campbell-Orde, until recently operations manager of British Airways, had just transferred to Imperial Airways in a similar capacity – much to the chagrin of Major Herbert Brackley, who was an old school friend and supporter of Imperial's axed manager, George Woods Humphery, whom he regarded as the lost leader. It seemed a bitter blow, but in effect Campbell-Orde would apply his long experience of civil aviation and testing to the technical side of operations formerly dealt with by Major Mayo who would now concentrate on research for the company, and 'Brackles' would remain manager in the field, with the pilots still under his care.

For the time being British Airways and Imperial Airways were combining to operate the London-Paris service. A White Paper on 20th April stated that from 16th April the maximum subsidies would be £67,000 for the London-Paris service and £22,000 for the London-Zürich service 'subject to subtraction of all amounts paid for carrying mails, and of half the amount by which the costs fall below, or the revenues rise above, the estimates on which the subsidy is based.' From 17th April, British Airways would also be responsible for the services to Brussels, Warsaw by way of Berlin, and Budapest by way of Frankfurt. The Brussels service would be subsidized by £19,500, with £53,500 for extension to Warsaw, and £53,000 for Budapest.

Three days later, despite all earlier contention and even recent denials, it was announced in the Commons that a Ministry of Supply would be created, though powers would at first be limited to Army supplies. In the following week, on 26th April, the Government published plans for compulsory military service and the limitation of arms profits. Men aged twenty and twenty-one would be called up for six months' military training – a tremendous break with the past, owing much to the pressure of Hore-Belisha at the War Office but bitterly opposed by Liberals and Labour.

Not only was the Civil Air Guard flourishing but the Air Defence cadets now numbered more than 10,000 and 100 squadrons had been formed. Clubs were busy teaching them to fly. To simplify tuition by reducing the difficulty of landing, General Aircraft hoped to put in production a tricycle version of the Cygnet design they had bought from CW Aircraft when that concern foundered through lack of capital. Hollis Williams had been careering around Hanworth aerodrome on a tricycle ground-rig towed by a car, and this resulted in several patents on the avoidance of 'shimmy'. He had redesigned the Cygnet with neatly moulded windscreen, twin fins and rudders, limited elevator, greater dihedral, and a modified wing-section with a carefully blended juncture to the fuselage, the ensemble giving almost foolproof characteristics – but it was the ability to drive around the field, using the rudder-bar controlled steerable nosewheel to manoeuvre in tight turns with the ability of a car which seemed so remarkable at that time. Applied to military aircraft it could have enabled inexperienced pilots to tackle war-time operations much earlier and with far greater safety, for no landing could be easier. I liked the Cygnet. It was comfortable, gave an outstanding view when taxi-ing, and nothing could be simpler to handle; even the rudder seemed almost unnecessary, for good turns could be made on ailerons alone though the customary slight use of elevator was needed, but there was no impression of wallowing such as that experienced with the two-control Stearman-Hammond from the USA which I tried at Farnborough. General Aircraft had ventured only on a production line of six, for the factory was busy with Hawker Hart airframes, wings for the AW Whitley, wings and tailplanes for the Spitfire, and a section was making fittings for the Blackburn Roc as well as shaping metal skins for the Cygnet.

Now that metal-skin construction was almost universal there was a general similarity in fabrication techniques, though recently a system of skin stretching to form a smoother surface had been imported from the USA where development still seemed to be outstripping Britain. The twenty-fourth annual Report of the NACA revealed that they now had a 500 mph 8-ft wind-tunnel and a 19-ft pressure wind-tunnel 'for investigations on large models in conditions more nearly those of free flight than in any existing tunnel.' By comparison all RAE and NPL equipment was much more limited, though in general covered similar ground. As Dr George Lewis, chairman of the NACA, stated in a lecture to the RAeS on 25th May: 'Speed, range, and efficiency are the most important aerodynamic problems of an aeroplane designer; consequently the main object of research is the reduction of drag.' A special low-turbulence tunnel had given illuminating results, indicating that the laminar boundary layers ahead of the transition point often accurately followed the laminar boundary layer theory. Said Dr Lewis: 'Aerofoils theoretically designed to take advantage of true low-drag laminar boundary layers over a major portion of their surfaces showed spectacular results from the standpoint of drag reduction when tested under suitable conditions in the two-dimensional flow tunnel.' This had led to a high aspect ratio wing designed by David Richard Davis, formerly of the Douglas Aircraft Company, who claimed 20 per cent reduction in drag compared with current designs, and it had been adopted by the Consolidated Aircraft Corp of San Diego for a twin-engined flying-boat and a proposed four-engined bomber, later known as the Liberator, which was currently being designed. Meanwhile surface roughness from protruding rivet heads and lap joints could account for 10 to 35 per cent of the profile drag of any wing.

Research at the NACA emphasized the importance of eliminating air leakage through joints, for even when fully retracted an undercarriage could account for 9 per cent of an aeroplane's total drag unless effectively sealed, and a leaking engine cowling might add 16 per cent. Even worse was the effect of ice formation, for it could reduce the lift coefficient from 1.32 to 0.9 and increase the profile drag coefficient by 90 per cent with a disastrous increase in stalling speed. Prevention of profile-distorting ice by distributing exhaust heat to the leading edge and even the entire wing chord was being studied. Strength of wings was also being re-assessed. A complete stress history had been obtained of the Martin flying-boats on the Pacific route giving valuable data on fatigue from stress variations in rough air, and it was found that gusts imposed far greater impact loads than previously taken into account, ranging from – 3.2 'g' to + 4.5 'g' with gust velocities of almost 60 ft/sec accelerating to maximum intensity within 150 ft.

Information on German progress was readily available in the 1,200 pages of their latest *Annual of Aeronautical Research*, and they similarly investigated the strength of airframes under fatigue and alternating loads, much of it in co-operation with Lufthansa on the frequency and extent of gusts. But, as with the Americans, investigations covered a great range of inquiry from aerodynamics, to structures, powerplants, and operational methods. Aircraft stresses were a major study, and a detailed method of calculation in set tabular form had been devised for engineers unfamiliar with theoretical presentation, enabling them 'to take careful account of the aileron-angle and flap-angle combined with whatever twist there may be in the wing as a whole.' Air loads caused by operating the elevator have been investigated, also variation of balance with Reynolds number, leading von Doepp of Junkers to warn that trim tabs tested in the wind-tunnel gave results for undisturbed flow, whereas in practice they operated in unsteady conditions giving a different answer. Although the Germans claimed that the direction of airscrew rotation had a major effect on stability, and could improve it by 2.3 per cent if oppositely rotating blades of a twin moved up near the fuselage, no measureable difference had been found with the Whirlwind between the first prototype with contra-rotation moving downward towards the fuselage, and the second with mutual rotation and the tips in opposed relation to the fuselage.

Lecturing to the RAeS student section of the RAeS, Dr Hilton of the NPL explained the problem of speeds approaching that of sound in air – 771 mph, decreasing inversely with height to a constant 660 mph after 36,000 ft – and showed that a 'compressibility stall' occurred by building up a forward-moving flow against the main direction of the air stream, which he pictured as a thick surface at which there was a rapid increase of pressure, density and temperature, and the component velocity at right angles to the surface changed from a speed greater than sound on the up-stream side to less than sound on the downstream side, the loss of energy involving a great increase in drag. He described how the ideal aspect ratio to avoid shock stalling became steadily lower as speed increased, until the ideal was a streamlined wingless 'body of revolution' of 0.482 thickness ratio with an added streamlined tail – yet it was evident that Britain had not yet realized the importance of sweep-back, for he only said that an aeroplane could currently be designed to attain 550 mph, which was 0.714 that of sound at sea level, but must have a smooth upper surface because the compressibility stall first appears where the air is speeded up, and he suggested an aspect ratio of 2, with touchdown at 95 mph as the fastest at which a pilot could reasonably be asked to land. Looking further, he predicted aeroplanes with normal propellers for take-off and alighting but at high speed they would feather and fly with some kind of reaction jet or by rocket propulsion.

Stirred by the success of the NACA high-speed wind-tunnel, Farnborough was building a 4,000 hp tunnel capable of testing 6 ft span models up to 600 mph. Studies of the effect of compressibility on aircraft performance and flying quality were presently initiated chiefly using high-speed fighters, and the characteristic increasing nose-heaviness to a point where the elevator became ineffective was eventually established. Basic principles for longitudinal stability had also been formulated by S. B. Gates and Miss H. M. Lyon, and a new and comprehensive system of basic stressing was presented by Dr P. B. Walker in R & M 1916 *Rationalized Stressing Cases* which was adopted for all RAF designs. Similarly Dr A. G. Pugsley wrote R & M 1839 *A Simplified Theory of Wing Flutter*, which combined earlier British work with more recent investigation by Kussner in Germany, and enabled a more accurate prediction to be made using a new form of 'resonance' ground test described by Dr P. B. Walker in R & M 2046.

Methods of calculating wing profile drag had been developed by H. B. Squire and A. D. Young, and investigation made of the effect of surface roughness on drag. There had also been much wind-tunnel research on engine installations to reduce interference drag. Radio, electrics, photography, armaments, mechanical testing, all had their large share at the RAE, and a much bigger version of the 'Temple' structural test frame, which had been in constant use since 1930, had been designed by I. J. Gerard with a length of 120 ft, 30 ft wide, and 15 ft high, to accomodate complete wings or fuselages of heavy bombers up to 120 ft span, and was known as the 'Cathedral'.

More futuristically, but unknown except to the few, design and construction was proceeding of a contra-flow turbine geared to a propeller shaft, in contrast to the gas turbine that Whittle was producing in which the efflux formed a propulsive jet. Meanwhile Pye, the DSR, was becoming 'irritated by Whittle's importunity in urging that an aircraft should be built to test his engine.' Initial running had been attempted in April 1937, but the engine screamed out of control into higher and higher revs. Changes to the fuel control and injector mechanism failed to make it fully controllable and the compressor seized in its casing and stopped with a shriek. Modified and rebuilt it had run in the late spring of 1938 for one hour forty-five minutes at 13,000 rmp with a thrust of

480 lb, but ended with another write-off. Officials decided it was useless to continue but BTH indicated that having gone thus far they were prepared to spend another £60,000. Whittle made a drastic redesign, replacing the single large combustion chamber with ten small ones, and received a new contract from the Air Ministry, supplemented by another twenty hours running at £200 per hour. In December 1938, tests began again, and by stages Whittle managed to get it up to 14,000 rpm in March, and now a new impeller was to be fitted to increase the speed further. Encouraged by Tizard, the DSR began to see that after all this engine might have valuable potential for high-speed aircraft, and presently George Carter, who had discussed the possibilities with Whittle, was called in to discuss a projected Gloster metal-skinned single-seat fighter powered with a jet of 750 lb thrust.

Unknown to British Intelligence was the advanced development of gas turbines in Germany. The Technische Amt had placed a contract with the Berlin-Spandau BMW Works, to develop Dr Ostrich's design of an axial-flow gas turbine, and in the autumn of 1938 awarded Messerschmitt AG a contract for an airframe powered with two of these engines. The result was about to be submitted as Projekt 1065 in the form of an all-metal, 30-ft span low-wing cantilever monoplane with retractable undercarriage, estimated as having a 560 mph potential at an available thrust of 2,323 lb.

In even greater secrecy, Dr Ernst Heinkel three years earlier had engaged a young graduate, Dr Pabst von Ohain, who was joined by Max Hahn, an automotive engineer, to develop a centrifugal turbojet, of which a miniature had run in September 1937, and currently a larger version of 992 lb thrust was being tested beneath the He 118 V2 dive-bomber prototype. Meanwhile Heinkel's technical director, Dipl-Ing Robert Lusser, had designed a 26 ft 8-in span research shoulder-winger, He 178, fitted with one of these engines, and it was rapidly taking shape. Britain, somewhere down the line, had once again been too conservative in rejecting Whittle's earlier proposals.

5

At Southampton a block of terminal buildings for Imperial Airways was being erected in the new Liner dock and would connect by covered way to nearby rafts for passenger disembarkation from the flying-boats. The first of the 24-ton Atlantic machines, *Connemara* and *Clyde*, had arrived at Hythe and were being tried by Capt Kelly-Rogers and Capt Bennett, but the prototype *Cabot* was still undergoing tests at Felixstowe, and *Caribou* was expected shortly. *En route* from its base at Baltimore via the Azores was the rival PAA Boeing 314 *Yankee Clipper* under the command of Capt Harold Gray with a crew of eleven. Each day there were rumours that it would arrive 'to-morrow'. On 4th April this impressively huge sponson flying-boat hove in sight, almost dwarfing *Connemara* flying in formation with it; then after several circuits the Boeing alighted some two miles further towards Southampton than the waterway carefully marked for it. After staying at moorings over Easter, it left on 13th April shortly after 7 am for Lisbon, where it arrived at 4.15 pm. Next day it took off just before 7 am for Horta, which was reached at 3.30 pm. On the 15th the Boeing left for Bermuda where it alighted for fuel, then proceeded to Baltimore, arriving in the afternoon of 16th April.

An interested onlooker at Southampton wrote: 'Many rumours are running round about the motors of the *Yankee Clipper*. Nobody has been allowed to see them, but the engineers worked almost day and night while the machine was at Hythe and I gather that two of the motors have given considerable trouble. To those used to the 20-second take-off of the Empire boats, the *Clipper* seemed a long time unsticking, for it was clocked at 60 seconds. This is understandable as she is 100 per cent heavier than the Empire boat but has only 50 per cent more power for take-off. The principle is sound enough as water is cheaper than motors if you can get enough of it.'

A few days before the *Clipper*'s departure, Juan Trippe, president of Pan American Airways, made a formal appearance before the US Civil Aeronautics Authority for permission to run an Atlantic service but stipulated an annual retainer for PAA of £500,000 for carrying mails, and additional payment on poundage bringing total government payments to £1 million. Without subsidy he estimated that the company would lose £700,000 on the first year's operation even if it carried the expected 2,500 passengers. Meanwhile in anticipation of British transatlantic landplanes such as the Albatross, Hattie's Camp in Newfoundland was ready as an airport after nearly three years' work in making four runways, of which the longest was 4,800 ft.

Showing what could be done in the way of long distance travel, General Vladimir Kokkinaki and Major Mikhail Gordienko flew a converted Russian twin-engined bomber from Moscow to Miscou Island off the northern tip of New Brunswick, Canada, a distance of 5,000 miles in twenty-two hours fifty-six minutes – but bad weather and lack of fuel caused a crash landing.

Germany too was all set for records, and on 26th April, using the Messerschmitt Me 109R, Flugkapitän Fritz Wendel, test pilot of the company, achieved 469,225 mph over the standard 3-kilometre course, thus barely beating by 5.28 mph the Heinkel record made in March. There had been hope that the special 'speed Spitfire' with a specially rated Merlin III could gain the world's landplane record, but trials revealed a speed well below that of the Messerschmitt.

As antidote to German publicity the Air Ministry began a series of Press demonstrations of Britain's latest aircraft. The first, by No. 2 (AC) Squadron with Lysanders at Hawkinge, was a literal wash-out in a display of fortitude by pilots and spectators alike while waiting for low rain cloud to move from the hills, but a brief demonstration of low-level bombing in line-astern proved possible, though marked with simulated smoke puffs which unfortunately appeared in the wrong places at the wrong times. A solo flight under difficult conditions by Major A. J. W. Geddes of the Royal Artillery caused the comment: 'The demonstration was nicely planned to show the many outstanding qualities of the Lysander when flown by a Service pilot – a method as convincing if not so spectacular as if the machine was shown off by the firm's pilot.'

The Press show for Spitfires at Duxford also suffered from weather, for an attack on a Blenheim by six Spitfires and air drill by twelve Spitfires was hidden by cloud. The best that reporters could say was that: 'When taking off in squadron formation, flying slowly or all out, doing moderately tight turns or drilling in precise close formation at speeds well over 200 mph, the Spitfire gave the impression that it possesses all the qualities of an extremely fast and highly efficient aggressor, with the manners of a gentlemanly aeroplane.'

Nor was the RAeS any luckier with its annual Garden Party on 14th May at the Great West Aerodrome of Fairey Aviation, where the thousand guests were received by Roy Fedden as president, helped by his niece Miss Betty Fedden. For those with technical intent the static exhibition in the big Fairey hangar afforded a wealth of interest with forty stands of equipment of every kind including twin three-bladed Fairey constant-speed propellers mounted for contra-rotation – the first seen in this country. 'The flying programme despite the shocking weather went perfectly to schedule', recorded *Flight*. 'Only one demonstration machine failed to appear, but fifty odd arrived during a disarmingly fine morning. The spectators either crowded the static show hangar, sheltered under wings, the Wellington proving popular and the sharply tapering Albatross less so, or remained battened in their cars, a process which involved switching on screen wipers or opening sun (*sic*) roofs if aerobatics were to be watched.' *The Aeroplane* added: 'The RAF won the honours of the occastion. The Spitfire formation was excellent. No. 601 Squadron AAF gave the assembly the kick of the afternoon when after some very accurate formation flying in their Blenheims, they passed from east to west in line-ahead and then dived off the line as if merely doing dive bombing in Harts, and howled eastward at not more than 200 feet over the crowd. It was a most shaking performance. No. 610 were evidently amateurs in the proper sense of the word.' Who could have imagined that the amateur airmen of England would soon save their Country?

There were several novel aircraft such as a curious twin-engined high-wing monoplane, the Willoughby Delta, with triangular aerofoil-section tail-booms, demonstrated by that test pilot of an earlier vintage at Beardmores, A. N. Kingwill – but less than two months hence it was fated to crash and kill the designer and his next pilot. Another private venture was the twin-engined advanced trainer built by the famous blind-flying instrument-makers, Reid and Sigrist Ltd, and it attracted much attention because of the beautifully streamlined monocoque fuselage and Heinkel-like elliptical wings which had flap and ailerons mounted with leading edge slightly ahead and below the wing trailing edge. In the hands of George Lowdell it showed spectacular agility despite a wing loading of 23 lb/sq ft, which was only 2.7 lb less than the Spitfire. Reggie Brie followed, industriously whirling his way with the new C 40 Autogiro on which the pylon structure for the autodynamic head was faired with windows to make a comfortable side-by-side two-seat cabin. The long, stalky, semi-cantilever undercarriage enabled almost vertical landings to match the jump-start, and as one spectator commented: 'She can rise and descend on her own shadow.' By contrast Stephen Appleby, former exponent of the Pou, cavorted with the Scheldemusch pusher biplane from Holland, and like the Cygnet displayed by Hollis Williams 'who lightened the gloom considerably by his lapses from orthodoxy', it had a tricycle undercarriage which was the object of contemplation by several designers.

Of the RAF's latest equipment, there was the Boulton Paul Defiant which its test pilot, Flt Lieut Feather, handled with aplomb, though tacticians were increasingly sceptical of the value of its turret, for even the beloved Bristol Fighter of the Great War only became effective when the pilot depended on his forward guns. Summers sedately flew the Vickers Wellington; F. H. Dixon displayed the Fairey P 4/34, and a Firefly biplane performed beautiful aerobatics in the hands of Staniland. The Lysander was demonstrated by Flt Lieut George Snarey, the new Westland assistant pilot, and a long-nosed Blenheim was impressively flown by Uwins's new assistant, Flt Lieut A. J. Pegg, recently of Martlesham, who made several slow rolls to show that a modern twin could take such things in its stride. But what interested so many old-timers was the show put on by the International Horseless Carriage Corporation with their Blériot IX, 1911 Deperdussin, and Sopwith Pup, respectively flown by Shuttleworth, Edmonds, and Wheeler. I watched them in company with Geoffrey de Havilland Junior, who disarmingly said: 'I wonder if they would let me fly that Pup?'

I told Geoffrey that the Pup was a reconstruction of the Sopwith Dove which Hollis Williams and Lowe-Wylde had made from scrapped units, including a Snipe tail, in the mid '20s, and after the latter's death in 1933 it was acquired by the firm of Ogilvie at Watford where a keen youngster, G. A. Chamberlain, found it in the early summer of 1935 and traded in his Parnall Pixie and £10 to meet the price of £45. As winner of the first *Daily Express* Flying Scholarship, Chamberlain had qualified for his 'A' licence early in February 1934 at Marshalls Flying School at Cambridge, but a few days later was severely injured in a road accident and on recovery was warned by his instructor to keep away from 'A' licence examining doctors for a few years! But he was determined to fly. The Certificate of Airworthiness of the Dove had expired. He decided to fly 'incognito', and kept it in a barn at Kempston on the outskirts of Bedford where he had to spend an hour rigging it before every flight and afterwards dismantle again. 'I had to drag the fuselage out of the farm shed, through the farmyard, and through two five-bar gates to get her into the field – so it was always a full day's job for one flight', he told me. 'With neither licence, C of A, nor insurance I didn't log anything for obvious reason, but I must have flown ten to fifteen hours in the Dove during the short time I had her. Inevitably the day came when in heavy rain the Le Rhône cut as I was making a low pass at the field down-wind prior to turning and landing. This left only a small corner available where we had previously driven the cattle, over which I had to play leap-frog. Jazzing the throttle got a few spluttering revs and I hopped the hedge to the next field, but it had deep furrows and on touching down they shot me up violently. The Dove stalled, nose well up at 50 feet with no more revs forthcoming. The port wing was far down and I had on full starboard stick. As we struck the second time the port tip virtually slithered along the grass as the V-struts of the undercarriage failed on one side, yet the only damage was tearing off the strut attachment lugs as the undercarriage disintegrated.'

The next six months were occupied in repair, and when the machine was ready in the summer of 1936, Shuttleworth heard of it and offered an Avro 504, G-ABSN, in exchange. Thus began the metamorphosis of the Dove into the Pup which still flies at Old Warden. At that time Ogilvie also had the crumbling remains of a BAT Bantam as well as a flyable Bristol Fighter on offer for £10 with spare engine, and both went to the Shuttleworth Collection.

The day after the RAeS Garden Party, Francis St Barbe's brother Sydney, one of the most colourful personalities of British civil aviation, spun into an orchard when the Lynx engine of his banner-carrying Avro 504 failed just after take-off from Hanworth. He scrambled from the wreck and walked back to the Club, but admitted feeling shaky though apparently uninjured. Two days later he died from an internal haemorrhage. 'His death will deprive a great many people of a lot of free entertainment,' wrote a friend. 'But his life was a tragedy. He had one of the most brilliant brains I have met. He was one of our best pilots. He was capable of unlimited hard work. He had immense artistic appreciation, was highly sensitive, and in a curious way very lovable. Yet he never made anything of his life and was essentially unhappy although he had more friends than most of us. A man of such talent ought to have been more than a mere pilot of advertising aeroplanes. But he seemed to have no objective in life, no ambition. Yet if Sydney had chosen to write he could have made a fortune, for his outrageous descriptions and phrases had a truth which bit deeply into the self-complacency of people for whose mental, spiritual or commercial good they were intended. I am sorry that he was not spared for many years to shame people with those witty extravagancies of devastating truth.'

Old or new, every aeroplane, irrespective of type and size, was now regulated by the Air Registration Board whose Report for 1938–39 issued on 11th May showed a great increase in activities. Nineteen civil aircraft firms were approved for design, some of which tended to resent supervision, but others depended on the Board for advice to such an extent that it amounted to assistance in design. Some 2,000 firms had been approved to supply materials for civil aircraft, and almost 2,000 ground engineers had been licensed. The Register of Aircraft contained 1,725 machines, though only 70 per cent were currently airworthy, and 3,836 visits had been made in connection with C of A renewals. Several prototypes were flying under 'B' conditions of which the latest was the pressure-cabin Monospar ST-25. It was the last of its breed employing the patented Monospar wing structure and had a heavily swept-up trailing edge at the fuselage. On 12th May it was shown to Sir Kingsley Wood and demonstrated in flight by Hollis Williams who had flown it for the first time the previous day but had not yet attempted pressurization produced by a supercharger fan driven by a 27 hp Douglas Sprite engine. When sufficient experience had been obtained it was hoped to secure a pressurization sub-contract from Faireys for the FC 1 medium-haul airliner, when instead of a separate auxiliary it was proposed to use two blowers from the flight engines to maintain an 8,000 ft cabin pressure at 15,000 ft. Shorts with their long-haul S 32 had stolen a march over Faireys and were already involved in pressure cabin design and had built a section for proof testing.

Meanwhile *Yankee Clipper* made another run from New York to Europe, having left Port Washington on 20th May, the twelfth anniversary of Lindbergh's solo flight from New York to Paris, and arrived at Lisbon on the 22nd using Horta in the Azores for topping up. Next day the *Clipper* continued to Marseilles, and on the 23rd arrived at Southampton. 'No mails were brought except a few special packages and Press photographs, for arrangements between the Post Offices of the two countries have not been finalized,' noted a newspaper. Meanwhile the British boats were held up by various technical difficulties, so there was no hope of initiating the Imperial service which had been announced for 1st June. Spat C. G. Grey: 'From the point of view of national prestige this is a disaster and a disgrace.' On 24th May the *Clipper* left on its return journey via Marseilles, Lisbon, Horta for refuelling, and arrived back at Port Washington, New York, on 27th May. Just after Lisbon it passed *Lieutenant de Vaisseau Paris*, a 33 ton six-engined Latécoère piloted by Capt Guillaumet, famous for his Andean flights, who was just ending a forty-eight-hour flight from New York to the French flying-boat base at Biscarosse, France, after refuelling at the Azores.

Britain was catching up. On 24th May Geoffrey Tyson with the Harrow tanker and Lankester Parker with *Cabot* successfully accomplished air-to-air refuelling above Southampton Water, and the *Caribou* was almost ready for similar trials – but the importance of this faster transatlantic project meant nothing to the average man. Even Munich seemed forgotten despite dramatic evidence of massed military aircraft of every type displayed at the sixty-three Air Force stations opened on Empire Air Day, 20th May. Sir Kingsley Wood with Air Chief Marshal Sir Cyril Newall flew a 435-mile tour of the stations in the Air Council's DH 86B from Biggin Hill, and Capt Harold Balfour made a separate tour, piloted by Grp Capt Orlebar, one-time commander of the Schneider team. Many a visitor remarked that officers and men no longer were wearing peaked forage caps but had the jaunty RFC type of soft cap mounted very right wing low – but for the more technical it was obvious that monoplanes at last had won the day, though many biplanes were still in yeoman service. Every station reported crowds double that of the previous year. Northolt drew the biggest with 56,000, but few were less than 10,000. Including the fifteen civil airfields, more than a million spectators participated that day.

An aeroplane which hitherto had received no previous publicity was the Martin-Baker fighter brilliantly demonstrated by Val Baker at Heston on 26th May during which he dived at 400 mph from 10,000 ft finishing his pull-out at

150 ft above the ground. The svelte MB 2 had been designed by James Martin, a plump, sandy-haired engineer, who came to earlier notice with a side-by-side low-winger with shaft-driven propeller intended for Amy Mollison's Australia flight in 1939 but was never completed at his small Elstree aircraft maintenance business. In 1934 he was joined by popular Val Baker in forming a company in their joint names backed by wealthy Francis Franncis, and in the following year produced the MB 1 – a neat low-winger of ingenious steel-tube construction; but the financial risk of production was too great, though Baker reported that it handled well.

Late in 1935, after seeing the prototype Hurricane, Martin decided he could design a better figher, but overtures to the Air Ministry revealed no interest. Nor was Rolls-Royce prepared to loan an engine, but an approach to Napiers resulted in the offer of a Dagger, for Martin could be remarkably persuasive, and his engineering skill was very evident. Francis Francis agreed to a private venture effort. From the outset Martin aimed at far easier maintenance than any other fighter, and made the metal covering panels of the steel tube fuselage frame instantly detachable. No existing aircraft could be so quickly armoured, for a turn of a small handle caused the upper surface of the wing to open for access to the four guns each side, and they could be taken out and replaced in five minutes. The wing span was only 34 ft compared with the 40 ft of the Hurricane, and it had a trousered undercarriage. Design and construction took more than two and a half years, so not until 3rd August the previous year had Baker made the first flight from the relatively long aerodrome at Harwell.

As might have been predicted, the rudder and miniature fin gave impossible directional characteristics, for Martin had assumed that the almost parallel profile of the rear fuselage afforded sufficient weather-cock stability – so Baker was lucky to make an accident-free circuit. A taller triangular fin and almost triangular rudder were quickly devised and fitted, but Martlesham pilots considered the machine a little brute compared with the Hurricane, for its stalling speed was high, the elevator too sensitive, ailerons too light at slow speed and too heavy at high, lateral stability negligible, and it was directionally unstable for small angles of yaw; so in turns it tended to slew instead of taking a natural bank. Nor was it easy to keep straight at take-off unless started across wind and using aileron, otherwise it pulled uncontrollably starboard against full rudder. The hard undercarriage on the rough surface of Martlesham also caused untidy landings. Nevertheless Sqn Ldr McKenna, before retirement, showed it to me with scarcely concealed malice as the best example of easy maintenance of any recent aeroplane submitted to the A & AEE, and he recommended to the Air Ministry that James Martin should be given a contract for an improved fighter with retractable undercarriage and Merlin engine, though eventually a Sabre was used. Currently this was being designed as the MB 3 of similar wing planform but with dihedral to overcome the directional-cum-rolling disability. A steel-tube fuselage skinned with detachable light alloy panels was again employed, and the wing spar had booms of laminated high-tensile steel strips decreasing in number spanwise using a D torsion nose. But what attracted the attention of

DOR was the unprecedented armament of six 20-mm cannon, fed with 200 rounds on a patented flat feed system. If it worked, this might even prove to be the Spitfire's successor.

Just before Empire Air Day, General Franco reviewed his entire Spanish Air Force and its German and Italian Auxiliaries at Barajas Aerodrome, near Madrid, where 600 machines were massed in squadrons of six, eight, and nine according to function, and included Junkers Ju 52 tri-motor bombers, Dornier Do 17 twin-engined bombers, Heinkel He 111 twin-engined bombers, Heinkel He 70 and Arado 45 single-engined bombers, and Messerschmitt Bf 109s. The Italians were represented by Savoia Marchetti S 79s and 81s, Caproni C 101s and Fiat CR 32s. After the review, the GOC of the Nationalist Air Force, General Kindelan, who had been taught to fly in 1911 by the Australian-born pioneer Bristol pilot Harry Busteed, announced that in recognition of the heroic services of the officers of the Kondor Legion and Italian Air Legion 'in the anti-Bolshevik crusade', General Franco would personally decorate them with the Spanish Military Medal.

On the day of the Kondor Legion's return to Germany there were further felicitations, and in his farewell address, General Franco said the Germans had shed their blood generously in Spain and it flowered again with poppies of peace which he offered to the great German nation and its Leader. Thereupon the squadrons, for the last time, started their engines and in massed flight headed for home and further commendation from Hitler as Heroes of the Air – for they had gained experience beyond price for the Luftwaffe.

That same day, in far less emotional manner, under a sun-filled sky glinting with light cloud, a large party from both Houses of Parliament attended an RAF demonstration at Northolt. Big formations impressively flew over while MPs and Lords were examining the many parked aircraft, and a miniature Hendon Display followed which opened with twenty-four Wellingtons formating at 1,000 ft, and as they swept away a squadron of Northolt Hurricanes took-off and joined an approaching armada of six Hurricane squadrons and a single Spitfire squadron while a broadcast commentary stressed that the formation could fire a quarter of a million rounds in little less than a minute. By contrast Sqn Ldr Buckle's war-time 'Horace' Farman pusher biplane flew past in stately fashion piloted by its burly owner. Watching with nostalgic interest was Marshal of the RAF Viscount Trenchard, who on a later occasion, looking at this same Farman, said to me: 'We used to get up early and fly them when the air was calm. There was a kind of music in the wires ...'

Target-towing followed, using a Hawker Henley which was attacked by a Harvard; Battles demonstrated ground-attack with smoke bombs; there was a touch of the past when a Gladiator formation looped and rolled in perfect station; three pilots of the Volunteer Reserve showed their skill with Hurricanes; two Spitfires revealed their versatility; three Sunderlands with deep diapason came low across the staring guests who marvelled at such size; the Northolt squadrons made a mass take-off and displayed the formation 'used by fighter squadrons for tactical need.' Then some of the newer types were demonstrated during which 'two mystery monoplanes flew across as fast as they could. One was a twin-engined high-wing monoplane with a gun turret well aft and the other a twin-engined monoplane with a Blenheim-looking nose and twin-rudder tail.' No names were given, but these were the Blackburn Botha and Gloster F 9/37 fighter which had made its initial flight at Hucclecote on 3rd April piloted by 'Gerry' Sayer.

A famous pilot of an earlier generation whose great talents had brought success was Rear-Admiral Richard Bell Davies, VC, DSO, AFC, selected after transfer of administrative control of the FAA to the Admiralty on 24th May as the first Commander-in-Chief of all Naval Air Stations, each of which as a Fleet Establishment would be independently commanded by a Captain. The first four Stations were now renamed – Lee-on-Solent, *HMS Daedalus*; Ford, *HMS Peregrine*; Worthy Down, *HMS Kestrel*; Donibristle, *HMS Merlin*.

In recent years Bell Davies commanded cruisers in the Mediterranean and China Fleets. He had entered the Navy in 1901, learning to fly at his own expense on a Farman at Hendon in 1911, and was with the first Naval unit to go to Dunkirk in 1914. In the following year he achieved the dramatic distinction of a Victoria Cross for the first air rescue after spotting an Henri Farman piloted by Flt Sub-Lieut G. F. Smylie which had forced landed badly damaged among Turkish troops. As they raced to capture the machine Smylie blew it up with a bomb, and Bell Davies, flying a single-seat Nieuport, landed alongside, beckoned him to jump on the lower wing root, and took-off for home. Laconically he reported: 'Returning saw H 5 burning in marshes. Picked up pilot.'

6

As a *divertissement* under the sombre clouds of rearmament and international strain the fourth Isle of Man Air Race took place at Whitsun on 26th May, but the weather did its best to make conditions difficult for the twenty-seven competitors which included such mem as Geoffrey de Havilland Junior, T. W. Brooke-Smith, Sqn Ldr H. R. A. Edwards, Alex Henshaw, Bill Humble, the Marquess of Londonderry, Tommy Rose and the redoubtable Edgar Percival – and it was young Geoffrey in the TK.2 who won, with Henshaw second, and Percival third. Sunday was a day of rest, but the Tynwald and Manx Air Derby circuits above the broad beaches followed on Whit Monday watched by a large crowd. 'Pop' Henshaw proposed piloting his Vega Gull for the Derby but Alex opposed it on the ground of filial anxiety, and though he greatly wanted to race his own Mew Gull he reluctantly and somewhat heatedly compromised by accompanying his father who insisted that he alone would still do the piloting – and he won the race, beating Tommy Rose and Edgar Percival by a few seconds. Afterwords Sqn Ldr and Mrs Edwards, flying his old Hermes Avian biplane, came first in the Tynwald Handicap during which one competing machine flopped on the sea after engine trouble and sank in five fathoms, though not before the pilot was rescued. The meeting was fun in its way, yet the heart seemed to have gone from such occasions when compared with the careless delights of the first half of the decade.

Meanwhile Germany was still playing poker. On 31st May a Non-Aggression Pact with Denmark was signed in Berlin, and a week later another with Latvia and Esthonia. Concurrently the Soviet made it clear that it would only adhere to the Pact of Mutual Assistance with Great Britain and France if Finland and Esthonia were included. After the Czechoslovak *anschluss* what could people believe? If far enough from the scene, neutrality could be a paying proposition. The USA did not officially recognize the war between Japan and China, nor the impact of European diplomatic manoeuvring, but in the past year had exported $68 million of aircraft, engines, and parts. Not all countries revealed their purchases, but Japan bought $11 million; the Netherlands 8^1/_2$ million; China 6^1/_4$ million; Argentina $6 million; the Soviet Union $5 million, and Britain $4 million. By the same token British aircraft exports had more than doubled since 1937, and from April 1938 to April 1939 amounted to £2,143,679.

Britain was far more concerned by headline news that the submarine *Thetis* had failed to surface while undergoing diving trials in Liverpool Bay. For a few days it was hoped that the ninety-nine men could be saved. The RAF was summoned for assistance and three Ansons were sent from Abbotsinch and four trainers from Sealand, the former sighting the submarine's marker buoy two hours after receiving the instruction to search. Meantime three flying-boats from No. 4 Armament Training Station, West Freugh, flew with doctors and salvage experts to the scene of the wreck – but nothing could be done and all the *Thetis* crew died.

The disaster obscured the announcement that fifty-one-year-old Sir Philip Sassoon died at his Park Lane house on 3rd June following an illness which doctors had been unable to diagnose. He had seemed the very epitome of an Etonian Englishman, but the Sassoons were a Parsee family who had intermarried with Jews. As Under-Secretary of State for Air in three terms of office he had been outstanding. So great was his interest that he had been a member of No. 601 (County of London) Bomber Squadron since 1929 with the rank of Hon Air Commodore. His obituary described him as having 'a brilliant intellect, keen judgment, great enthusiasm, and enormous wealth with which to put his ideas into execution, and though he could be hard and even cruel, his natural impulse was to be generous and kind.'

A man of similar penetrative ability but with great humility was Henry Self of the Committee of Supply who received a KBE in the King's Birthday Honours List in which Grp Capt HRH The Duke of Kent was promoted to Air Vice-Marshal, Air Marshal Christopher Courtney was awarded a KCB, and William Lindsay Everard became a Knight Bachelor.

Sir Henry Self was responsible for a Bill currently setting up the British Overseas Airways Corporation with measures to take over Imperial Airways Ltd and British Airways Ltd on terms already agreed by the parties. There had been considerable sniping in Parliament over the valuations. Early in May, Sir

Kingsley Wood stated that the Government was prepared to take over Imperial Airways retrospectively as from 31st March 1938 for the sum of £2,659,086, representing a payment of 32*s*. 9*d*. for fully paid £1 ordinary shares. British Airways would similarly be retrospectively taken over as from 30th September for £262,500 plus repayment of £311,000 advanced by the principal shareholders. In both cases interest at 4 per cent from the take-over to the date of payment would be allowed. Inevitably some Imperial Airways shareholders argued that the shares were being given away, though Sir John Reith pointed out that the Air Minister had already been criticized for paying as much as 32*s*. 9*d*. As from 5th June all business was transferred to the new headquarters in Buckingham Palace Road, from which passengers would be taken to Croydon by bus, and those for the Empire route would go to Southampton by special train. The BOAC Bill still had to receive Royal Assent.

C. G. Grey angrily wrote: 'The word Imperial has been dropped in accord with the new principles of government afflicting this country for the past twenty years with a series of politicians afraid to call a spade even an agriculture implement, and have demonstrated in Ireland and Palestine and elsewhere that they are afraid to govern a couple of million people let alone an Empire of hundreds of millions.'

Imperial Airways had not yet caught up with their transatlantic arrangements. On 11th June, Pan American announced that their service between Port Washington and Marseilles would commence on 28th June, with two round trips a week, of which one was for mail and the other for passengers who would be charged $375 single or $675 return.

At Croydon, the Ensigns, whose withdrawal in January for modifications had caused considerable loss of revenue, were being redelivered at one a week. They now had up-rated Tiger engines giving 935 hp for take-off, and constant-speed propellers had improved the full load climb to 800 ft/min. The sixth production machine with these engines, *Eddystone*, had been evaluated at Martlesham in April, but though the rudder and elevators no longer snatched, the machine was judged longitudinally and laterally unstable, and the ailerons were heavier than the prototype, so the machine was unpleasant in bad weather. As a result its C of A was reduce by 500 lb from the original 49,000. There was still disquiet over the unreliability of the Tiger engines, and Imperial Airways had tentatively ordered forty-eight Curtiss-Wright Cyclones for use on the Eastern route, otherwise take-off weight would have to be massively reduced. Even now, despite many miscellaneous improvements, there were signs that the oil coolers were inadequate and there was trouble with the constant-speed unit of the propellers. The burning question was why the Bristol Hercules was not selected instead of the Cyclone, but Imperial Airways said there had been no commercial experience of the British engine and they could not afford further delays.

Extensions to Croydon were suffering a delay. Significantly on 6th June Capt Harold Balfour moved the second reading of a Bill to provide extensions to Heston Airport which was still managed by Airwork although owned by the Air Ministry, and was designated as one of the ring of London airports to be ready in September. Runways of 6,000 ft were envisaged by 1941, so many thought that Heston would become London's Air Terminal.

Currently its co-founder, Nigel St Valery Norman, succeeded to the baronetcy on the death of his father the Rt Hon Sir Henry Norman on 5th June. Sir Henry, an MP from 1900 to 1923, had long been associated with radio communication, for he was chairman of the War Office Committee on Wireless Telegraphy in 1912 and then of its successor the Imperial Wireless Telegraphy Committee. As *The Times* said: 'Though not an inventor in the field of wireless, he was among the first public men to foresee the enormous scope, utility and popularity of this means of communication.' He had also been interested in flying from the early days, and had taken Lloyd George to see Wilbur Wright at Le Mans in 1908.

The once gay Nick Comper became the next aviation casualty. Bad luck had dogged him, and he had taken to drink. Near midnight on June 17 he had been in a brawl at Hythe and shouted: 'I am an IRA man' as he threw a firework. Believing it was a bomb, a man rushed across and struck him on the jaw while a policeman put the firework out. Nick fell on the roadway and was knocked unconscious. Next day he died.

Costly, but less lethal, were Imperial Airways' troubles with their Empire flying-boats. On 12th June *Centurion* nosed into the water while alighting on the Hooghly River at Calcutta and was partially submerged, but only one passenger was injured. Newsmen were told that a sudden gust lifted the tail at a critical moment – but as it was a very restricted fairway, the machine more probably stalled in the anxiety of the moment. This was followed by the night-time burn-up of *Connemara* at moorings at Hythe after the refuelling petrol barge alongside caught fire, lighting up the entire port. So furious was the blaze that the flying-boat's wings fell off and she sank within twenty minutes. One of the bargemen was drowned and another badly burned. *The Times* computed that the three latest losses of these boats would cost the underwriters £150,000, and warned that the London Insurance market was finding aviation so expensive that higher rates were bound to follow.

Even the attempts to salvage *Corsair* at its remote location in the Belgian Congo seemed doomed to failure. By the end of June repairs had been completed under the most difficult circumstances, and the boat had been relaunched while the rainy season was at its height and the river in full spate, but it was only twelve yards wider than the wings. Capt Kelly-Rogers attempted a take-off despite a dangerous bend half way along the narrow fairway, but realizing that the unwieldy machine was not unsticking in a safe distance, throttled back, turned with the aid of the ground crew pushing the wings from the bank, and began motoring back but hit a rock and holed the hull. The only way to save it was to lighten by removing the engines and raising the airframe on petrol drums lashed together. Repairs started again. Boldly it was decided to dam the river and form a lake big enough for take-off when the next wet season came. Gangs of tribesmen were recruited to make it, build a village for the workers, and a road for transport of supplies from Juba. The end of this epic is beyond the time-scale of this book, but success attended the next attempt after ten months of tremendous endeavour.

At Hamble, flight refuelling of *Caribou* on 3rd July showed that all snags had been finally overcome. That the transatlantic service was within grasp seemed even more obvious when the majestic Short G-class *Golden Hind* was launched at Rochester on 17th June and towed to the barge yard for finishing and engine runs. Increase of price with size tends to follow the cube law, though cost per pound of structure weight was the usual guide, and although the deeper hull and 20 ft increase of span in relation to the Empire boat might seem relatively small, the volumetrically increased size resulted in a selling price more than twice the Empire boat's £41,000 – but the Air Ministry had subsidized the three purchased for Imperial Airways on condition that they were transferred to the RAF in time of war, entailing the later installation of gun turrets, armour protection for the crew, and bomb stowage under the wings for eight 500 lb bombs. Meanwhile further Sunderlands were under construction.

With the rival Supermarine shops full of Sea Otter, Walrus, and Spitfire production there was no time for so manpower-consuming a project as another big flying-boat, but Joe Smith was busy not only with modifications for all three types but also the design of the variable-incidence Dumbo replacement for the Albacore. Production of the Walrus and Sea Otter was also sub-contracted to Saunders-Roe where Henry Knowler and his directors were still interested in the possibility of bigger flying-boats. To that end, pursuing the line of thought which prompted development of the Empire boats from the Scion Senior, a 50 ft half-scale model flying-boat with four Pobjoy engines was being designed as a guide to a large reconnaissance project to meet specification R 3/38.

Ciphered A 37, the model was known as the Shrimp and its hull lines were based on the A 36 Lerwick which since its first flight had been fitted with a new planing bottom to overcome porpoising. Since the early part of the year three Lerwicks had been used for prototype trials in an attempt to overcome the all-too-often-encountered problem of combined roll and yaw deficiencies, and the elevator was too heavy. Auxiliary fins were fitted and the tailplane raised, but this gave no cure. A taller rudder and fin eventually improved the directional characteristics, but lateral wallowing remained, and the machine was difficult to trim because of a slight longitudinal instability, though pleasant enough until a stall was attempted, when there was a quick loss of height but no spin, and it was even more uncomfortable with flaps lowered. When stalled with engine on, a most stringent condition, not only did the machine get out of control but the centre-section cockled. Despite the bigger fin, the machine was difficult to fly straight on one engine unless side-slipped; and if the tail turret was rotated abeam, the nose pitched down. However ten had been ordered off the board at a cost of almost £55,000 each. It was clear that something better would be required, but hopes for the equivalent Blackburn B-20 with patented retracting hull bottom had vanished because the Rolls-Royce Vultures were still beset by development problems. However, the MAEE was testing the first Service aeroplane to be delivered from the USA to England by air – the Consolidated

PBY-4 flying-boat, later known as the Catalina, which had flown non-stop from Botwood, Newfoundland, to the MAEE at Felixstowe, a distance of 2,450 miles, in thirteen and a half hours, and then spent two hours cruising over the east coast keeping clear of Service exercises. The Air Ministry stated that this boat had been bought to keep in touch with design developments in other countries. Meanwhile it was making a favourable impression as a possible replacement for the Lerwick, though 20 mph slower.

It was a Consolidated flying-boat of this type, the *Guba*, with which the Australian airman, Capt P. G. Taylor, an associate of the late Kingsford Smith, had just completed a survey of the Imperial Reserve Air Route across the Pacific. With Capt L. A. Yancey as navigator, R. Rogers as second pilot, and two flight engineers, their first stage on 4th June was a nonstop 2,600 miles from Sydney to Port Headland, then 1,400 miles expecting to find the Cocos Islands, but the stars were obscured by cloud preventing sextant readings, and daylight visibility was bad, so after twenty-two hours flying they turned away and flew 700 miles to Batavia. On 7th June they left for the Cocos Islands by way of Christmas Island half way, reaching their destination after eight and a half hours flying. On the 13th the *Guba* departed for the Chagos Archipelago, alighting next day in Diego Garcia Harbour alongside *HMS Manchester* with which they had been communicating by radio throughout the 1,500 miles. On 21st June Capt Taylor concluded his flight at Mombassa in Kenya. In unusually generous spirit the British Government had contributed £3,500 of the cost, but it was through a small group of Australian sportsmen in conjuction with Major R. G. Casey, the Australian Minister of Supply and Development, that the Federal Government had made a meagre grant and secured the backing of a sporting American millionaire who offered his flying-boat with which to make the survey. Except for *The Times* the flight secured little reference in Britain, but the ability of a Consolidated flying-boat to make so great a journey and alight safely in strange seas had not gone unnoticed by the Air Ministry, and reinforced interest in ordering the Catalina.

7

The thirtieth anniversary of Handley Page Ltd was on 17th June. A further order for 200 Hampdens had just been received and that emboldened its omniscient founder to reserve eight trains on a Works outing for 4,500 employees from Waterloo to Southsea for a feast in what was claimed to be the world's biggest marquee. Two days later Frederick Handley Page threw a great party for 400 guests at Grosvenor House to celebrate the anniversary. Every one whom he considered important was there, including every available pioneer of 1909–1914 vintage and the now high-ranking officers who had formed the nucleus of the RFC and RNAS, for not only was he a great technician but also a great tactician and overwhelming personality. When it came to interesting the RAF and the Air Ministry: 'By a liberal display of hospitality and his own considerable powers of persuasion, Handley Page would help the authorities to appreciate to the full every merit of his aircraft on test,' recorded his PRO. To that end it was established policy for his chief test pilot to 'go down to Martlesham Heath and entertain the Air Force people there regardless of cost. Expenses in this connection were never challenged.' Perhaps that led Sir Kingsley Wood to make a sly dig in proposing the toast of 'The Firm', for he said that Mr Worley, as chairman, was fortunate in being a chartered accountant! More seriously he continued: 'The anxious moments through which we are passing demand a continuous flow of first rank aircraft such as these of Handley Page – yet there are more and better types to come from him and others which will soon be flying and carry the remarkable development of British aircraft production a stage further.' He added that the country was now spending almost £2 million a week on aeroplanes.

The Marquess of Londonderry added his laudations and commented that: 'The RAF is the greatest factor for preserving peace. Turning the other cheek has failed as a policy. The British Empire must have the strongest armed force in the world so that it can speak with authority.' Perhaps because of the presence of important guests from Germany and Italy he added: 'We are tired of being browbeaten by totalitarian powers, and are determined to lead the world in the way we believe to be right. I suggest that the Government should say now that if those countries want to fight we can fight too. We need a more full-blooded doctrine.'

Smiling benevolently Handley Page rose to the occasion, saying: 'I have never before found such words of commendation in official circles for our aeroplanes. They always seemed to prefer other firms than mine. But I hope that officialdom will now see the light in the sky and find the street which was called straight and leads to my Works.' He could hardly complain that the way was unknown, and his new Rolls-Royce car reflected the results. With quick wit he touched on the outstanding features of aircraft constructed by other pioneers present that evening, but sadly admitted that modesty prevented him from referring to outstanding developments of his own company, such as slots. 'Speaking as the earliest employee of Handley Page Ltd, I thank Mr Walter Dunkels for providing money for the firm in its early days. I thank George Volkert who has had the work of designing all the HP aeroplanes while I take the credit. I thank Mr Worley who has steered the company through difficult times, and Mr Easey, the secretary, who can always tell me just how much the Air Ministry owe the firm. I thank James Hamilton, who with Mr Volkert devised the system of split construction. But chiefly I express thanks to the foremen, charge hands, and all the men of the bench. Like the NCOs in the Army, they are the real backbone of the firm. In a larger way, the aircraft industry are the foremen and workmen of the Air Ministry. I assure the country that the Trade will not batten on the public purse but it must not be pilloried by politicians. It is we who have kept our staff together in lean times.'

Admiral Sir Murray Sueter in his toast to 'The Pioneers' said that he knew all about their misdeeds, but proceeded to recount their great activities, finally quoting the late Lord Montagu of Beaulieu as saying that the penalty of all pioneers was to be jeered and scoffed at, but when at last the harvest came it was they who had the satisfaction of reaping.

Moore-Brabazon, one of the greatest after dinner speakers of that epoch, made reply: 'H P pretends to be a realist, but in fact he is an idealist. There is an impishness about him that suggests he was the originator of H.P. Sauce. But some pioneers such as I have sunk so low as to become politicians. Others have become millionaires. In any case the original pioneers are not played out, but I wish the War Office had been right in its judgment of the uselessness of flying. Think of all the things that could have been avoided – even the awful tragedy of the establishment of the Air Ministry!' Referring to the present, Brabazon remarked how the gods must laugh at the foolishness of the human race. He hoped that if he escaped motor-buses and did not fly too much he might celebrate the fiftieth year of aviation and see flying as something which would knit all nations together. That was what all pioneers had fought for and died for.

On that note the party turned to less histrionic discussion over coffee and brandy. Never before, or since, was there so great a gathering of the British pioneers. Had the Germans dropped a bomb on Grosvenor House that evening they would have considerably improved their chances of winning a war with Britain. Somewhat cryptically, C. G. Grey merely reported: 'At no time in its existence have the public relations of Handley Page Ltd with people outside the firm been happier than they are at present.'

Public relations had become a catch phrase, used by manufacturer and government alike, but it could not hide the fact that Britain had been beaten at the post by the Americans in establishing the first regular North Atlantic passenger air service when *Yankee Clipper*, piloted by Capt Gray, arrived at Southampton Water on 28th June, this time with nineteen passengers, a crew of twelve, and 60,000 letters weighing three-quarters of a ton. The passengers included representatives of both Houses of Congress, various Government Departments, the Civil Aeronautics Authority, and members of Pan American Airways. The *Clipper* had left Port Washington, Long Island, two days earlier, but alighted at New Brunswick, Canada, due to low cloud and fog and could not leave until the evening of 27th June, but reached Botwood just before 11 pm and took 3,485 gallons aboard, immediately leaving for the 1,970-mile crossing which took thirteen and a half hours at an average of 144 mph, and alighted at Foynes at 3 pm on 28th June, still with 1,880 gallons left. A little over three hours later the *Clipper* departed and touched down at Southampton that night at 8.37 pm.

Imperial Airways had chartered the *SS Calshot* to convey newspapermen and a welcoming deputation, including Sir Francis Shelmerdine, to the alighting area. A reporter wrote: 'After about an hour in a fresh wind and under complete overcast at 500 feet, the motors of the *Clipper* were heard overhead, but passed above the clouds, and the noise faded. We heard that Capt Gray had been advised by radio to make for a patch of broken cloud further east through which to come out below the overcast. Fifteen minutes later we saw the *Clipper* approaching up Southampton Water from the Isle of Wight. It circled twice and came in to a very slow and clean alighting. The US Flag and Red Ensign were run up, and the *Clipper* taxied in on its outboard motors to a mooring buoy alongside the Short Empire flying-boats near the Netley shore.' Two Power Boats took off the passengers and mail and brought them to the *Calshot* where the Press and newsreel operators fell upon them.

Next day at a Savoy lunch given by the Government to the passengers of the *Yankee Clipper*, Sir Kingsley Wood said that Imperial Airways hoped to inaugurate their British transatlantic service from the beginning of August, having been delayed 'by the need to concentrate on the rearmament programme which had to take first place.' In fact the *Caribou* had only been successfully flight-refuelled on 3rd July, and there would be another nine days before it was cleared for take-off at the requisite 55,000 lb with permissable alighting at 48,000 lb. 'By next year,' said Sir Kingsley Wood, 'Great Britain should have equipment comparable with the American flying-boats when the new Short *Golden Hind* class will be operating along with the American Clippers.' He could have mentioned that she had been launched on 17th June and was almost ready for flight.

Some forty-three hours after arrival, the *Yankee Clipper* commenced its return to New York which it reached on 2nd July having taken thirty-one hours fifty-one minutes. Meanwhile the *Dixie Clipper* had left Port Washington on the Southern route by way of the Azores and Lisbon for Marseilles, carrying twenty-two fare-paying passengers at £135 return. These activities stirred the Post Office to announce that air mail for the USA, Canada and Newfoundland could now be accepted for transmission if posted in London not later than 7.30 am on Saturdays at a cost of 1*s.* 3*d.* per half-ounce for letters and 7*d.* for post cards. Not to be outdone the French Government concluded an agreement with the British and Newfoundland Governments for the use of Newfoundland Airport during a series of experimental flights across the North Atlantic to the USA, and intended to make six crossings with flying-boats alternating with six by landplane.

In the USA another brilliant airliner, the Douglas DC-4 powered with four 1,400 hp P & W twin Hornets, had made the re-powered and slightly larger Armstrong Whitworth Ensign seem outdated in design and performance, for it cruised 50 mph faster, had better climb, far greater range, and carried fifteen more passengers. The all-up weight of 73,000 lb compared with 49,000 lb for the Ensign until extended to 55,500 lb when fitted with Wright Cyclone engines.

Since the spring of 1930 the airlines of the United States had featured women stewardesses, and the lead had been followed by Swissair and KLM but women in Britain were now offered a more martial vocation. On 1st July the formation of the Women's Auxiliary Air Force, for war-time duty with the RAF was approved by HM The King. This was a revision of the Auxiliary Territorial Service for Women formed in September 1938 for non-combatant duties with the Army and RAF, the latter having forty-eight Companies mustering 250 officers and 2,500 other ranks. All transferred to the WAAF, and Miss J. Trefusis Forbes was appointed Director with the rank of Senior Controller. The intention was to draft many of the young women to the Barrage Balloon Units in each industrial area. In London some eighty sites had been approved in parks and open spaces for training in handling and winching these elephant-like silver monsters. The Under-Secretary of Air described the balloon barrage as a 'minefield of the air,' which was misleading, as that implied explosive charges, whereas it was a static defence with the psychological intent to force hostile aircraft to fly above the balloons lest they hit the invisible cables – but almost as important was the psychological effect on the population below who now felt they were protected from direct attack and that any high-altitude bombing probably would miss them! Germany was using balloons to the same end.

A very notable German aeronautical engineer, Dr Adolf K. Rohrbach, died at the age of fifty-one from heart failure on 6th July. His long pioneering work on metal aircraft construction had boldly set the pattern of future 'giant' airliners when he designed the great Zeppelin-Staaken four-engined high-winger which the Allied Commission of Control had been prompt to destroy at the end of the Great War. That did not expunge his dream of big, metal-plated, aerodynamically clean monoplanes. In July 1922 he founded the Rohrbach Metall-Flugzeugbau GmbH and soon produced the twin-engined all-metal flying-boats with the high-aspect-ratio rectangular wing of unique design for which Beardmores took a licence and sponsored his Roma three-engined flying-boat for the Air Ministry. Next came the Roland three-engined, high-wing, slab-sided transport monoplane, of which several were built for the Deutsche Luft Hansa. Best known in England was the vastly impressive, stark 150 ft span, 17 ton, three-engined Beardmore Inflexible powered with Rolls-Royce Condors and constructed in a manner reminiscent of bridge-building, so massive were the plates and so extensive the riveting. But Rohrbach was before his time, and his firm got into financial difficulties and disappeared, but in 1934 he joined the Weser Flugzeugbau-Gesellschaft where he had been working on a paddle-wheel 'cyclogiro' equivalent of the Autogiro.

8

Whatever the complications and pressures of producing more and more aircraft for the Air Ministry, export possibilities still had attractions, and nine British firms found time to build aircraft for display at the Brussels Second International Salon of Aeronautics which opened on Saturday 9th July. There were an Airspeed Oxford, de Havilland Moth Minor, GAC Cygnet, Miles Magister, Reid & Sigrist Trainer; but more spectacular were the Fairey P 4/34 two-seat fighter, a Hawker Hurricane with Merlin II, the high-speed research Spitfire with Merlin S of unspecified but enormous power, and biggest of all, a Vickers-Armstrongs Wellington I with Pegasus XVIII engines.

The Germans similarly had nine aircraft, but modestly limited themselves to tourers and trainers except for the compelling Ju 87 dive-bomber which attracted enormous attention because of its reputation in the Spanish war. Belgium showed fighters and bombers as well as trainers; France had two fighters, a fighter-reconnaissance type and a trainer; the Netherlands displayed the twin-engined Fokker torpedo-bomber and a Koolhoven trainer. There was a full spate of every country's aero engines including the direct-injection 1,200 hp Junkers Jumo 211 twelve-cylinder inverted vee on which Germany now heavily depended for all its new aircraft.

At the official banquet for 200 guests on the opening day so great was the spirit of accord, not least in relation to the Germans, that the Toast to the King of the Belgians was preceded by drinking to the combined Emperors, Kings, Princes, Regents, Duces, Führers, Presidents, Mayors and so forth of the nations represented at the tables.

On Sunday a great International Military Air Meeting was held at Brussels' Evère Airport, with King Leopold III and his handsome daughter Princess Josephine heading the 80,000 spectators. The participants were multi-national. Belgium performed splendidly flown aerobatics with Fairey Fireflies and displayed its nine Battles and three Hurricanes as well as its Guantlets and Foxes. The French performed with nine Morane Type 225 parasols eloquently described as the *Patrouille de l'École de l'Air*, and its bombers were represented by flights of Breguet 691 and Lioré et Olivier 45s; Britain produced three flights of Wellingtons, and Flt Lieut R. C. Reynell, a Hawker assistant pilot, managed to stray from England into the programme with a Hurricane which performed vertical climbs while slowly revolving on its longitudinal axis. The Swiss displayed a dignified formation of Hispano-powered Fokker CVEs built by EKW at Thun and designated C 35. Nor were the Germans out of the picture, but they encountered disaster when the leader of one of three flights of aerobatic Bücker Jungmeisters, while making high-speed flick rolls, dived vertically into the ground behind the Sabena hangar. In true tradition of military air forces the German team continued with a brilliant display of aerobatics, ending breathtakingly with their star flying extremely low down while performing double flick rolls and what was described as a horizontal upside-down spin.

Watching were representative senior officers of many countries, such as Air Chief Marshal Sir Cyril Newall, General Milch for Germany, General Vuillemin for France, General Ryski for Poland, and their host General Gillieaux for Belgium – all of whom chatted fraternally. A visitor remarked: 'Some of the English who do not often go abroad were astonished at the obvious friendship existing between air forces of the various nations. Nobody who has seen them together could doubt that the big chiefs of the French, German and British Air Forces are personal friends. They were all experienced war pilots. They all know what war means.'

Not to be outdone, Italy held a great air display at Guidonia, their equivalent of the RAE, A & AEE, and NPL rolled into one. Il Duce himself piloted a Savoia bomber, and demonstrated his high competence as a pilot. Said one of his passengers: 'The landing was just as good as his take-off. It was one of the smoothest I have ever had. He just put the machine on the ground and kept it there. The only thing I saw the second pilot do, beside using the radio, was after we landed he used the engine throttles while Il Duce taxied back into position.' Of the general standard of Italian flying there was no doubt either that it equalled the RAF and even revealed a new event, the Fantasia, when a massed formation of Fiat CR 32 biplanes disported each flight of five with the leader flying level and the machines each side displaced and inclined to form a letter 'W'.

Both as a threatening growl to Germany and an expression of *camaraderie* with France, twelve RAF squadrons of light and heavy bombers made flights over France on 11th July as part of the training programme of Home Command, and they were to be reciprocated by similar flights over England by the French Air Force following agreement between the two governments. Most of the flights were at 6,000 ft, but for better effect, as the formations of Battles, Blenheims, Whitleys and four squadrons of long-range Wellingtons passed over the larger towns and cities in France, they came lower so that the inhabitants might see the massed might. All landed at Le Bourget to participate in the Review of the French Armed Forces during the Fête Nationale in Paris on 14th July at which fifty-two British aircraft were programmed for the fly past. Air Chief Marshal Sir Cyril Newall flew to Paris for the occasion as the guest of M. Daladier. The day after his return he cut the first turf on the site of the new RAF Staff College at Weyhill, Andover.

On 19th July another series of flights were made over France, among which five squadrons of Wellingtons flew some 1,500 miles to Marseilles and back, with Capt Harold Balfour piloting one of the machines; two squadrons of Whitleys flew to Avalon, 120 miles south of Paris, and back, totalling 750 miles, and several Blenheim squadrons crossed the French coast at Treport for a circuit of Paris and nearby cities, returning by way of Dieppe. Ninety-five aeroplanes were involved. Cuttingly, Arthur Henderson in the Commons asked whether HM Government proposed practice flights to Poland, Turkey, Rumania, and other countries. Tauntingly Sir Kingsley Wood replied that such flights might be possible. Meanwhile General Sir Edmund Ironside, Inspector-General of Overseas Forces, visited Warsaw for Anglo-Polish Staff Conversations.

The skies of England echoed to the hum of many aircraft these summer months. Continental touring in no way diminished, despite uneasy politics. Aero clubs were holding their flying displays, and Birmingham, with the aid of the Duchess of Kent, opened the City's new Airport at Elmdon amid pouring rain.

At Rochester, Lankester Parker had been waiting for a quiet day with the wind blowing along his habitual take-off strip on the Medway, and on 21st July accomplished a 33-minute maiden flight of the impressive *Golden Hind*. Four days later flew it for seventy-two minutes and substantially established that the machine could be handed over to Imperial Airways, though two further half-hour tests were also made. In the light of later years this minimal testing by Shorts of an untried and complex passenger machine seems astounding. As long as engines performed with good response and without overheating, then merely reasonable handling in normal cruising flight and at take-off and landing were all that was expected. Yet the *Golden Hind* had many novel features such as the heating system in which steam was raised in boilers around the exhaust pipes and led to a bank of heat-exchangers through which fresh cold air was led from openings in the wings; there was the mechanism of the big flaps driven by a Rotax motor to wind them up and down, with automatic cut-off as soon as they reached closed or open position; there were the involved starting system, electrical generators, fuel transfer system, the galley and its heating and water supply – in fact scores of ancillary devices which could go wrong, yet were only ground-tested. Instead, the rush was to fit the passenger accomodation. Meanwhile there was mounting contention as to whether to enclose the waters of Langstone Harbour for the future Imperial Airways flying-boat base, or adapt Pagham Harbour a few miles further east, or continue with the hazards of shipping in Southampton Water.

Though the glow of publicity focused on the *Golden Hind*, far more important to the Air Ministry, the DOR, and Fighter Command was the first flight of Frise's conversion of the Beaufort to the powerful and fast Beaufighter which for the time being was stripped, but would be armed with four cannon in the nose and a dorsal Vickers K on a pillar mounting, and might later carry additional .303 guns in the wings, or a ventral torpedo, and bombs under the wings, or rockets. Initially flown on 17th July by the now white-haired, venerable-looking Cyril Uwins, it revealed every promise of becoming a formidable fighter, for it was as fast as the Hurricane, and by fitting Mk VI engines might prove almost as fast as a Spitfire – which the Air Ministry announced as 367 mph in the battle of intimidation with the Germans, though production Spitfire Is were in fact 12 mph slower when fitted with armament. As with all prototypes, the Beaufighter had minor snags despite evolution from an established design, and it was somewhat heavy laterally and fore and aft, soon resulting in elevators altered to an inset instead of horn balance. Directional stability was marginal, but production standardization made it undesirable to change the fin and rudder areas when a dihedral tail might afford a solution, particularly as this would reduce changes in the trim between power 'on' and 'off'. For night landings a longer stroke oleo leg was an improvement. So convinced had been the Air Ministry on seeing the almost completed prototype at the beginning of the month that this private venture fighter was immediately covered by an official order for four prototypes and a production contract for 300 capable of being re-powered with the Rolls-Royce Griffon if necessary. Nevertheless Fedden's vigorous intention was that the Hercules would be steadily stepped up in power, perhaps to almost 2,000 hp, and would remain the standard.

In the week that the Beaufighter was making its first flights at Filton the twin-engined Avro Manchester rival of the unfortunate four-engined Stirling had been wheeled out at Ringway for engine running and final flight clearance, and on 25th July Sam Brown and his assistant, Flt Lieut S. A. 'Bill' Thorne, made a successful first test to the relief of Roy Chadwick, who like other designers, became more tense than the pilots on such occasions. The fact that the all-up weight of this massive but clean-looking monoplane was four times as great as anything which Sam Brown had flown created no difficulty: seated well ahead of the wings in a cockpit long familiar from the mock-up and later stages of construction, he would have no particular impression of big span, but would be wholly involved with the instruments, the broad vista ahead, and the general response to his instinctive control – for the aim is to ensure within broad limits of physical strength that an aeroplane answers its three controls in the standard fashion with which every pilot is familiar. The test pilot's art was to diagnose the differences, and as hydraulically assisted controls were not yet practicable, appropriate control balance was discovered through small changes of shape or servo; similarly the tail areas and combination of fin and dihedral had to give sufficient stability on all axes through the range of design centres of gravity which trials had to show were permissable.

Although wind-tunnel tests had established the size of the tail units, the Manchester proved directionally unstable, but to avoid the complication of increasing the size of the fins and rudders on the tailplane extremities it was decided to add a small central tail fin, but even this was inadequate, as I discovered a year later when I flew the increased span second prototype after the A & AEE had been transferred to Boscombe Down near Salisbury. The Manchester nevertheless handled beautifully in its massive way, sweeping easily from one grandiloquent turn to another, and seemed to have neutral longitudinal stability, holding whatever speed was set, though I felt that at high speed it might have an increasing nose-down divergence. The field of view was tremendous: the feeling of power from the big 1,760 hp Vulture engines seemed enormous – but therein was the snag common to all twin-engined aircraft, and particularly marked in the Manchester with such great power, for if at slow speed one engine was cut, then an uncontrollable swing followed, and the outer wing would lift and the machine drop like a brick, for the loading was 40 lb/sq ft, and would be higher in the production version. The risk was made worse because the Vulture was still far from a developed engine, and there were many teething troubles, some as serious as fatigue failure of connecting-rod bolts. Within a fortnight of my flight, an engine failure occurred on take-off, and the very able young Australian pilot who had captained the machine I flew, was killed when it dived into the ground just beyond the aerodrome.

Major Bulman recorded: 'Air Marshal Tedder and Sir Wilfrid Freeman went to Avros just after this prototype had spun in, and Dobbie Dobson the irrepressible, with a model Manchester in his hand, said none of them were happy with results to date and that Rolls-Royce seemed luke-warm about the Vultures. Whereupon he slipped off the model's wing and replaced it with another mounting four Merlins. Hence the Lancaster which was destined to become the best British bomber of the war. No doubt Hives had warning of this vital change in talks with Dobbie, and may even have inspired the thought in Dobbie's mercurial brain.'

For all their great prestige, Rolls-Royce were involved in problems with the Merlin as well as the Vulture, and in any case there was unending need to achieve greater power. To cope with the enormously increased requirements, Rolls-Royce had their newly opened £1 1/2 million factory at Crewe which had been built and equipped in only twelve months. Another was being constructed at Glasgow – but Derby remained the hub. Arthur Sidgreaves, the managing director, in welcoming Sir Kingsley Wood to the new Crewe factory, accompanied by Sir Wilfrid Freeman, Sir Charles Bruce-Gardner, and Frederick Handley Page, pointed out that unlike shadow factories, his company had also spent £700,000 on expansion of their own Works, £250,000 on development of Rotol Airscrews Ltd in alliance with the Bristol Aeroplane Co Ltd, and £150,000 on Phillips & Powis Aircraft Ltd. But as he said: 'Constructors of aircraft and aero engines in these days are well looked after by the Government. Just in case the price agreed is too favourable there is the National Defence contribution, the Income Tax authorities, and the Armament Profit's Duty to be satisfied!'

Despite Sidgreaves' antipathy to shadow factories, Sir Wilfrid Freeman insisted that Fords of Dagenham, with their enormous resources, must contribute to Merlin production, and A. R. Smith, their managing director under the chairmanship of Lord Perry, agreed that 400 a month was feasible. Said Major Bulman: 'Smith was a great man in every sense, and realized that his team must steep themselves in all the know-how of building aero engines as distinct from Ford cars. With the complete support of Hives, he therefore sent ninety of his best men to Derby for training and they were there for nine months. The Ford estimate of production cost was £5 million, and it was Freeman's view that the matter was so urgent that he side-stepped the usual approval of the ACCS and sent me to the Treasury to get the necessary authority, and incredible though it seems, I received it within five minutes due to the perception of our administrative Civil Servants of those days.' Smith decided to build an entirely new factory at Trafford Park, outside Manchester, compromising on the true Ford method of production in order to begin delivery in June 1940 rather than his initially estimated two and a half years. 'The result was a machine shop containing single-purpose tools tailored to the Merlin's design, though less capable of being switched to different engines, fundamentally far more productive of Merlins than the Rolls-Royce line, and making a cheaper engine of no less superb quality.'

The pressure for more and more engines was paramount. At Bulman's instigation, it was decided that the Standard and Daimler companies in conjunction, and similarly Rovers and Rootes, would build complete Bristol aero engines as a second Bristol shadow group, producing 400 per month at a capital cost of £9 million; but it was stipulated that none of the four new works should be adjacent to existing shadow factories less they all be put out of action through bombing.

9

Since the Great War the number of people who were paid during holidays had steadily increased, but now with the passage of the Holidays with Pay Act of 1938, some eleven million families had the opportunity of a fortnight by the sea, and Billy Butlin's Holiday Camps at Skegness and Clacton at £4 a week per head were proving gold mines. Good train services, buses, and two million private cars were making seaside holidays attractive everywhere. Participation in tennis, golf, and cricket was booming, and in addition to the steadily increasing number of sailing clubs and flying clubs, the gliding enthusiasts were becoming ambitious. At the National Gliding Contest at Great Hucklow nothing counted under two hours duration, or height less than 1,500 ft, for the fifty pilots and their twenty-five sailplanes, and cross-country flights earned extra points if across wind or to a named destination. Philip Wills made daily jaunts of 100 miles or so to stated goals, efficiently retrieved by his wife whose technique was to drive down-wind as soon as her husband took off, and at intervals she telephoned to the launching ground for news, otherwise assuming that he was continuing down-wind. Nobody yet had a radio inter-com. Rivals were Christopher Nicholson with a flight of 162 miles, and apparently Joan Meakin who astounded officials by phoning that she had landed at New York – but it was found that this was a small village sixty eight miles east of her launching point! Sadly, the carefree atmosphere was marred by two fatal accidents, for British sailplaning still had an amateur touch compared with the Germans at the Wasserkuppe where thermal soaring was the order of the day. Their permanent quarters on the mountain top were roomy and delightful, and numerous enthusiasts could be accommodated; but all were State-subsidized members of the Nationale Sozialistisches Flieger Korps. Nevertheless, the quasi-military background was not flaunted; the young girls and boys seemed like their British equivalents, but with more obvious patriotism and discipline.

When British pilots visited Germany there was always a friendly welcome and what seemed heartfelt expressions of goodwill. The long-feared enemy was Russia. Though Hitler in earlier years had visualized England as a potential ally, the case now was very different. He realized that Chamberlain's commitment to help Poland was unlikely to be practicable if Germany struck swiftly this year. Had not Lloyd George told Chamberlain in the Commons: 'If war occurred tomorrow you could not sent a single battalion to Poland. I cannot understand why before committing ourselves to this tremendous enterprise that we did not secure beforehand the adhesion of Russia.'

In the summer's dalliance, Britain was making time-consuming overtures to Russia, despite the springtime rejection of Stalin's proposals for a conference. Now that Chamberlain had guaranteed help to the hereditary enemy Poland, Stalin preferred closer relations with Germany with whom he was prepared to discuss a trade agreement. Nevertheless a negotiating British Mission arrived at Moscow on 11th August, but it had not the military significance which Stalin required, and next day he notified Germany that he was ready for political talks.

Russia now played a temporizing game, protracting the discussions with both countries; but after a message from Hitler urging that Ribbentrop be invited to visit Moscow for political negotiations, Stalin conciliated by signing a Russo-German commercial agreement on the 19th.

Clearly events were marching towards a crisis, yet to the mass of the population it seemed unreal and was discounted, for these were prosperous days. With rearmement, unemployment had steadily dropped and was a mere 9 per cent of insured persons. Almost everyone was well off compared with the 1920s. Civil aviation was pursuing its steady way. Croydon had never been so busy; most of the modified Ensigns had returned, and there were four operational Albatrosses. Said an awed spectator: 'Beside those parked outside the main shed I have seen six big airliners on the operational area at the same time and as many as thirty or forty others parked in front of the hotel or by the hangars. Two Ensigns are now running regularly on the Paris service with an elapsed time from concrete to concrete of 1 hour 25 minutes, for taxi-ing out, arrival circuit, and taxi-ing in take far more time than people think because the big liners now in use have to make a very wide arrival circuit especially if other aeroplanes are about. The cross-Channel Ensigns take a few minutes longer than the Albatrosses and French machines because the rpm are being kept down well below the maker's figures until the pilots and engineers get used to them. Because the operational height of the Ensign is 7,000 ft and the Frobisher is 8,000 ft, they climb slowly and steadily for the first half of the run and coast slowly downhill for the second half as the short Paris haul does not allow their operational height for more than a few minutes. I hear that Imperial Airways are proposing to start the Empire services in September with the Ensigns, so presumably they will not wait for the Cyclones which will not be delivered until that month.'

In fact the Ensigns would be operated by British Overseas Airways Corporation, for despite harassment in committee the Bill for the Amalgamation of Imperial Airways and British Airways received Royal Assent on 4th August, following the second reading in the Upper House on 1st August when Their Lordships unanimously agreed to it, and took the opportunity of congratulating the Air Minister on his escape four days earlier when the Air Council's DH 86B forced landed in fog on Kirby Moor and nosed over, injuring the pilot and two of its complement of senior officials.

The amalgamation did not mean that the Corporation was immediately effective. A Board had to be elected and management appointed, but it was assumed that Sir John Reith as chairman of Imperial Airways would continue with the Corporation in that position. A few days later, Imperial Airways announced the curtailment of passenger bookings on its Empire routes due to increased mail loads and insufficient numbers of aircraft, and it was thought likely that the passenger situation would not be much better until January 1940.

Imperial Airways' troubles with the Empire boats seemed unending. In the first two years of operation eight had been lost in major accidents, and in the current year there had been three more, and *Corsair* was still being repaired in the Belgian Congo. On 9th August there was more trouble when the *Australia*, taking off from Basra for Karachi on 9th August, ran aground on a submerged sandbank and was badly damaged. However flying-boat problems were not confined to Imperial Airways, for on the 13th one of the PAA Sikorsky flying-boats hit a floating log when alighting at Rio de Janeiro, crashed into a dry dock and caught fire, and of the sixteen aboard only two were saved.

On 11th August British Airways had another near disaster with a Lockheed 14 on the London-Zürich service. One of the engines gave trouble at 10,000 ft and the machine failed to hold height. At 3,000 ft the carburettor caught fire, and the flames spread. Captain David Prowse put the machine down in a field near Luxeuil and the undercarriage collapsed; the crew, though suffering slight burns, were able to get the nine passengers out – but the machine was destroyed by the fire. Four days later British Airways had another disastrous set-back when one of the Electras on the Stockholm run caught fire in the air, and the only recourse was to ditch in the Storstroem Straits sixty miles south of Copenhagen, but the impact and nose-over killed all four passengers and the radio officer, though the pilot, Capt C. F. C. Wright survived after being found floating unconscious when the machine had sunk in nine fathoms.

The day after the Royal Assent, *Caribou*, commanded by Capt J. C. Kelly-Rogers, was ready for the opening transatlantic service, and left Southampton for Foynes in formation with *Maia* carrying Sir Francis Shelmerdine, Marshal of the RAF Sir John Salmond, now a director of Imperial Airways, Leslie Runciman as deputy chairman, Major McCrindle chairman of British Airways, Mr J. Duranty as High Commissioner for South Ireland, and Capt A. S. Wilcockson as manager of the Atlantic Division of Imperial Airways. Arriving at Foynes at 3.30, having covered the intervening 355 miles at 156 mph, Kelly-Rogers was met by a delegation from Foynes Regatta Committee, of which he was a member, who presented him with a silver tray while *Caribou* was being filled to take-off weight. At just after 6 pm he took off in thirty six seconds, contacted the Harrow tanker which Geoffrey Tyson flew from nearby Rineanna (Shannon Airport), and topped up to maximum permissable tankage of 2,000 gallons while circling the area for the benefit of cameramen and spectators and those in *Maia* who now included Mr de Valera.

With practice it took three to seven minutes to contact and link the hoses, and eight minutes to pass 800 gallons. Refuelling was completed in sixteen minutes from take-off and a course set for Botwood, butting into a strong head wind, with the result that the crossing took three hours twenty-two minutes longer than the sixteen hours expected. *Caribou* alighted at Botwood at 1.30 pm GMT on 6th August, after encountering heavy rain and fog off the Newfoundland coast, having covered the 1,980 miles at a mere 102.5 mph due to the wind. She was then quickly refuelled and 2 hours 20 minutes later left for Boucherville on the St Lawrence, seven miles from Montreal, arriving mid-afternoon, having speeded the 900 miles at 142.5 mph. Capt Kelly-Rogers went ashore and was received by Mr G. D. Howe, Canadian Minister of Transport, then reboarded and within an hour of arrival took off for Port Washington which was reached two and a half hours later – the entire journey from Ireland having taken thirty six and a quarter hours. On 9th August the return was commenced, and three days later *Cabot*, piloted by Capt Donald Bennett, took-off on the second outbound scheduled service.

Not to be outdone, the French had already made their third trip across the North Atlantic using the six-engined Latécoère flying-boat, with far-famed M. Guillaumet, chief pilot of Air France-Transatlantique, in command – but lacking air refuelling it was necessary to top-up at Lisbon and the Azores, so the overall 4,334 miles took thirty-nine hours thirty-two minutes, of which thirty-six hours thirty-two minutes were airborne.

From peaceful air transport intended to bring the nations closer, Britain turned to warlike Exercises in which some 1,300 RAF aircraft took part to test air and ground defence of SE England, commencing on 8th August and ending with a practice black-out of all that area and Greater London on the 11th. In addition to the aircraft 53,000 men participated, with 110 guns, 700 searchlights, and 100 barrage balloons. Defenders Westland had 500 fighters, 50 general reconnaissance aircraft and 250 bombers, four anti-aircraft divisions, ten London balloon barrage squadrons, and fifteen groups of Observer Corps, all under the supreme command of Air Marshal Sir Hugh Dowding, C-in-C Fighter Command. Eastland attacked with 500 bombers under the direction of Air Chief Marshal Sir Edgar Ludlow-Hewitt, AOC Bomber Command, comprising Blenheims, Hampdens, Wellingtons, and Whitleys which raided military objectives from 100 miles off shore, and the single-engined Battles from fifty miles. The essence was to discover the effectiveness of interceptions, this time aided by the full range of some two dozen mostly coastal radar stations covering an area from Portsmouth along the French coast to Holland and 150 miles off shore from the entire coastline of England and Scotland to the Orkneys. Radio signals from their 350-ft stacks of aerials when reflected back by an intruding aeroplane were received on 250 ft aerial systems, and by measuring the fractional elapsed time between transmission and reception on a calibrated cathode-ray tube, the distance, angular bearing in altitude and direction, and therefore the flight path could be assessed and marked on a large table map by a WAAF plotter using small plaques and coloured arrows, a teller passing the information to Fighter Command and Sector Operations Room, who in turn transmitted interception instructions to defending fighter pilots. Meanwhile the Observer posts, of which there were fifty in each of the sixteen Observer Groups, reported the visual position and direction of all aircraft spotted over their areas.

Weather, as usual, played an important part, and was particularly bad with low cloud, rain, wind, and ground fog, hampering attack and defence with uncalculable impartiality, even suspending activities between midnight and 5 am on 10th August – yet it was the acid test of the success of radar. Nearly every attack was intercepted.

There was special commendation by the Air Ministry for all the participating amateurs – the outstanding skill of the spare-time boys who had graduated to

Hurricanes, Spitfires, and Blenheims, as well as the determined enthusiasts of the Observer Corps and Anti-Aircraft and Searchlight Units of the Territorial Army. A journalist cryptically reported: 'The dangers of air attack have been much magnified. This country is protected by stretches of sea too wide for the enemy to have an effective escort of fighters. Our own fighter pilots are protected by the motors of their machines and in these days by other methods, and our bombers incorporate certain features which give them an advantage over the bombers of other countries.' But what were those devices? Some people murmured the magic word 'oboe'. Could the pilots of British bombers now actually see their target through cloud, and were fighter pilots brought visually to their targets in the dark?

To convince the nation of its safety from air attack, Air Chief Marshal Sir Hugh Dowding broadcast a short summary of the Exercise from the BBC on 12th August, and even admitted that various new methods had been tried with most satisfactory results. 'It only remains for us to see that our technical equipment keeps ahead of that of our potential enemy. What we have been doing is to work at increasing interception towards the 100 per cent which is our goal. I am satisfied with our progress, and I confidently believe that serious attack on these Islands would be brought to a standstill within a short space of time.'

Next day the Vickers Warwick, powered with two Vultures at last made a brief initial circuit at Brooklands, piloted by Summers with Westbrook as passenger, uncertainties of engine response necessitating an immediate landing. Six days later, with an improved carburettor linkage, it performed satisfactorily and trials began.

In the following week seven Wings of bombers, fighters, and reconnaissance aircraft of the French Army of the Air flew over England in the course of a training and technical Exercise in which Fighter Command took part. Each Wing consisted of thirty to forty aircraft guarded by fighters, and the first raid crossed the coast at Harwich before dawn to attack Manchester, Liverpool and other northern towns. Later there were a series of unheralded raids over industrial areas in the Midlands and West Country during which the French formations were attacked by British fighters who found them despite thick haze over the Channel. London had the full force of the third raid during the afternoon when bombers came fairly low with escorting fighters above, but RAF Hurricanes, Blenheim fighters, and Spitfires intercepted them over Surrey, Sussex, and Kent and followed them over London in continuous battle. Said the official communiqué: 'Coming after the raids of British planes over France last month, the Exercise gave an equal opportunity for intervention by the British and French detector services, working in co-operation, and these manoeuvres form a contribution to air co-operation between France and Britain which each day thus becomes closer and more efficacious.' Their serious purpose became even more apparent when the French High Command sought assurances that not less than six British fighter squadrons would be despatched to France immediately war broke out.

Concurrently Air Marshal Sir Charles Burnett, Inspector-General of the RAF, and other members of an Anglo-French Staff Delegation were in Moscow discussing Soviet co-operation. At a great flying programme emphasizing Russian prestige, they inspected the astounding 210-ft span six-engined ANT 20 bis, successor to the crashed *Maxim Gorky* monoplane. At an all-up weight of 45 tons it lifted a disposable load of 14 tons compared with a contemplated 7 tons for the Short Stirling. As an airliner the ANT had seating for sixty-four passengers and crew of eight. Next day they learnt that the Russo-German commercial agreement had been signed, and the Embassy warned that a Russo-German non-aggression pact was imminent. On 22nd August the British Cabinet met to discuss reports of the menacing movements of German troops. That same day Hitler gave orders indicating zero-hour for an attack on Poland.

Parliament was summoned on 24th August to enact an Emergency Powers (Defence) Bill, and Reserves of the Army, Navy, and Air Force were called up and ARP services alerted. Chamberlain, still convinced of his personal influence, wrote privately to Hitler warning him that no greater mistake could be made than thinking that British intervention on behalf of Poland could be ignored as the Government was determined to fulfil its obligations so that there should be 'no such tragic misunderstanding, as was in 1914.'

Hitler, who had remained uncertain of Britain's determination to fight over Poland, seemed taken aback and summoned Sir Nevile Henderson next day, pledging himself to the continued existence of the British Empire, even promising the power of the Reich to British defence if it was threatened anywhere in the world. But as to Poland, he offered nothing. That same afternoon, after hearing from Henderson, Chamberlain signed an Anglo-Polish Treaty of Alliance, reaffirming the April agreement. Daladier immediately announced France's determination to uphold their pledges to Poland.

On Saturday the 26th and all Sunday, the British Cabinet was painfully arguing the crucial instructions that must be sent to Henderson 'to safeguard Poland's essential interests' and thus open the way for 'a wider and more complete understanding' between Britain and Germany. This was delivered by Henderson to Hitler at 10.30 pm on 28th August. Twenty hours later he was informed that Germany was ready to negotiate with Poland if an emissary arrived in Berlin with full powers next day and was ready to accept whatever the Germans proposed.

While the hurrying consultations and passage of messages was proceeding, Germany made a different mark in history with the first-ever flight of a turbojet-propelled aeroplane – the Heinkel He 178 with a single He S 3B unit giving 834 lb thrust. This little machine, much like the Gloster-Whittle design, was briefly flown in the greatest secrecy on 24th August at Marienehe by Heinkel's test pilot Erich Warsitz, and for a longer flight three days later. Here was a major break through in technology offering boundless possibilities. Heinkel's technical director had already schemed a clean single-seat fighter powered with two of von Ohain's bigger turbojets, each delivering 1,587 lb thrust, and following the successful flight of the research machine, work on the mock-up was hastened in order to show it, together with the He 178, to Udet and Milch at the earliest possible date, unaware that the two Generals had already studied the proposal for a twin axial-flow turbojet fighter which Messerschmitt had submitted early in June.

In England, Glosters were working in close collaboration with Whittle on the installation problems of his new form of engine in which the single large combustion chamber had been replaced with ten small ones. Whittle, on extended leave for a post-graduate research year, was likely to be instructed to return for Service duty with the rank of Squadron Leader, but his anomalous position as a serving officer would then produce complications because the Air Ministry could not place contracts with its own employees, yet his full-time attention in developing the engine was essential if its snags were to be overcome. By June, a new impeller had enabled speeds to be increased to 16,000 rpm for short periods, but there was trouble with the combustion and problems of steel behaviour at the high temperatures of operation. Major Bulman recorded: 'Towards the end of August, Pye assured Whittle that the development of both engine and airframe would continue if war occurred. A new and more powerful W 2 engine was to be made giving 1,600 lb thrust compared with a hopeful 1,200 lb of the W 1 flight engine.' But Whittle suspected that the Air Ministry regarded his engine only as an interesting piece of research and feared the aero engine industry would snatch the result of his work and 'resurrect it to create enormous profits and kudos for their firms, in league with Air Ministry officials'. Everyone seemed to find him extremely difficult. Nevertheless Roxbee Cox in later years told the Royal Commission on Inventions that Britain would have been two years longer in bringing fruition to the jet engine had it not been for Whittle's fiery crusade. Yet who remembers von Ohain, Ostrich, or Helmut Schelp, those contemporaries of Whittle who had seen the value of Professor Betz's work on axial compressors at Göttingen and had argued the case for jet propulsion of German fighters in the late mid-'30s?

While the ominous future was beginning to reveal itself, a forgotten British aviation pioneer died on 14th August – the hunchback Leonard F. Howard Flanders, graduate of Emmanuel, who had helped A. V. Roe at Lea Marshes in 1908. He had subsequently designed and built several early biplanes and monoplanes which were highly regarded by such pilots as Fred Raynham and E. V. B. Fisher, and so successful that the War Office ordered the then vast number of four of his F 4 monoplanes, but their few weeks of RFC use abruptly ended with the War Office ban in 1913 on all monoplanes. Before he could fully develop his B 2 biplane, Flanders crashed on a motor-cycle and was convalescent for a year, during which his firm collapsed, but in 1914 he joined Vickers Ltd as chief designer. Such were the hazards of working under the difficult managing director, Capt Herbert Wood, that when his twin-engined FB 7 precursor of the Vimy proved dangerous laterally due to the then unrealized phenomena of aileron reversal, he lost favour and was transferred to development of the unsuccessful Hart radial engine for which a licence had been obtained by Vickers from the USA. At the end of the War he joined the Cosmos Engineering Co Ltd, successors to the Brazil-Straker car company where he had gained his earliest engineering experience, but when it went into liquidation in 1921 he became a lecturer at the Regent Street Polytechnic until 1923, and then joined English Electric. In 1925, he started his own business with the then novel idea of producing portable radio receivers – but he was too early in that field and reverted to employment with an engineering firm. As his friend C. G. Grey said: 'Because of an accident in infancy he had a deformed spine which always affected his health, and his motor-cycle accident permanently damaged his eyesight. In spite of this he was a determined worker, with mathematics as his favourite occupation, and as he was good at secretarial work, he gave much time to the British Gliding Association in the early days of the movement and also to the Institution of Aeronautical Engineers until amalgamation with the RAeS.' Recently he had rejoined the Bristol Aeroplane Co as Fedden's personal assistant, but again became ill and died at the age of fifty-six, after a spinal operation. With him went yet another link with the old flying days in which he had participated so eagerly.

10

A further sign of the imminence of war was the cancellation of the King's Cup Race which was to have been run on 2nd November, but only twelve machines had been entered – such was the absorption of members of civil aviation in the all-embracing rearmament programme.

The signature of the Nazi-Soviet Non-Aggression Pact had brought a storm of protest from all over the world, as well as peace appeals to Hitler from President Roosevelt, the Pope, the King of the Belgians. But Mussolini temporized, despite his assertion to the British Ambassador at the beginning of July: 'If England fights in defence of Poland, Italy will take up arms with its ally Germany.'

At midnight on 30th August Henderson was summoned by Ribbentrop, who tersely defined Germany's offer to Poland. But exaggerated stories of Polish atrocities, coupled with the obdurate attitude of Britain and France, had already prompted Hitler, whatever Poland's reply might be, to confirm instructions for an invasion at dawn on 1st September.

And so it began. Britain had no chance of going to Poland's help. Germany attacked Poland without declaration, opening fire with artillery and raiding the four nearest towns and a dozen Polish aerodromes, later bombing Warsaw several times. 'Most of the places bombed seem to have been of definite strategic importance,' ran one report. 'Most of the bombers came from East Prussia, but there seems to have been little long-range bombing and no escorting of bombers by fighters.' The Poles alleged that 1,500 people were killed or wounded in these raids, including women and children.

Complete mobilization of the RAF, Army, and Royal Navy was ordered, and a black-out throughout Britain from sunset every evening. Legislation was announced that morning for the extension of military service to all physically fit men between the ages of eighteen and forty-one. The full ARP organization was put into operation. Evacuations from London of small children and their mothers immediately began. The French Government ordered general mobilization. Canada announced that she would stand on Britains's side in a conflict with Germany. President Roosevelt appealed to European nations not to bomb civilian populations or unfortified towns.

The German plan was to attack simultaneously from north and south across the great plain of Poland, which had indeterminate boundaries and few natural obstacles, in order to cut off the Danzig corridor and force the way to Warsaw. Had the Poles concentrated their main defences instead of attempting to hold the entire length of German frontier, hostilities might have been prolonged enough for some support to reach them from Britain and France – though the latter would have been a reluctant starter, for French military thinking was defensive and relied almost entirely on the great Maginot Line of fortresses on the German frontier. As it was, the grey-clad German troops found it easy to crush the opposition of the scattered Poles.

Everybody waited, tense and uneasy. The British Cabinet on that afternoon of the German assault agreed on an instruction to Henderson declaring that Britain would stand by her obligations unless the German forces were withdrawn, otherwise Britain 'was resolved to oppose them by force'.

Next day, ten Fairey Battle squadrons of the Advanced Air Striking Force flew to France. At midnight the eastern half of England became a prohibited area for civil aircraft. Most air transport companies cancelled all services except to the Channel Islands and the West and those from Scotland to the Northern Isles, Hebrides, and Northern Ireland. Imperial Airways announced a reduction of their air mail services, and with other companies operating from London, was evacuating to headquarters as yet unnamed. The Prime Minister of Eire announced that his country would endeavour to maintain her neutrality in a European conflict.

When Parliament met that afternoon, members were angered by the dilatory tactics, as they thought, of Chamberlain. Arthur Greenwood, Deputy Leader of the Labour Party, protested that Germany had been at war against Poland for thirty-eight hours and nothing had happened because the Treaty obligations towards Poland had not automatically been put into effect. 'I wonder how long we are prepared to vacillate at a time when Britain and all that Britain stands for, and human civilization, are in peril?' But Chamberlain denied any weakening on the part of the Government and said there was need to synchronize action with the French.

That evening the Cabinet instructed Lord Halifax to telephone the French Foreign Secretary, Georges Bonnet, that a declaration of war would have to be brought to Parliament next day. At dawn on 3rd September Henderson was duly ordered to present an ultimatum to the German Government that if no satisfactory reply to the note of 1st September was received by 11 am BST, a

State of War would exist between Britain and Germany from that hour. This was delivered at 9 am, and it is said that Hitler, Göring, and Göebbels received it with stunned surprise.

Big Ben struck at 11 am. A despondent Chamberlain was sitting alone in the Cabinet Room with a microphone before him and a BBC announcer standing by. At 11.15 listeners all over Great Britain, all over the world, heard a tired old man say: 'This morning the British Ambassador in Berlin handed the German Government a final note stating that unless we heard from them by 11 o'clock that they were prepared at once to withdraw their troops from Poland a State of War would exist between us. I have to tell you now that no such undertaking has been received, and consequently this country is at war with Germany.'

Ten minutes after he had finished, the first air-raid sirens began their ominous wailing. Many people were too stunned to move. Others rushed to their home-built Anderson air-raid shelters, the tubes, and even the trenches which had been cut in Hyde Park. The blue sky was dotted with silver barrage balloons around London and most big cities. Again everyone waited expectantly. But it was a false alarm.

Five hours later, their ultimatum having expired, the French were also at war as allies of Britain. In the House of Commons that noon a deeply moved Chamberlain told silent MPs: 'Every thing that I have worked for, every thing that I have hoped for, every thing that I have believed in during my public life, has crashed in ruins.' But it was Winston Churchill, who had for so long warned the Government of Germany's increasing armament and its dangerous intent, who now expressed the new, resolute attitude of Great Britain: 'Outside the storms of war may glow and the land may be lashed with the fury of its gale, but in our own hearts this Sunday morning there is peace. This is no question of fighting for Danzig or fighting for Poland. We are fighting to save the world from the pestilence of Nazi tyranny. A war to establish and revive the stature of man.' When Chamberlain re-formed his Cabinet that evening he could no longer overlook Churchill, and it was almost with relief that he brought him in as First Lord of the Admiralty – the same post he had held in 1914. Headed by the Prime Minister, the War Cabinet of nine was set up, comprising Simon, Halifax, Hoare, Lord Chatfield, Churchill, Hore-Belisha, Kingsley Wood and Lord Hankey as secretary. Said Churchill: 'Six of them average over sixty-four years old – only one year short of the old age pension!'

That night ten Whiteley night bombers from Nos. 51 and 58 Squadrons flew across the North Sea to Hamburg, Bremen and the Ruhr, and dropped bundles of printed leaflets of anti-war propaganda to the German people telling of the iniquities of Hitler. Except for searchlights the British raiders met no opposition, for Germany was too engrossed with Poland to spare aircraft for the Western Front – but two hundred miles out in the Atlantic the Cunarder *Athenia*, bound for Canada, was sunk by a U-boat and 120 of the crew and passengers were drowned; yet it helped to establish a spirit of grim determination when those in Britain learnt that everyone was singing the current popular song *Roll out the Barrel* as they were marshalled to the life-boats. On the radio that night a jeering artificial voice, soon to be known as Lord Haw Haw, announced not only the sinking of the *Athenia* but declaimed from Berlin Radio that the British Air Ministry was to be immediately transferred to Code location ZA, which was Harrogate – thus revealing that there were effective spies in Britain, for it was true.

In Fighter Command everyone was ready. Under Dowding it was divided into No. 11 and 12 Groups respectively commanded by Air Vice-Marshal Ernest Leslie Gossage and Air Vice-Marshal Trafford Leigh-Mallory. All other operational squadrons were set to go to France, and already a Blenheim IV from No. 139 Squadron at Wyton, flown by Flg Off A. Macpherson, had been the first to make a reconnaissance across the actual German frontier. Preparations were in hand for ten Blenheim IVs from Nos. 107 and 110 Squadrons at Wattisham and fourteen Wellingtons from Nos. 9 and 149 Squadrons to attack the German Fleet off Brunsbüttel.

The number of first-line aircraft was a little under 2,000, though the RAF had a total of 10,208 aircraft on charge, of which more than a quarter were in store, 2,000 with training units, 918 in non-operational units, and 397 with manufacturers and testing establishments, and the balance was overseas. The German first-line strength was twice as great, and that of France about 250 less than Britain.

Of individual types, the RAF had 1,089 Blenheims and 1,014 Fairey Battles, but 240 Blenheims were in the Middle and Far East Commands, and over 300 in store. What mattered was the total Squadron strength in Britain, of which the more important comprised 530 Fairey Battles, 350 Bristol Blenheims, 347 Hawker Hurricanes, 187 Supermarine Spitfires, 169 Handley Page Hampdens, 160 Vickers Wellingtons, 140 Armstrong Whitwork Whitleys, and 300 Avro Ansons. Apart from the many machines in store at home and a scattering of flying-boats, the rest made a mixed bag of Audaxes, Gauntlets, Gladiators, Hinds, Lysanders, Magisters, Tutors, and a scattering of others of which the imported American Hudsons were the most numerous.

Of the ability of the aircraft industry to provide the tools of war there was no doubt. The pioneers were still actively playing an important part in the same or derivative firms which had helped to save Britain in the previous war. Robert Blackburn was fifty-four, Geoffrey de Havilland fifty-seven, Richard Fairey fifty-two, Frederick Handley Page fifty-four, Tom Sopwith fifty-one, Oswald Short fifty-six, Charles Walker sixty-one. There were the great designers who similarly had played their part in that war – George Carter, Roy Chadwick, Arthur Davenport, Henry Knowler, John Lloyd, John North, Rex Pierson, George Volkert – and others such as Arthur Gouge, Sydney Camm, Frank Bumpus and Hollis Williams who had gained extensive experience in the twenty-one years between the two wars. There were also stalwarts like Ron Bishop, Joe Smith, George Petty, Leslie Frise, recently appointed as chief designers, and that uprising but unworldly and difficult thirty-one-year-old genius, Teddy Petter. Behind all these men were senior assistants, often of great calibre, some destined as future leaders.

Even more important had been the consolidation of the engine industry – the Bristol radials through the untiring zeal of Roy Fedden; Rolls-Royce and the skill of Arthur Elliott; Frank Halford with de Havilland and Napier engines; and there was thirty-two-year-old Whittle and his revolutionary turbine-jet.

Behind the swiftly increasing production of established types now being built at a rate almost matching the German effort were still more powerful fighters and bombers. The second prototype Short Stirling would be ready for flight trials in another three months, and their great production line was already far advanced, with deliveries scheduled from May 1940 onwards. The equivalent Handley Page Halifax prototype would be ready for flight within two months, and a great line of production machines was matching those of the Stirling. Of new fighters, hope was pinned on the twelve-gun 1,760 hp Vulture-powered Hawker Tornado interceptor which was scheduled to fly within a month, and the similar Hawker Typhoon was almost ready, though delayed by its intended 2,100 hp Sabre. Had Churchill's warnings been acted upon several years earlier those bombers and fighters would already have been established in squadron use, and Germany could have been reduced to submission while engaged on its Eastern Front. Nevertheless this ultimate armada was likely to extract a great reckoning if in the meantime existing machines could hold the expected assault. What mattered now was the swift availability of more and yet more of every existing type of fighter and bomber and engine, and to the basic 174,000 officers and men must be added a huge input for training in the big establishments which the Treasury in early years had so unwillingly agreed.

On this first day of war the King sent his encouragement to the rank and file, to his novice pilots and veterans alike, auxiliaries or regulars: 'The Royal Air Force has behind it a tradition no less inspiring than those of the older Services, and in the campaign which we now have been compelled to undertake you will have to assume responsibilities far greater than those which your Service had to shoulder in the last War. One of the gravest will be the safeguarding of these Islands from the menace of the air. I can assure all ranks of the Air Force of my supreme confidence in their skill and courage and in their ability to meet whatever calls may be made upon them. George R. I.'

The determination and courage were there, and the voice of Shakespeare seemed to echo through the land:

'Come the three corners of the world in arms
And we shall shock them. Naught shall make us rue
If England to itself do rest but true.'

INDEX

BRITISH CONSTRUCTORS

BRITISH ENGINE MANUFACTURERS

OVERSEAS AIRCRAFT CONSTRUCTORS

MISCELLANEOUS AIRCRAFT

AIRSHIPS

AIRLINES

GENERAL

Printed in England for Her Majesty's Stationery Office
by Billings & Sons Limited, Worcester
Dd587534 K40